Revealed Sciences

Demonstrating the vibrancy of an Early Modern Muslim society through a study of the natural sciences in seventeenth-century Morocco, *Revealed Sciences* examines how these sciences flourished during this period, without developing in a similar way to the those in Europe. Offering an innovative analysis of the relationship between religious thought and the natural sciences, Justin K. Stearns shows how nineteenth and twentieth-century European and Middle Eastern scholars jointly developed a narrative of the decline of post-formative Islamic thought, including the fate of the natural sciences in the Muslim world. Challenging these depictions, Stearns uses numerous close readings of legal, biographical, and classificatory texts – alongside medical, astronomical, and alchemical works – to establish a detailed overview of the place of the natural sciences in the scholarly and educational landscapes of the Early Modern Maghreb, and considers non-teleological possibilities for understanding a persistent engagement with the natural sciences in Early Modern Morocco.

Justin K. Stearns is Associate Professor of Arab Crossroads Studies at New York University Abu Dhabi, where his research interests focus on the intersection of law, science, and theology in the pre-modern Muslim Middle East. He is the author of *Infectious Ideas: Contagion in Premodern Islamic and Christian Thought in the Western Mediterranean* (2011) and an edition and translation of al-Hasan al-Yusi's *The Discourses, Vol. 1* (2020).

Cambridge Studies in Islamic Civilization

Other titles in the series are listed at the back of the book.

Revealed Sciences

The Natural Sciences in Islam in Seventeenth-Century Morocco

JUSTIN K. STEARNS
New York University Abu Dhabi

CAMBRIDGE
UNIVERSITY PRESS

CAMBRIDGE
UNIVERSITY PRESS

Shaftesbury Road, Cambridge CB2 8EA, United Kingdom

One Liberty Plaza, 20th Floor, New York, NY 10006, USA

477 Williamstown Road, Port Melbourne, VIC 3207, Australia

314–321, 3rd Floor, Plot 3, Splendor Forum, Jasola District Centre, New Delhi – 110025, India

103 Penang Road, #05–06/07, Visioncrest Commercial, Singapore 238467

Cambridge University Press is part of Cambridge University Press & Assessment, a department of the University of Cambridge.

We share the University's mission to contribute to society through the pursuit of education, learning and research at the highest international levels of excellence.

www.cambridge.org
Information on this title: www.cambridge.org/9781107638709

DOI: 10.1017/9781107588523

First published 2021
First paperback edition 2023

A catalogue record for this publication is available from the British Library

ISBN 978-1-107-06557-4 Hardback
ISBN 978-1-107-63870-9 Paperback

For my parents, Bev and Steve

Contents

Figures

Tables

Preface: Paths not Taken

When some years ago I began writing a dissertation on how Muslims and Christians responded to the Black Death in Iberia in the fourteenth century, I thought that I was writing a comparative study of social and intellectual responses to a devastating epidemic. And, to some degree, this is what the resulting book became.[1] Yet, working on the subject of contagion in the pre-modern Muslim world also confronted me with the extent to which nineteenth to twenty-first century debates on the tension between modern science and religion on the one hand, and the marginal position of the Muslim world in modern scientific production on the other, had distorted historians' understanding of Muslims writing about the natural sciences in the pre-modern world. Thus, scholars such as the Granadan Lisān al-Dīn Ibn al-Khaṭīb (d. 776/1374), who affirmed the phenomenon of contagion, had in much of the previous scholarship on Muslim responses to plague been described as exceptions proving the rule of general Muslim fatalism and anti-empiricism when faced with epidemic disease.[2] The sources I found suggested, instead, that while innovative and creative, his response and his defense of the transmission of disease was not unique and was part of a larger body of writings by Muslim authors that drew in various and

[1] Justin Stearns, *Infectious Ideas: Contagion in Premodern Islamic and Christian Thought in the Western Mediterranean*.

[2] See Michael Dols, *The Black Death in the Middle East*, 92–94; Ullmann, *Die Medizin im Islam*, 246–47; Ullmann, *Islamic Medicine*, 92–96; Rofail Farag, "The Muslims' Medical Achievements," 303; Vásquez de Benito, "La materia médica de Ibn al-Jaṭīb," 140–41; Arjona Castro, "Las epidemias de peste bubónica en Andalucía en el siglo XIV," 58; Calero Secall, "La peste en Málaga, según el malagueño al-Nubāhī," 58; al-Bazzāz, *Tārīkh al-awbi'a wa-l-majā'at bi-l-maghrib fī l-qarnayn al-thāmin wa- l- tāsi' 'ashara*, 393; Congourdeau and Melhaoui, "La perception de la peste en pays chrétien byzantine et musulman," 110.

diverging ways on religious, medical, and empirical evidence and authorities to address the challenge posed by the transmission of disease.[3] The historiography surrounding contagion had been substantially distorted by our present day understandings of disease transmission and of the necessity of distinguishing between religious and medical discourses. As I finished this first project, a figure marginal to one of its later chapters, the eleventh/seventeenth-century Moroccan polymath al-Ḥasan al-Yūsī (d. 1102/1691), increasingly preoccupied me and became the impetus for writing the book the reader holds now. I had initially become interested in al-Yūsī for his views on contagion, eloquent but not particularly innovative.[4] Yet, reading al-Yūsī pushed me to broaden my interest in the contemporary framings and concomitant distortions of Islamic intellectual history from contagion to the natural sciences in general. Specifically, I became interested in how contemporary teleologies of the rise of modern science had encouraged historians to either ignore developments in Islamic thought that had no place in its genealogies or to more starkly dismiss them as pseudoscience or intellectual decline. There were two central issues here, the first more clearly related to what I had observed with contagion and which could be summarized as the reduction of past intellectual thought to those elements that had a role in shaping our current understandings of science, or at least represented a parallel with them. The second was distinct, if related, and consisted of clarifying the social and intellectual context of the natural sciences and the ways Muslims drew on them within the religious discourses of law, theology, and Sufism – aspects largely unexplored for the post-formative period. This engagement with the writings of al-Yūsī and the scholarly dynamism of his age, along with the accident of my own specialization in the Islamic West, led to my pursuing these interests through the prism of Morocco in the eleventh/seventeenth century, a period of intellectual and political ferment that I describe in Chapter 1.

Much of what I have addressed above could be summarized under the rubric of Whig History or writing the story of the past with the values of the present, something that all students of history and especially of the history of science are repeatedly warned against. Yet this belongs to the class of imprecations that are especially poignant because we sense that they are in vain: while in the following I have done my best to establish the nature and

[3] Not all my readers were convinced. See the passing remarks in Gotthard Strohmaier, "Galenism Caught between Faith in God and 'Prophetic Medicine'," 38, and in the same volume the warning of Lutz Richter-Bernburg, "Medicine in Islam: Contested Autonomy," 49.

[4] See Stearns, *Infectious Ideas*, 137–38, and al-Yūsī, *The Discourses*, vol. 1, 245–51.

significance of the natural sciences for the scholars of early modern Morocco, my own interest in doing so is a profoundly modern one that situates itself in conversation with a host of other modern voices. Many of these are scholarly ones whose names the reader will find in the body of the book itself. But there are also two conversations taking place in the broader public sphere that have played a role in my choosing to write *Revealed Sciences*, although neither finds much support among students of Islamic intellectual history today.

The first of these is the argument that Islam or the Islamic world has opposed the development and study of philosophy and the natural sciences since a purported Golden Age that ended sometime in the (European) Middle Ages after European scholars translated what Muslims had preserved of Greek philosophy.[5] Once widespread among historians and some orientalists, this view is today especially prevalent among the so-called New Atheists, including Richard Dawkins and Sam Harris, but also among other prominent scientists who have turned to writing histories of science, such as Steven Weinberg, and populizers of science such as Neil DeGrasse Tyson. At their most benign, these figures depict all religion as inimical to scientific inquiry; more perniciously they can represent Islam as particularly opposed to science and to modernity in general.[6] In the Introduction below, I will discuss how at the end of the nineteenth century two narratives became widespread: 1) the story of Western European history containing a struggle between religion and science culminating in science's victory; and 2) the description of the Muslim world having been in a state of intellectual decadence and decline since the Middle Ages. Here, I would like to draw attention to how the simplistic civilizational discourse that underlies these narratives and locates both science and modernity in a uniquely European rationality adds to the force of the Islamophobia that both Europe and the United States are witnessing in the first decades of the twenty-first century.

[5] See my comments in "Writing the History of the Natural Sciences in the Pre-modern Muslim World: Historiography, Religion, and the Importance of the Early Modern Period," *History Compass*, vol. 9 (2011), 923–51, and at 926–28 for my discussion of the great Orientalist Ignaz Goldziher's confusing and influential views on the fate of the rational sciences in post-formative Islamdom.

[6] For Dawkins' remarks, which are part of his more extensive views of Islam on social media, see the overview in Abby Ohlheiser, "A Short History of Richard Dawkins vs. The Internet"; for Sam Harris' views, see the opening to his chapter "The Problem of Islam" in *The End of Faith: Religion, Terror, and the Future of Reason*, 108–09; for Steven Weinberg, see *To Explain the World: The Discovery of Modern Science*, 116–23; Neil DeGrasse Tyson's views on the influence of al-Ghazālī in bringing about the intellectual collapse of Islam are expressed in his reboot of the series *Cosmos* and in this lecture (www.youtube.com/watch?v=Fl1nJC3lvFs).

An opposing view that is nonetheless based on the similar premise of the Islamic world having witnessed its last moment of vibrancy and contribution to world history during the European Middle Ages is found in the exhibition *1001 Inventions* and the associated publications that have proceeded from the work of the British-based Foundation for Science, Technology, and Civilization.[7] Here we have a narrative that not only celebrates the achievements of Muslim scholars during a Golden Age that stretched from roughly the third/ninth to sixth/twelfth century celebrated, but they are described as the origin of modern science and modernity in general. As with the first group, the intellectual production of the Muslim world during the Early Modern period is passed over entirely and assumed to be either irrelevant or in decline.[8] This narrative also originated in the late nineteenth century, this time in the works of Muslim reformers who responded to the writings of European orientalists and colonial administrators by stressing the past glory of Muslim achievements and the need to return to that glory through dramatic reforms. Here I note that such attempts to emphasize the importance of the work of scholars working in the Muslim world during the European Middle Ages – based on the same Hegelian civilizational approach shared by many of the first group – not only distort the past, but also lead Muslims and non-Muslims alike to a lack of interest in those aspects of Islamic intellectual history that were not valued by European scholars or which did not lead to developments in European intellectual history.[9]

The misleading relay-race narrative in which the Muslims receive the baton of science from the Greeks and then pass it on to the Europeans is shared with varying emphases by both groups, and contributes to a general lack of interest in what happened intellectually in the Muslim world after Europeans stopped translating Arabic texts.[10] Even more important for

[7] See www.fstc.org.uk/, www.1001inventions.com, as well as the associated short film *1001 Inventions and the Library of Secrets* (www.youtube.com/watch?v=JZDe9DCx7Wk).

[8] The *1001 Inventions* and associated FSTC events and publications have been incisively critiqued in Sonja Brentjes et al. (eds.), *1001 Distortions: How (Not) to Narrative History of Science, Medicine, and Technology in Non-Western Cultures*. See also Sonja Brentjes' review of Salim T. S. al-Hassani (ed.), *1001 Inventions: The Enduring Legacy of Muslim Civilization*.

[9] A point I made in an abbreviated fashion in *The National* after visiting the *1001 Inventions* exhibition when it came to Abu Dhabi in 2011 (www.thenational.ae/1001-innovations-and-the-living-heritage-of-islamic-science-1.375391).

[10] The impression that translations into Latin from Arabic ended in the Middle Ages is widespread among historians as well. For a nuanced and comprehensive discussion of the Renaissance translations of Arabic philosophical works, which are rarely invoked in these narratives, see Dag Nikolaus Hasse, *Success and Suppression: Arabic Sciences and Philosophy in the Renaissance*. Shifting the period of translation from the Middle Ages to

this book, however, is that the bulk of the scholarly research on the natural sciences in the Muslim world over the past century has focused on the so-called "Golden Age" associated with the ʿAbbasid caliphate from the third/nineth to seventh/thirteenth centuries and subsequent Mongol and Timurid states into the nineth/fifteenth centuries, with less attention being paid to the Islamic West, much less Morocco, following the division of much of the Islamic world into the Mughal, Safavid, and Ottoman Empires in the sixteenth century.[11] This generalization should clearly not be pushed too far, for as Chapters 2–4 will show, there is a growing body of work addressing the natural sciences during what we can tentatively call the post-classical period of Islamic intellectual history – chronology being a constant bugbear for historians – which stretched from the thirteenth to eighteenth centuries.[12] Yet the balance of research and interest is still clearly in the Eastern Mediterranean and Central Asia during the eleventh–fourteenth centuries and that is where much of the best work is being done.[13] The richness of secondary scholarship on this region and period has no parallel in either the later centuries or in the Islamic West. An exception to this last statement is the study of the natural sciences in al-Andalus, which has received considerably more attention than North Africa, but which also ended, at the most generous estimate, with the expulsion of the Moriscos at the beginning of the seventeenth century.[14]

the Renaissance does little, of course, to increase interest in the intellectual production of the Muslim world following this.

[11] For a recent nuanced discussion of some of the problems of identifying the Abbasid caliphate with a Golden Age of Islamic civilization, see Michael Cooperson, "The Abbasid 'Golden Age': An Excavation."

[12] The "post" here refers to a general agreement in Islamic studies that by the fifth/twelfth century the classical intellectual and institutional structures for Islamic jurisprudence, theology, and mysticism (a poor gloss for Sufism) had crystalized and would form the basis for subsequent developments. For a concise overview differentiating between processes that could be considered formative and classical in terms of the development of Islamic civilization, see Chase Robinson, "Conclusion: From Formative Islam to Classical Islam." For insightful remarks on the problem of chronology in the historiography on the Muslim world, see Shahzad Bashir, "On Islamic Time: Rethinking Chronology in the Historiography of Muslim Societies," 519–44.

[13] To take just four important works of the past two decades, this is as true of the biographical monographs of Robert Morrison and Nahyan Fancy on, respectively, Nīẓām al-Dīn al-Nīsābūrī (d. ca. 1330) and Ibn al-Nafīs (d. 1288), as it is of François Charette's study and translation of Najm al-Dīn al-Miṣrī's fourteenth century treatise on mathematical instrumentation or Sally Ragep's recent translation of Jaghmīnī's thirteenth century introduction to contemporary reformulations of Ptolemaic Astronomy.

[14] Consider only two comparatively recent works, Miquel Forcada's 2011 study of Ibn Bājja (d. 1139), and Robert Morrison's 2016 study and translation of Joseph ibn Naḥmias' (fl.

Revealed Sciences aims to address this imbalance by focusing on the natural sciences in seventeenth century Morocco, but the lacuna that it mainly attempts to fill is not principally geographic but methodological. Instead of focusing on the works of a scholar who specialized in one or more of the natural sciences, or on a specific site of scientific production, it seeks to contextualize the role played by the natural sciences within the broader scholarly production of the time. Doing so entails the reading of biographical dictionaries, scholarly autobiographies, legal opinions, theological asides, mystical and moral reflections, as well as works on astronomy, medicine, and alchemy. I am interested here not so much in intellectual influence or progress, terms with their own connotations, as much as I am in establishing the presence and range of the natural sciences in knowledge production and transmission.[15] Accurately carrying out this task necessitates first sidestepping the teleological questions of progress and development that characterize so much writing in the history of science in order to fully establish the range and nature of the natural sciences during this period. Only once we have described an epistemological field that included Prophetic and humoral medicine, astronomy, alchemy, astronomy, magic and mathematics, and the various debates in which scholars discussed the categorization and the value of these sciences, can we turn to the broader historical question of change over time. Put differently: we should describe what Muslim scholars of seventeenth century Morocco perceived as natural sciences and the uses to which they put them before we can begin to move to tracing how their understandings differed from those of their colleagues from previous or subsequent centuries. Finally, if we wish to understand the broader social impact of the natural sciences, we need to move our focus from the work of the exceptional scholar of one of these sciences to the more general reception of these disciplines by both nonspecialists and their more ordinary representatives.

I am well aware that the approach taken here will not appeal to a number of groups. For those who believe in a singular teleology of science that can be traced from Aristotle through the European Middle Ages to the European Scientific revolutions of the seventeenth–nineteenth centuries, and then to the present day, the story told here will have the value of a historical curiosity at best, proof of civilizational

1400) Judeo-Arabic work on theoretical astronomy, *The Light of the World: Astronomy in al-Andalus*.

15 For a nuanced discussion of terminological pitfalls related to the spread of knowledge between distinct cultural spheres, see James E. Montgomery, "Islamic Crosspollinations."

decline at worst.[16] Many historians of science in the Muslim world, for their part, who tend to focus on the work of a scholar specializing in one of the natural sciences, may look askance at the marginal place of the discussion of texts in the natural sciences, and note that there is little of scientific interest here in general. While more sympathetic to the second than the first group's concerns, *Revealed Sciences* represents an argument for shifting our focus from exceptional scholarly production to the context of what might be called the more pedestrian types of Kuhnian normal science.[17] I do not subscribe to a "strong" version of Kuhn's evocative, influential, and thoroughly critiqued theory of scientific paradigms setting out research agendas for normal science that experiences crises leading to scientific revolutions, nor do I think that it readily applies to the material under consideration. Yet, Kuhn's description of normal science as "mop-up work" dealing with "puzzle-solving" the problems set out by an initial paradigm does go some way toward explaining the types of scholarship produced in the natural sciences in seventeenth-century Morocco.[18] The reference is imprecise, and not only because Kuhn himself understood science and its revolutions, regardless of how one understands the much debated term "incommensurability" that he used to describe distinct paradigms, as decidedly a story of Western European modern science.[19] Nevertheless, it was a passage at the end of Thomas Kuhn's classic *The Structure of Scientific Revolutions* that precipitated my thinking at an early stage of this book. Describing Darwin's greatest challenge in gaining acceptance for the theory of evolution, Kuhn notes that it was Darwin's arguing that evolution had no goal where he encountered the greatest resistance: "What could 'evolution,' 'development,' and 'progress' mean in the absence of a specified goal? To many people, such terms suddenly seemed

[16] For one unapologetic defense of a teleological reading of the history of science, see Steven Weinberg's defense of Whig readings of the history of science in "*Eye on the Present – The Whig History of Science*," The New York Review of Books, December 17, 2015 (www.nybooks.com/articles/2015/12/17/eye-present-whig-history-science/). But see also the comments of the scholar of Avicenna, Dimitri Gutas, who argues for a clear differentiation between "science" and "religion" in the formative period of Islamic thought, with all philosophy tinged by religious dogma being paraphilosophy (Dimitri Gutas, "Ibn al-Nafīs's Scientific Method").

[17] See Thomas Kuhn, *The Structure of Scientific Revolutions*, chapter 3 "The Nature of Normal Science."

[18] See ibid., 24 and chapter 4 "Normal Science as Puzzle-Solving."

[19] This aspect of Kuhn's theory received an especially great deal of criticism. For some of his later reflections on the subject, see Kuhn, "Commensurability, Comparability, Communicability."

self-contradictory."[20] Darwin's famous tree of life, ever branching outwards in many directions but not moving toward any single goal, provided Kuhn with a powerful analogy to his own description of scientific progress – one that offered a sustained critique of modern science as an asymptotic movement toward absolute truth. Insofar as the analogy remains useful, it helps us understand the story of the natural sciences in early modern Morocco not as one of retardation or decline, but of divergence from the path taken in Europe during the same centuries – distinct, but no less rational or curious.

If the main impulses of this book are, first, to draw attention to the nature and place of the natural sciences in the Muslim world in the Early Modern period, and, second, to focus on engagement with these sciences beyond the writings of exceptional thinkers, the third is to add nuance to the now rather stale debate about the compatibility between science and religion. Proceeding from the assumption that both this debate and the two central terms assumed their current significance at the end of the nineteenth century, *Revealed Sciences* dwells on the porous nature of the natural and transmitted sciences during this period, tracing how they lent their authority to each other and the ways that individual scholars saw them as compatible. While these and related questions emerge from a decidedly twenty-first century context and are those of a scholar writing from within a modern, secular, institution that is distinct from those in which the scholars discussed here lived and wrote, I do not think that these questions would have been wholly unintelligible to them (or incommensurable with their own).

[20] Kuhn, *The Structure*, 171. See also the excursus on Kuhn after Chapter 3.

Acknowledgments

It has been over a decade since I first began working on this project, and it has acquired many debts along the way. It is a great pleasure to be able to acknowledge them here, even as doing so leaves me chastened to contemplate how long it took me to get to this point. I need to start by thanking Cambridge University Press for its patience. The book began with a conversation with Marigold Acland at MESA in 2010, and following Marigold's retirement it went through several hands, resting now with the very capable and encouraging Maria Marsh. I am deeply grateful to everyone at Cambridge for giving me the time I needed to complete this project. Finally, I am indebted to Martin Grosch for his work on the wonderful map of Morocco that accompanies the text. A good map is worth far more than a thousand words.

Along the way, my work on this book has received support from a number of fellowships, beginning with a residency at the Rockefeller Center at Bellagio in the fall of 2010, where I completed my first reading of al-Yūsī's writings and presented my still inchoate thoughts to the other fellows. It was a gift to be able to return to Bellagio in the spring of 2019 as my work on a first draft of this book was nearly done, this time accompanying my wife Nathalie Peutz on her residency there, in something of a full circle. An American Institute for Maghrebi Studies Grant allowed me to visit the Ḥamziyya-ʿAyyāshiyya lodge in June of 2015 and to consult the manuscripts there, and an NYUAD Research Enhancement Fund supported research on the project between 2015 and 2017, including visits to the Nāṣiriyya lodge in Tamgrout in 2016 and to Taroudant in 2018. The bulk of the writing of the first draft took place with the support of an American Council for Learned Societies grant in 2017 and 2018, and I had an opportunity to revise and rethink this initial draft with the support of a residential fellowship at the Institute for Advanced Study (IAS) in Princeton in 2019–20. I am deeply grateful all this support and the opportunity to focus on this work. Finally, I am indebted to the NYU Abu Dhabi Grants for Publication Program, which covered the indexing costs of this book (and deeply thankful to Pilar Wyman for the indexing itself).

At IAS, I benefited a great deal from the hospitality of Suzanne Akbari, Sabine Schmidtke, and Francesca Trivellato. Along with the seminars they ran there, I enjoyed many conversations with them and many others, including with Laurie Benton, Godefroid de Callatay, Daniel Hershenzon, Piet Hut, Webb Keane, Daniel Varisco, Cord Whitaker, and Alden Young. As I write these lines, I only regret that the coronavirus brought that year to an early end. Finally, I would like to say thanks to those scholars and mentors who wrote letters for me over the years and without whose support I would not have been able to do this work: Michael Cook, Maribel Fierro, and David Powers, you have been wonderful.

Early versions of parts of material that made its way into this book were presented at talks or workshops at Princeton University, New York University Abu Dhabi, Columbia University, Yale University, the Sorbonne Abu Dhabi, Washington University in St. Louis, the University of Colorado at Boulder, Ibn Khaldoun University in Tiaret, Algeria, the American University in Dubai, Georgetown University in Qatar, the American Legation in Morocco in Tangiers (TALIM), the Consejo Superior de Investigaciones Científicas (CSIC), McGill University, the Middle East Studies Association meetings in 2010, 2013, 2014, the History of Science Society annual meeting of 2018, and at the Institute for Advanced Study in 2019. I am especially indebted to Najam Haider for bringing me to Columbia, Asad Ahmed for inviting me to take part in a workshop at Washington University, Aun Hasan Ali for hosting me in Boulder, Sonja Brentjes for suggesting I join her in Tiaret, Maribel Fierro for running the workshop in Madrid, and Aslıhan Gürbüzel for organizing and hosting the Montreal workshop.

Sections of Chapters 2 and 3 are drawn from "'All Beneficial Knowledge is Revealed': The Rational Sciences in the Maghrib in the Age of al-Yūsī (d. 1102/1691)," *Islamic Law and Society*, vol. 21 (2014), 49–80, and "The Legal Status of Science in the Muslim World in the Early Modern Period: An Initial Consideration of *fatwa*s from Three Maghribi Sources," in Ahmed, Sadeghi, Bonner (eds.), *The Islamic Scholarly Tradition: Studies in Islamic History, Law, and Thought in Honor of Professor Michael Allan Cook on His Seventieth Birthday* (Leiden: Brill, 2011), 265–90, and a version of the section on al-Ṣāliḥī's medical writings – here in Chapter 4 – should appear in a special issue of *Early Science and Medicine* in 2021.

Along the way a small number of colleagues were generous enough to read one or more draft chapters and to offer their insight. These included Robert Allen, Nora Barakat, Maya Kesrouany, Caitlyn Olson, and Erin Pettigrew. Caitlyn Olson deserves special thanks for having read all of the main chapters and given extended comments. I am deeply grateful to her for her time and for giving me the chance to watch her dissertation on seventeenth-century theological debates in Morocco take shape over the past few years. Michael Cook generously agreed to read a full draft of the book. An earlier form of part of what would become Chapter 3 had been presented at a conference in his honor in 2010 (and had then appeared in the resulting book, cited above). Michael's comments, insights and suggestions made this a much stronger book. Throughout my career, his example as a scholar and advisor has influenced me profoundly. Finally, my thanks go to Cambridge's anonymous readers, whose comments pushed me to make the manuscript better on multiple fronts. I am deeply aware that I was not able to address all of the suggestions and critiques made by these readers.

The number of colleagues and friends who provided intellectual support and conversation was much greater in number. At NYUAD I am grateful to, again, Robert Allen, Nora Barakat, Maya Kesrouany, and Erin Pettigrew, but also Kevin Coffey, Phil Kennedy, Taneli Kukkonen, Nathalie Peutz, Melina Platas, Maurice Pomerantz, and Mark Swislocki. After hearing me give a talk, it was Mark who suggested that I simply make my working title – *Revealed Science* – a plural to address my qualms about giving the impression I was writing about a singular science. Further afield, on separate occasions Sajjad Rizvi and Nahyan Fancy co-organized workshops with me that helped me think through the historiographical implications of the project. It was at the latter workshop with Nahyan in December, 2019, that we were able to bring together many of the scholars whose work influenced me in the writing of this book, including Sonja Brentjes, Noah Gardiner, and Matthew Melvin-Koushki. Conversations with Robert Morrison over the years have always been fruitful, and I am grateful for his support and scholarship. On several occasions, David King answered questions and supplied references, for which I'm deeply indebted. In Morocco, where I carried out several trips to consult and acquire copies of manuscripts, I received hospitality and support from Hamid Lahmar, and enjoyed conversations with Fouad Ben Ahmed. My heartfelt thanks to them both.

The researching and writing of this book took place during a transformative decade in my life, one in which Nathalie and I moved to Abu Dhabi to join a community of academics and administrators to start a new university (NYUAD), were blessed with triplet girls, and in which I received tenure. The support and love that I received from family and friends during these years was crucial to both this book and my overall happiness and sanity. There is no way for me to convey my gratitude to them. A special thanks goes to Apsara Perera Gangodage, whose work taking care of our children during these years was crucial to ensuring the craziness of a two-career household did not get too out of control, and whose warmth and presence has provided our children with so much. Finally, my family: I owe so much to my children Mattheus, Anahita, Clio, and Makeda. Thank you. Thank you. I love you so much. And to Nathalie, my partner through some truly amazing and difficult times: all of this means so much more because of you. I love you. You are home.

This one is for you, Mom and Dad. As time goes past, I see more and more how much you have given me and how far your love has taken me.

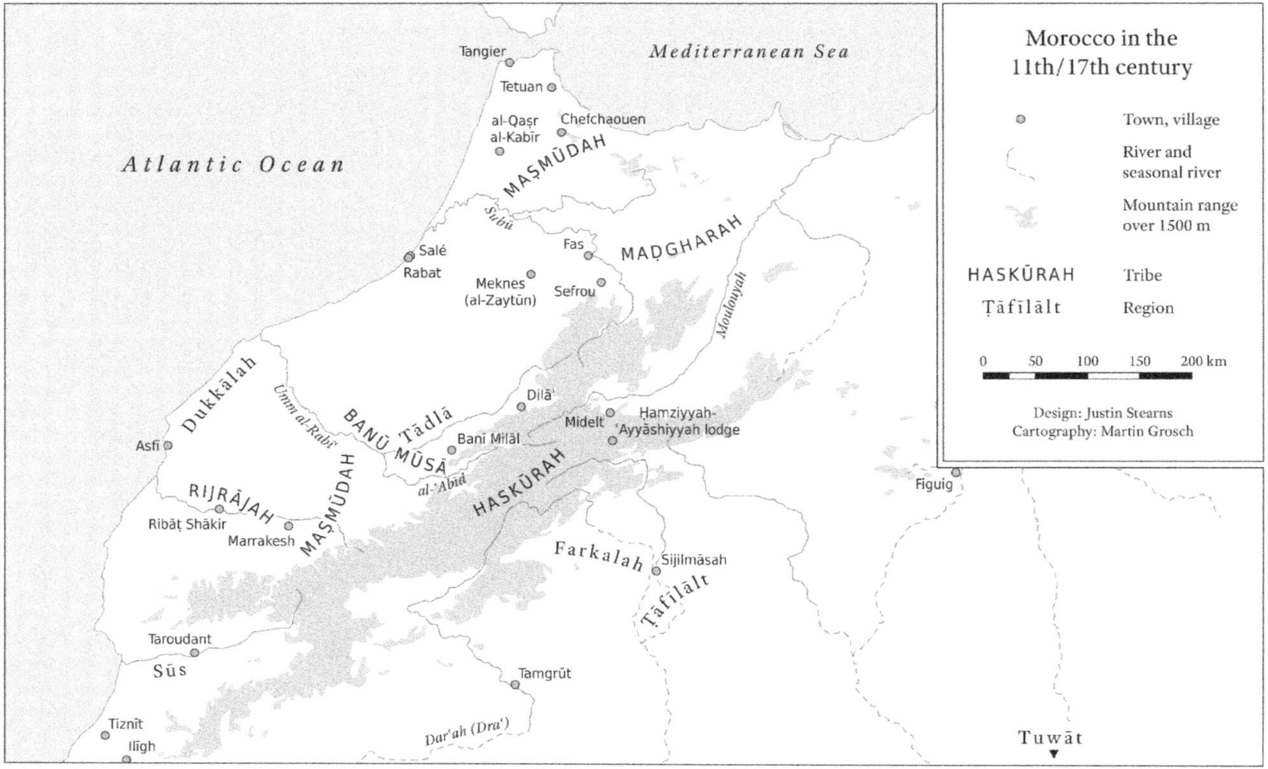
Morocco in the
11th/17th century
Town, village
River and
seasonal river
Mountain range
over 1500 m
HASKŪRAH
Tribe
Ṭāfīlālt
Region
0
50
100
150
200 km
Design: Justin Stearns
Cartography: Martin Grosch
Atlantic Ocean
Mediterranean Sea
Tangier
Tetuan
al-Qaṣr
al-Kabīr
Chefchaouen
MAṢMŪDAH
Subū
MADGHARAH
Fas
Salé
Rabat
Meknes
(al-Zaytūn)
Sefrou
Moulouyah
Dukkālah
Umm al-Rabī'
BANŪ MŪSĀ
Tādlā
Dilā'
Midelt
Hamziyyah-
'Ayyāshiyyah lodge
Banī Milāl
Asfī
al-'Abīd
HASKŪRAH
RIJRĀJAH
Ribāṭ Shākir
Marrakesh
MAṢMŪDAH
Figuig
Farkalah
Sijilmāsah
Ṭāfīlālt
Taroudant
Sūs
Tamgrūt
Tiznīt
Ilīgh
Dar'ah (Dra')
Tuwāt

INTRODUCTION

Narratives of Science, Old and New

> Hence the humanistic historian must concern himself with the great commitments and loyalties that human beings have borne, with which every sort of norm and ideal has been made explicit; and he must concern himself with the interactions and dialogues in which these commitments have been expressed. Hence, for an "exceptionalizing" historian with such intentions, it is Islamdom as a morally, humanly relevant complex of traditions, unique and irreversible, that can form his canvas. Whether it "led to" anything evident in modern times must be less important that the quality of its excellence as a vital human response and an irreplaceable human endeavor. In this capacity, it would challenge our human respect and recognition even if it had played a far less great role than, in fact, it did play in articulating the human cultural nexus in time and space and in producing the world as we find it now.
>
> Marshall Hodgson, *The Venture of Islam*

OVERVIEW

For this book to tell its story another needs to be untold. This is because the seismic political and intellectual changes that took place globally in the nineteenth to twentieth centuries during successive stages of European colonial and economic expansion and the subsequent periods of decolonization and globalization have profoundly shaped our understandings of the preceding centuries.[1] It has frequently been argued that modernity

[1] As one eloquent example of how our understanding of the Middle Ages came into focus during European colonial expansion, see Kathleen Davis, *Periodization and Sovereignty: How Ideas of Feudalism and Secularization Govern the Politics of Time*. For two masterful

itself emerged in connection with industrialization in the long European century that began in the eighteenth century with European colonial adventures into the Middle East and Africa and ended with a world war in 1914–18, although it is far less clear what this actually means.[2] Today, to attempt to reconstruct the premodern, preindustrial societies before this century requires a considerable feat of imagination.[3] Our interest here is less the political and economic changes witnessed by this century than the intellectual and cultural shifts that accompanied them, and specifically the ways in which both European and Middle Eastern scholars adopted new definitions of science and religion during the long nineteenth century. Concurrently, and in the context of increasing European colonization of the Middle East, many European Orientalists and traditionally educated scholars in the Middle East came to view the intellectual landscape of the region in the pre-nineteenth century as largely static, and in stark contrast to an earlier period of intellectual fertility.[4] Unsurprisingly, the lessons that colonial administrators, Middle Eastern intellectuals, and Western Orientalists, drew from this insight differed. For many of the first group, Eastern decadence and weakness justified if not necessitated colonial tutelage. For their part, scholars in the region were divided between those who believed in the necessity of defending the traditional educational institutions and their curricula in order to resist the cultural imperialism of the colonial powers, and reformers who argued for a radical break with the recent past in order to restore the scholarly creativity and vigor of

syntheses on the importance and nature of the changes brought about by the long nineteenth century see C. A. Bayly, *The Birth of the Modern World 1780–1914* and Jürgen Osterhammel, *The Transformation of the World: A Global History of the Nineteenth Century.*

[2] For a valuable discussion of the problematic fashion in which modernity in the Middle East has traditionally been linked to Napoleon's invasion of Egypt in 1798 see Dror Ze'evi, "Back to Napoleon? Thoughts on the Beginning of the Modern Era in the Middle East." On the confusion surrounding the term "modernity," see Dipesh Chakrabarty, "AHR Roundtable: The Muddle of Modernity."

[3] For one lucid attempt to lay out the differences between our world and the preindustrial one, see Patricia Crone, *Pre-Industrial Societies: Anatomy of the Pre-Modern World.* The story as I present it contains a number of crude generalizations, which need to be nuanced: different parts of the world experienced industrialization at different times, Britain, notably, in the eighteenth and not the nineteenth century. For this and much more, see Robert Allen, *The British Industrial Revolution in Global Perspective.*

[4] See Indira Gesink, "'Chaos on the Earth': Subjective Truths versus Communal Unity in Islamic Law and the Rise of Militant Islam," and ibid., *Islamic Reform and Conservatism: Al-Azhar and the Evolution of Modern Sunni Islam.* This story is laid out now in Ahmed Shamsy in a fashion that complements Gesink's analysis, even as it comes to some distinct conclusions, in his *Rediscovering the Islamic Classics: How Editors and Print Culture Transformed an Intellectual Tradition.*

a distant Golden Age. Both groups of local scholars agreed that, in terms of scholarly production, the centuries preceding the arrival of colonial powers were characterized by adherence to tradition, though they differed on its nature and desirability. At the beginning of the twentieth century, and at times in conversation with scholars in the Middle East, Western Orientalists, a group with a diverse set of relations to the colonial project, developed an increasingly consistent argument for the region having long been intellectually dormant.[5]

These narratives that developed in Europe and the Middle East coincided with the emergence among European historians of science in the first half of the twentieth century of the concept of a Scientific Revolution that had taken place in Northern Europe in the seventeenth century and which set Europe alone on a path toward modern science and modernity itself.[6] The story of the Scientific Revolution drew on the nineteenth-century belief in a historical European exceptionalism and the argument that modernization entailed secularization – the Weberian "disenchantment of the world" – which itself built on a late nineteenth argument that Protestantism – in stark contrast to Catholicism – had helped birth modern science.[7] The notion that the wrong kind of religion blocked rational thought and historical progress – the latter a notion that acquired greater currency due to the work of nineteenth-century thinkers such as Hegel, Marx, and Burkhardt – was transferred from Catholicism to Islam in greatly divergent ways by the Muslim reformers mentioned above and many of their Orientalist contemporaries.[8]

5 Edward Said's *Orientalism* presented a distorted (if influential) account of European scholars working on the Middle East and North Africa in the nineteenth and twentieth centuries. For an excellent example of a more accurate and productive analysis, see Suzanne L. Marchand, *German Orientalism in the Age of Empire: Religion, Race, and Scholarship*. For Morocco, see the insightful work of Edmund Burke III, *The Ethnographic State: France and the Invention of Moroccan Islam* and Manuela Marín, "Los estudios árabes y el colonialism español en Marruecos (siglos XIX–XX)," and *Testigos coloniales: españoles en Marruecos [1860–1956]*.

6 While generally separate conversations, both narratives emerged from a conviction in European exceptionalism that took on a new character in the nineteenth century. The assumption of the conflation of the Scientific Revolution and modernity is widespread, but see especially Herbert Butterfield, *The Origins of Modern Science: 1300–1800* and then Michael Adas, *Machines as the Measure of Men: Science, Technology, and Ideologies of Western Dominance*.

7 The *locus classicus* being Andrew D. White, *A History of the Warfare of Science with Theology in Christendom*, but see also Robert K. Merton, *Science, Technology and Society in Seventeenth Century England* and Peter Harrison, *The Bible, Protestantism, and the Rise of Modern Science*. For a favorable Turkish reading of John W. Draper's 1874 volume on the *History of the Conflict between Religion and Science*, see M. Alper Yalcinkaya, "Science as an Ally of Religion: A Muslim Appropriation of 'the Conflict Thesis'."

8 On the broad historical framings of nineteenth century historical thought, see Hayden White, *Metahistory: The Historical Imagination of the Nineteenth Century*. Protestant authors

Taken together, these narratives suggested that there was little for historians to study when it came to intellectual production, much less the natural sciences, in the Middle East (often conflated with the Muslim world as a whole) following a Golden Age that had ended at some point in the Middle Ages. Over the past decades, historians have readily provided a series of compelling critiques of this story. One lies in exploring how modern science emerged through the interaction of Europeans with their colonial worlds, thus partially de-centering if not provincializing Europe and drawing attention to the ways in which the production of science occurred globally.[9] Another, related, corrective is to push the date of the importance of the intellectual production of the Middle East for modern science forward from the Abbasid period (750–1258) to the beginning of the Scientific Revolution itself.[10] Still another has been to question the Protestant nature of the Scientific Revolution, with a special emphasis on the intellection production of Spain and its colonies in the New World.[11] As suggested in the Preface, *Revealed Sciences* charts a different path and looks instead at a history of science that is marginal in the genealogy of modern science. In this, it is distinct, but has parallels with recent efforts to recenter the importance of esoteric works in post-formative Islamic thought: in both cases, the aim is to explore and describe histories of rational thought within the category of natural philosophy broadly defined, which have fallen out of the teleological narratives that dominate contemporary histories of science. The most important difference between this book and those who have been writing on esotericism is not only geography, period, or subject matter. Instead of a focus on the natural sciences themselves, here I trace their presence and role in the hegemonic

had used Islam as a foil for criticizing both the Pope and Catholicism since the sixteenth century. For one especially relevant example, see Sonja Brentjes, "Pride and Prejudice: The Invention of a 'Historiography of Science' in the Ottoman and Safavid Empires by European Travellers and Writers in the Sixteenth and Seventeenth Centuries."

9 Two examples are provided by Kapil Raj, *Relocating Modern Science: Circulation and Construction of Knowledge in South Asia and Europe, 1650–1900* and Harold Cook, *Matters of Exchange: Commerce, Medicine, and Science in the Dutch Golden Age*. But see now the collection of articles with the excellent introduction of J. B. Shank, "Special Issue: After the Scientific Revolution: Thinking Globally about the Histories of the Modern Sciences."

10 See the contributions to Riva Feldhay and F. Jamil Ragep (eds.), *Before Copernicus: The Cultures and Contexts of Scientific Learning in the Fifteenth Century* and Robert Morrison, "A Scholarly Intermediary between the Ottoman Empire and Renaissance Europe."

11 See Victor Navarro Brotóns and William Eamon (eds.), *Más allá de la Leyenda Negra: España y la Revolución Científica* and Jorge Cañizares-Esguerra, *Nature, Empire, and Nation: Explorations of the History of Science in the Iberian World.*

Islamic religious discourses of their time: jurisprudence, theology, and, to a lesser extent, Sufism. Writings in the natural sciences themselves, principally astronomy and medicine, while playing an important role in the chapters that follow, are not the primary focus until Chapter 4. I am more interested in following the ways in which the natural sciences and the natural world were inextricably woven into Islamic thought as a whole, in challenging the assertion that Muslim scholars compartmentalized religious and philosophical questions, and in exploring how genre and subject matter were only partially successful in disciplining the natural sciences.[12] In this manner, *Revealed Sciences* moves past the now tired question of the compatibility of science and religion – and especially of Islam and science – as well as the question of the degree and nature of the influence of Muslims and Islam on modernity, to examine the significance of the natural sciences for scholarly individuals and networks that were profoundly religious.

CREATING SCIENCE AND ISLAM: REVISITING TERMINOLOGICAL ANXIETIES

During the nineteenth and early twentieth centuries, the disciplines of history and history of science were professionalized in Europe, and subsequently in the United States.[13] This process went hand in hand with the professionalization of science itself, the term "scientist" famously being coined in 1833 by William Whewell of Cambridge University, who became a pioneer in the field of the history of science by writing both a history and a philosophy of the inductive sciences.[14] Some work was involved in separating science from the natural sciences and natural philosophy, but by the end of the nineteenth century both historians and scientists could speak of science as an intellectual pursuit that had played a central role in Europe's past and which would continue to drive mankind's progress forward. Soon after, in the early twentieth century, you could even get a job doing it and not have to rely on private wealth or patronage.[15] All of this is to say that our understanding of science today is decidedly different

[12] The compartmentalization thesis has been eloquently advanced by Ahmed Dallal in *Islam, Science, and the Challenge of History.*

[13] The literature is extensive. Two places to begin regarding England and the United States are Steven Shapin, *The Scientific Life: A Moral History of a Late Modern Vocation* and Richard Yeo, *Defining Science: William Whewall, Natural Knowledge and Public Debate in Early Victorian England.*

[14] John F. M. Clark, "Intellectual History and the History of Science," 157.

[15] For this shift from vocation to profession, see Steven Shapin, *The Scientific Life.*

from that of the natural philosophers of the early nineteenth century, much less those who studied and wrote natural philosophy in the preceding centuries before science acquired its current meaning. The differences are so substantial that they urge us to question the degree to which the classic notion of scientific progression is still sufficient to explain attitudes toward the natural sciences over the centuries immediately preceding the nineteenth century.

Looking at science and its history in this fashion is possible in large part due to the cultural turn in historical studies of the 1960s and 1970s, and what came to be known as the externalist critique of a history of science that limited itself to the internal developments of scientific thought. The practice and results of science were, in this view, constructed, and not facts merely to be discovered.[16] Credit for this shift also lies with Thomas S. Kuhn's 1962 *The Structure of Scientific Revolutions* – which itself drew considerably on Ludwig Fleck's remarkable if neglected 1936 *Genesis and Development of a Scientific Fact* – that offered a sustained critique of the teleological notion of scientific progression that had characterized the nineteenth century and which carries through until today.[17] While Kuhn's theory of successive and incompatible scientific paradigms in which long periods of normal science were interrupted by crisis and revolutionary science that in turn ushered in new periods of normal science remains evocative and is rhetorically impressive, it was more effective in stimulating discussion around the social transmission of science than in producing disciples.[18] Despite these destabilizing and critical interventions from the 1960s onwards that came to occupy a field at times called science and technology studies, the older teleological understanding of the history of science persisted, albeit in some tension with the former.

An example may help clarify what this tension looks like in scholarship. In a series of exchanges in the 1990s and culminating with a debate in the 2000 volume of *Early Science and Medicine,* the historians of Medieval and Early Modern European science Edward Grant and Andrew Cunningham argued passionately over whether Isaac Newton's claim in the *Principia* that he was engaged in natural philosophy meant that he was

[16] See Jan Golinski, *Making Natural Knowledge: Constructivism and the History of Science.*

[17] Ludwik Fleck, *Genesis and Development of a Scientific Fact.*

[18] For an insightful discussion of the nature and influence of Kuhn's argument, see Ian Hacking, "Introductory Essay," in *The Structure of Scientific Revolutions*, and for Kuhn's own regrets with the relativistic ways some historians and philosophers used his work, see "The Trouble with the Historical Philosophy of Science." My reading of Kuhn has been influenced by Bojana Mladenović, *Kuhn's Legacy: Epistemology, Metaphilosophy, and Pragmatism.*

doing science.[19] Why did this matter? Grant believed strongly in the steady progression of scientific thought from the Medieval European universities to Newton as father of modern science, and then until today.[20] Cunningham, for his part, stressed that precisely because natural philosophy was about explaining the workings of God in the world, and not predicated on establishing natural laws, that the medieval study of natural philosophy up to and including Newton's own work needed to be understood as qualitatively distinct from modern science.[21] While my own scholarly sympathies are with Cunningham and his consistent focus on refraining from describing the writings of scholars with presentist categories, more important here is that Grant and Cunningham's differences derive in great part from the nature of their questions, which are as incommensurable as their conclusions. Grant was invested in exploring the vibrancy of the rational philosophical heritage of the Middle Ages and in stressing the continuities between the scholarship of that period and of the Scientific Revolution. Cunningham, for his part, although he seldom refers to Kuhn explicitly, comes close to positing the types of epistemological ruptures Kuhn memorably termed paradigm shifts in his contextualization of the conceptual worldviews of European scholars of the seventeenth–nineteenth centuries.

Rehearsing the Grant-Cunningham debate in the context of the emergence of modern science in the nineteenth century helps clarify what is meant in this book by science – a term that will more often appear in the plural, and which simply refers to a discrete body of knowledge that can equally refer to the natural as the religious sciences. I will spend some time in Chapter 2 examining taxonomies of knowledge of seventeenth–eighteenth century Moroccan scholars and will expound there on semantic range of the term science, but wish here to emphasize the term's historical contingency.

What is true for science is also true for religion: the word existed before the nineteenth century but it was during that century that building on changes

[19] I previously referred to this argument in "Writing the History of the Natural Sciences," 938.

[20] Along with the references to Grant's work given in the above-cited article, see the long chapter on the teaching of natural philosophy in the Medieval University in Grant's *God and Reason in the Middle Ages*, 148–206. Grant argues here that the institutionalization of natural philosophy in a separate faculty in Medieval European universities that marginalized theological questions laid the groundwork for the later emergence of Modern Science. See his comments on Newton in ibid., 204–05.

[21] Cunningham's views on this question are laid out fully in "Getting the Game Right: Some Plain Words of the Identity and Invention of Science." For his debate with Grant, see the references in Stearns "Writing the History of the Natural Sciences."

beginning in the sixteenth–seventeenth centuries, it acquired the meaning we associate with it today.[22] In the modern period, the concept of religion comes to entail a series of inner beliefs and external actions as well as implying the existence of a plurality of religions – in this it is not dissimilar from the Arabic word *dīn* in Islamic writings in the premodern period.[23] It was during the sixteenth and seventeenth centuries, however, and as a direct result of a concurrent European and (initially) largely Catholic expansion into the so-called New World on the one hand, and Reformation and Counter-Reformation polemics on the other, that religion became a universal category in European thought, one that could be applied to all peoples. Different from contemporary Western understandings of religion, European scholars of these centuries considered religion most legible in the ritual activities of its practitioners.[24] This early modern shift to religion becoming a universal category allowed European thinkers to place the worlds' peoples into an admittedly changing understanding of a universal history. This growing conceptualization of religion as both universal and historically contingent was accentuated in the Iberian context from the fourteenth to seventeenth centuries by the forced conversion of Jews in 1391, the expulsion of Jews and Muslims in 1492 and 1501 respectively, the concurrent discovery of the "New World," and then the wave of expulsions of Moriscos (descendants of converted Muslims, many of whom historians now believe were sincere Catholics) to North Africa between 1609 and 1614.[25]

For Early Modern European scholars, then, religion was first a comparative category, coming into focus at moments of categorization and comparison.[26] Its equation with rituals should be contrasted with the

[22] For much of this discussion I have benefitted greatly from Guy Stroumsa, *A New Science: The Discovery of Religion in the Age of Reason*, and have also drawn on J. Z. Smith's ever useful "Religion, Religious, Religions . . .." in Mark C. Taylor, *Critical Terms for Religious Studies*, and Peter Harrison's eloquent *The Territories of Science and Religion.*

[23] For nineteenth century developments in European thinking on religion, see Tomoko Masuzawa, *The Invention of World Religions: Or, How European Universalism was Preserved in the Language of Pluralism.* A good and for our purposes chronologically relevant overview of the meanings of *dīn* in early modern Muslim scholarship is found in Stefan Reichmuth, "The Arabic Concept of *Dīn* and Islamic Religious Sciences in the 18th Century: The Case of Murtaḍā al-Zabīdī (d. 1791)."

[24] See Stroumsa, *A New Science*, chapter One: "Paradigm Shift: Exploring the World's Religions," especially at 29.

[25] For the ways in which these events relativized religious truth for some, see Stuart B. Schwartz, *All Can Be Saved: Religious Tolerance and Salvation in the Iberian Atlantic World.*

[26] In this comparative sense, religion has more in common with the Arabic *milla,* pl. *milal* in the Islamic tradition, usually used to refer to religious communities. For a brief overview of its use in the heresiographic tradition, see D. Gimaret, "al-Milal wa 'l-niḥal."

interiority of belief, which later came to characterize the concept of religion during the nineteenth century. Within the European context in which this change occurred, it is related to theological debates between Protestants and Catholics during and following the Reformation in the sixteenth century, in which the former critiqued the importance the Catholic Church gave to acts and stressed the radical primacy of the individual's faith in order to attain salvation. The debates continued through the counterreformation into the wars of the seventeenth century and down to the writings of the great social theorists of the nineteenth century mentioned above (Marx, Hegel, Weber).

The point here is not that Enlightenment thinkers in Europe posited the universality of religion for the first time, much less that there was no concept of religion before this period – here I follow Daniel Boyarin's argument that the separation of Christianity from Judaism during Late Antiquity involved conceiving of a multiplicity of religions, including Hellenism.[27] Instead, I am offering the weaker argument that during the nineteenth century in Europe scholars came to understand religion much more than previously as a category of beliefs and attitudes that could be attributed a role in supporting or retarding other values or systems of belief such as the newly emerging category of science. And while this new understanding of religion had certainly had much to do with internal arguments within the Christian tradition, it was easily if not readily applied to other religious traditions, Islam being the one we are primarily concerned with here.

The tradition of Western European scholarship that preceded the transformation of the concepts of science and religion in the long nineteenth century had dealt with Islam in a range of ways.[28] Whereas Christian scholars in Late Antiquity had successfully (to themselves in any case) explained Judaism as a superseded revelation the true import of which had never been fully understood by Jews themselves, the early Muslim community emerged into a monotheistic Middle East that, for Christian writers, had already witnessed the last true prophets.[29] For European Christian observers of Late Antiquity Islam was idolatry, among Catholic scholars it attained the status of heresy in roughly the twelfth century, and for the Protestants of the

[27] See Daniel Boyarin, *Borderlines: The Partition of Judaeo-Christianity*, 202–11.

[28] The scholarship here is substantial. A good place to start is with Suzanne Conklin Akbari, *Idols in the East: European Representations of Islam and the Orient, 1100–1450* and John Tolan, *Saracens: Islam in the Medieval European Imagination*.

[29] For a stimulating overview of the seventh century world of the early Muslim community, see Garth Fowden, *Empire to Commonwealth: Consequences of Monotheism in Late Antiquity* and Fowden, *Before and After Muḥammad: The First Millennium Refocused*. For the indispensable survey and analysis of how others viewed the emergence and early

sixteenth century it served in part as metaphor for the error of Catholicism.[30] During the nineteenth century, with the ascendency of a historicist philology, and the awareness that the existential threat posed by the Ottoman Empire in the sixteenth–seventeenth centuries had been tamed, European scholars of comparative religion began to stress how Islam was essentially an Arab faith and not a full-fledged religion.[31] Among those European scholars who had established their professional credentials through their philological expertise in the languages of the Islamic world – Orientalists – the matter was different. Here, Islam did not represent only a religion (in the newly defined sense encompassing both external acts and professions of faith as well as internal convictions and beliefs) but a way of life, if not a civilization.[32] Indeed, many of the most innovative and creative approaches to the concept of Islam at the beginning of the twenty-first century have involved critiquing the many scholars who posited a singular Islamic civilization or culture in the late nineteenth and early twentieth centuries.[33] As European scholars increasingly described themselves as living in a modern, disenchanted world in which religious and secular spheres were neatly divided, they stressed the absence of this distinction in the Muslim world, where everything was subsumed under the rubric of religion.[34]

Recent trends in the study of Islam as a religious tradition have stressed the diversity of approaches Muslims have taken over time to understanding and practicing their faith. Taking this diversity seriously has involved on the one hand accepting an older anthropological critique of religious studies for privileging elite literate discourses over popular practices when defining orthodoxies, but more pertinently here in a related move it has entailed decentering the privileged place occupied by Islamic law and

expansion of this community, see Robert Hoyland, *Seeing Islam as Others Saw It: A Survey of Christian, Jewish and Zoroastrian Writings on Early Islam*.

30 Such a generalization has value as a heuristic, although it obscures the complex richness of Christian representations of Islam. See for example Tolan, *Saracens*, 51–55, for his discussion of John of Damascus' (d. 749) depiction of Islam as both Christian heresy and idolatry, and compare with Hoyland's discussion of seventh-century Christian authors seeing Islam as a primitive Abrahamic faith in *Seeing Islam as Others Saw It*, 535–38.

31 See Masuzawa, *The Invention of World Religions*, 179, and compare with the discussion in Alexander Bevilacqua, *The Republic of Arabic Letters: Islam and the European Enlightenment*.

32 For one example of such an approach, see G. E. Grunebaum, "The Problem: Unity in Diversity."

33 For two recent, sustained and eloquent critiques of essentializing visions of Islam, see Thomas Bauer, *Die Kultur der Ambiguität: Eine andere Geschichte des Islams* and Shahab Ahmed, *What is Islam? The Importance of Being Islamic*.

34 See chapter 6 of Bauer's *Die Kultur der Ambiguität*, entitled "Die Islamisierung des Islams."

restoring Sufism to a central place in the history of premodern Muslim societies.[35] The view in European and Middle Eastern scholarship that Sufism was a supplement or variant of a variously defined authentic Islam can similarly be dated to the long nineteenth century and complicated interactions between Islamic reformers and European Orientalists.[36] The narrative presented both by recent surveys and more focused studies has shown, instead, that not only was Sufism central to the societies of the premodern Middle East in terms of the institution of the Sufi lodge and practices such as saint visitation, but also that the majority of religious scholars in the postclassical period who wrote on the religious and rational sciences were affiliated in one way or another with Sufism.[37]

The above sketch suggests that the political, economic, and social effects of European colonialism in the nineteenth century, along with Muslim reform efforts from the eighteenth century onwards, and finally the emergence of modern science in nineteenth-century Europe, substantially influenced the historian's ability to evaluate the intellectual landscape of the Muslim Middle East in the preceding centuries. This was not only due to the essentializing tendencies of Western scholars or their own cultural chauvinisms – these certainly played a role – but also due to reform efforts within the Muslim world, with which European Orientalists interacted, and which shaped their own views of the intellectual history of Muslims.[38]

THE REFORMATION OF ISLAM AND THE BREAK WITH THE CLASSICAL TRADITION

Several decades ago, it was common to argue that in the nineteenth and twentieth centuries European scholars (and later their American

[35] The tendency in a colonial context to equate the religion of Islam with a civilization went hand in hand with the equation of an Islamic normativity with Islamic law. See the discussion in Chapter 3, and Léon Buskens and Baudouin Dupret, "The Invention of Islamic Law: A History of Western Studies of Islamic Normativity and Their Spread in the Orient," and David Powers, "Orientalism, Colonialism, and Legal History: The Attack on Muslim Family Endowments in Algeria and India."

[36] For three recent, excellent overviews of Sufism with different approaches, see Alexander Knysh, *Sufism: A New History of Islamic Mysticism*; Arthur F. Buehler, *Recognizing Sufism: Contemplation in the Islamic Tradition*; Nile Green, *Sufism: A Global History.* The emergence of the various Salafi criticisms is discussed in Knysh, *Sufism*, chapter 6.

[37] Along with the above-mentioned surveys, see Nathan Hofer, *The Popularisation of Sufism in Ayyubid and Mamluk Egypt,* especially the book's argument as laid out in its introduction.

[38] See here also the introduction of Abraham Marcus regarding writing the history of Aleppo in the eighteenth century *The Middle East on the Eve of Modernity*, 1–12.

counterparts) developed a narrative of Islamic decline that justified a paternalistic European colonialism and reassured Western audiences of their own inherent superiority. This is, after all, a central argument in Edward Said's influential 1978 *Orientalism*, a book that was instrumental in the development of the field of postcolonial studies and which initiated a sea change in Middle East studies in the United States.[39] Yet *Orientalism* was a polemic (an arguably necessary one), and as such it generalized broadly, suffered at times from conceptual incoherence, and, as Said admitted himself, did not address Middle East history, restricting itself to studying how Europeans and Americans wrote about the region.[40] A corollary to the focus on Western scholarship was a marginalization of Middle Eastern actors and the role that they played in establishing a narrative of decline to forward their own ends. To understand how both Muslim reformers and Orientalists came to advance this narrative at the end of the nineteenth century, Jonathan Wyrtzen's approach in the Moroccan colonial context to talk of a colonial intellectual field in which knowledge production of both foreign and local actors interacted is instructive.[41] The two groups clearly enjoyed differentiated access to power in a variety of colonial contexts, but the role played by Middle Eastern scholars in recasting their own intellectual heritage is vital to understanding how deeply entrenched aspects of the decline narrative became in both Middle Eastern scholarship and scholarship on the Middle East.

The story of modernist reform movements in the Middle East – often glossed in Arabic as *al-nahḍah* – has traditionally been told as an Egyptian one with a Levantine subplot. Egypt's outsized importance in the historiography of the modern Middle East as well as Islamic reformist movements can be linked to its demographic importance as the single largest Arab country in the Middle East, its strategic importance, reflected in Napoleon's misguided and brief occupation between 1798 and 1801, the subsequent reforms it experienced under its largely independent Ottoman governor Muḥammad ʿAlī (d. 1849), and above all to it being the home of al-Azhar, the most influential educational institution in the Sunni Muslim

[39] See Zachary Lockman, *Contending Visions of the Middle East: The History and Politics of Orientalism* (2nd ed.), chapter 6 "Said's *Orientalism*: a book and its aftermath".

[40] For a comprehensive study of what has been written about *Orientalism*, and a critique of Said's approach to his sources, see Daniel Martin Varisco, *Reading Orientalism: Said and the Unsaid*.

[41] See Jonathan Wyrtzen, *Making Morocco: Colonial Intervention and the Politics of Identity*, introduction.

world. So, to sketch the depiction of premodern Islamic intellectual history that emerged out of Islamic modernist thought – an essential task for the untelling of this narrative of decline – a detour through developments in the eastern Mediterranean is required.

The nineteenth century was of central importance to an array of reform projects in Egypt and the Ottoman Empire, most famously perhaps the administrative, military and educational reforms in the Ottoman Empire glossed as the Tanzimat Reforms (1839–76), and in Egypt Mehmet Ali's efforts to reform the economy, create a modern army, and a secular educational system during the first decades of the nineteenth century. When it comes to intellectual trends in Arab Muslim thought, Western scholars often refer metonymically to the trio of Jamāl al-Dīn al-Afghānī (d. 1897), his student Muḥammad ʿAbduh (d. 1905), and the latter's student Rashīd Riḍa (d. 1935) as representative of a reformist wave that decentered classical centers of religious education such as al-Azhar and broadened popular participation in redefining the significance of religious traditions.[42] These three scholars self-consciously drew on the Protestant narrative of European modernity – al-Afghānī called explicitly for an Islamic Reformation, and both he and ʿAbduh adopted the Protestant narrative that religious conservatism had distorted their respective tradition's early dynamism.[43] With ʿAbduh's student Riḍa the reform movement shifted from a liberal to a more neotraditional bent, but the rupture with the classical scholarly system was complete, due to ʿAbduh's emphasis on juridical reinterpretation, his efforts to reorganize al-Azhar and the overall use by the reformers of newspapers to promote their religious views. Islam had become modern.

There is much to take issue with the above narrative, of course, not to mention that there are other equally simplistic narratives, such as the one in which intellectual reform in the Arab Middle East in the nineteenth century began with its Christian minority, itself influenced by European and North American Protestant missionaries.[44] Recent scholarship, including notably

[42] The classic source being Albert Hourani, *Arabic Thought in the Liberal Age 1798–1939*, but see now Jans Hanssen and Max Weiss (eds.), *Arabic Thought beyond the Liberal Age: Towards an Intellectual History of the Nahda*.

[43] For al-Afghānī's explicit comparisons of Islamic reform with Protestantism and his implicit comparisons of himself with Martin Luther, see Gesink, *Islamic Reform*, 72–73. Compare with Nikki R. Keddie, *An Islamic Response to Imperialism: Political and Religious Writings of Sayyid Jamāl ad-Dīn al-Afghānī*, 171–72. For the impression that ʿAbduh undertook the "Protestantization of Islam," see Gesink, *Islamic Reform*, 228.

[44] This is, also the classic narrative for the beginnings of Arab nationalism, as found in George Antonius, *The Arab Awakening*. See chapter 3 ("The Start: 1847–68") for the importance of French and American missionaries in Beirut in sowing the seeds of Arab nationalism.

the work of Indira Gesink and Marwa Elshakry, has offered a valuable corrective to such diffusionist narratives where Europeans and Americans imported various aspects of modernity into the Arab Middle East.[45] Instead, this scholarship either depicts modernity coming into being at the encounter of European nations with their future colonies, or posits the existence of multiple modernities occurring in parallel within an increasingly globalized world.[46] When it comes to the Egyptian case, Gesink has provided us with a nuanced rereading of the reform aspirations of Muḥammad ʿAbduh, demonstrating that the utilitarian reforms pushed through the Azhar at the end of the nineteenth and the beginning of the twentieth centuries were due more to the conservative forces than to ʿAbduh and other liberal reformers.[47] The latter, an outsider who was viewed with suspicion by scholars representing the Azhar establishment, also did himself few favors by his close relationship with leading British colonial figures (despite his own anti-colonial views) and following his trips to Europe was subjected to a virulent smear campaign in the Egyptian press. When ʿAbduh died of liver cancer in 1905 he became a martyr for the liberal reformist cause, but the actual reforms that would go on to modernize al-Azhar were carried out by the conservative establishment that was intensely wary of ʿAbduh's perceived infatuation with European thinkers such as the educational reformer Herbert Spencer, and who had also internalized the need to reform the Azhar along utilitarian lines to meet the challenges of the times. What ʿAbduh himself had been aiming to accomplish with his championing of intellectual exertion (*ijtihād*) has similarly been largely misunderstood as his advocating for nonspecialists to play a role in reinterpreting Islamic law, when he was actually calling for a broad social reengagement with reforming Egyptian society.[48] Gesink's reading of the reform movement at al-Azhar argues that the vision of the intellectual history of the Middle East presented by ʿAbduh in his writings, and seized on by a generation of European

Tellingly, Mehmet ʿAli's efforts, which Antonius described in chapter 2, were given the title, "A False Start."

45 Indira Gesink, *Islamic Reform*; Marwa Elshakry, *Reading Darwin in Arabic, 1860–1950.*

46 For an overview of the misguided quest for the origins of Islamic reform movements in the nineteenth century and limitations of the Afghani/ʿAbduh/Riḍa triumvirate, see the exhaustive and insightful survey of Catherine Mayeur-Jaouen, "'À la poursuite de la réforme': Renouveaux et débats historiographiques de l'histoire religeuse et intellectuelle de l'islam, XVe–XXIe siècle." Much of the most innovative work when it comes to the intellectual significance of modernity in the Arab world has been carried out by scholars of literature. See Tarek El-Ariss, *Trials of Arab Modernity: Literary Affects and the New Political* and Maya Kesrouany, *Prophetic Translation: The Promise of European Literature in the Egyptian Imaginary.*

47 Gesink, *Islamic Reform*, chapter 9. 48 Ibid., chapter 8.

Orientalists, was at best an incomplete rendering of nineteenth-century Muslim reform, and at worst it contributed to a substantial misrepresentation of the vibrancy and creativity of premodern Islamic thought.[49] This representation continued and shifted in the thought of ʿAbduh's most famous student and collaborator, Rashīd Riḍa.

Henri Lauzière's careful study of the origins and significance of the term Salafism (*al-salafiyya*) has helped nuance our understanding of Riḍa's role in the decades following ʿAbduh's death in coordinating and shaping intellectual and political reform in the Middle East.[50] Relevant here is Lauzière's thoughtful tracing of how a series of careless translations and lazy historical narratives of European Orientalists – beginning with Louis Massignon (d. 1962) in the early 1920s – created a narrative of reformers harkening back to an early pure Islam that was initially linked to the efforts of Afghani and ʿAbduh and which was then through the efforts of Riḍa linked to the Hanbali school of Wahhabism that appeared in Arabia in the eighteenth century and which formed the intellectual basis for the third Saudi state at the beginning of the twentieth century.[51] Not only was this narrative incorrect – one main argument of Lauzière's book is that Salafism as a coherent school did not emerge until the middle of the twentieth century – it also simplifies the diverse intellectual landscape of the Middle East in the late nineteenth and early twentieth centuries. Yet, despite the shared confusion between Western and Middle Eastern scholars about the genealogies of various Islamic reform movements, Lauzière's analysis confirms that the increased importance of Salafi reformist thought throughout the Muslim world during decolonization and into the postcolonial period went hand in hand with a critique of certain Sufi practices as

[49] There is much more to say regarding how individual Orientalists drew on the arguments presented by Egyptian and Syrian reformers in the nineteenth century. For a particularly important case, Lawrence Conrad, "The Pilgrim from Pest: Goldziher's Study Tour to the Near East (1873–74)." Conrad shows how Goldziher's views of Islam and the Middle East profoundly shaped by his own participation in the *Haskala* Jewish reform movement in Hungary and Germany as well as his conversations with both al-Afghānī and Muḥammad al-ʿAbbāsī, who was both Grand Mufti of Egypt and Rector of al-Azhar in 1873 (see also Conrad, "The Dervish's Disciple: On the Personality and Intellectual Milieu of the Young Ignaz Goldziher"). I have previously commented on Goldziher's, at times, incoherent representation of Islamic intellectual history in "Writing the History of the Natural Sciences," 926–28.

[50] See Henri Lauzière, *The Making of Salafism: Islamic Reform in the Twentieth Century*, especially the introduction and chapter 3. Lauzière argues that Riḍa's support of Saudi Arabia and Wahhabism during the 1920s was closely tied to his belief that King ʿAbd al-ʿAzīz was the strongest Arab leader of the time and the best positioned to resist colonial powers and to help other Arab countries to do the same.

[51] On the influence of Massignon, see Lauzière, *The Making of Salafism*, 37–40.

well as much of the post-formative Islamic intellectual tradition.[52] This critique, which in scope addressed much of the premodern Islamic intellectual production, should not be understood to have entailed a rejection of all of it, much as Afghani and ʿAbduh, while criticizing numerous popular Sufi practices, did not reject Sufism itself.[53] Still, in conjunction with Orientalist depictions of intellectual stasis during the post-Abbasid period, these Salafi narratives have been effective in distracting attention from the period preceding nineteenth-century reforms.[54]

THE MOROCCAN CONTEXT

The narrative of Islamic modernism sketched out here, and its account of premodern Islamic intellectual history had a significant impact in the Moroccan context, where, however, political and social circumstances differed markedly from the Egyptian and Levantine landscape. Unlike Egypt, where the British intervened militarily in 1882 to consolidate their economic and political influence, or Algeria, which the French had incorporated into France from the 1830s onward at the cost of hundreds of thousands of Algerian lives, Moroccan political and intellectual elites were able to entertain hopes of continued independence through the end of the nineteenth century. This hope continued despite Morocco's military defeats by France in 1844,

[52] See Lauzière, *The Making of Salafism*, 51–58, 119. For an important critique of Lauzière's argument, see Frank Griffel, "What Do We Mean by 'Salafi'? Connecting Muḥammad ʿAbduh with Egypt's Nūr Party in Islam's Contemporary Intellectual History." Unlike Lauzière, who argues that a coherent Salafi movement didn't emerge until the second half of the twentieth century and that Orientalist observers in the 1920–30s confused two intellectual genealogies with each other, Griffel believes that there was general confusion among Muslim scholars as well as Western ones regarding the meaning of the term Salafi, and that it is not possible to untangle two (or more) genealogies cleanly. Like Gesink, whose work he does not engage with, Griffel thus links the liberal reformers of the late nineteenth century such as ʿAbduh, with more conservative rejections of the authority of established legal and theological schools in the late twentieth century. Lauzière provided a persuasive rebuttal to Griffel's arguments in "What We Mean Versus What They Meant by 'Salafi': A Reply to Frank Griffel."

[53] See, for example, Oliver Scharbrodt, "The Salafiyya and Sufism: Muḥammad ʿAbduh and his *Risālat al-Wāridāt* (Treatise on Mystical Interpretations)."

[54] The same was not true of the various reform movements of the eighteenth century, which, with the exception of the writings of Ibn ʿAbd al-Wahhāb, largely did not have the same dismissive attitudes towards the intellectual contributions of the preceding centuries. For an overview, see Ahmad Dallal, "The Origins and Early Development of Islamic Reform"; for an additional view on the importance of Rashid Riḍa in evaluating the intellectual contributions of earlier reformers, in this case, the Yemeni al-Shawkānī, see Dallal, "Appropriating the Past: Twentieth Century Reconstruction of Pre-Modern Islamic Thought." See also the survey of Mayeur-Jaouen, "À la poursuite de la réforme," 342–47.

by Spain in 1859–60, and the ongoing encroachment of both European powers on Moroccan territory.[55] It was only with the substantial erosion of Moroccan sovereignty that took place through the Moroccan crown's increased indebtment to France at the end of the nineteenth century and France's growing influence at the beginning of the twentieth century in part through its military advisors at the court of the Moroccan ruler ʿAbd al-ʿAzīz (rl. [1894–1908]), that Islamic modernism found a positive reception through the influence of Moroccans such as Abū Shuʿayb al-Dukkālī (d. 1937) who had studied in Egypt.[56] The story of the competing strands of Moroccan political thought at this time, and how Moroccan nationalism emerged out of Islamic modernism, competing Sherifian claims, and the social networks supplied by Sufi orders is a fascinating one and has been told in detail by a number of scholars.[57] More pertinent here is that a number of the central figures in Morocco who struggled to articulate a way forward for the country in face of European power, from the prominent Sufi scholar and rebel Muḥammad b. ʿAbd al-Kabīr al-Kattānī (d. 1909) to the Islamic modernist and nationalist ʿAllāl al-Fāsī (d. 1974) argued for the importance of renewing Islam and correcting the errors of the past.[58] For ʿAllāl al-Fāsī, who founded the Independence (*Istiqlāl*) Party, which played a central role in Morocco during its first years of independence, this process of renewal involved stripping Islam of fancies and accretions and returning to the pure faith of the forefathers, while also studying the modern sciences in European languages to strengthen the Moroccan nation.[59] While defending the

[55] Morocco's relationship to Europe, and its internal politics during the nineteenth century has been the subject of a number of excellent studies. See Edmund Burke, *Prelude to Protectorate in Morocco: Precolonial Protest and Resistance, 1860–1912* and Eric Calderwood, *Colonial al-Andalus: Spain and the Making of Modern Moroccan Culture*, but also Jean-Louis Miège, *Le Maroc et l'Europe (1830–1894)*.

[56] See Ann Wainscott, "Islamic Modernism, Political Reform and the Arabisation of Education: The Relationship between Moroccan Nationalists and al-Azhar University," 158. See also Mohamed El Mansour, "Salafis and Modernists in the Moroccan Nationalist Movement," and Edmund Burke, "Pan-Islam and Moroccan Resistance to French Colonial Penetration, 1900–1912."

[57] In addition to the sources mentioned in the previous footnotes, see Sahar Bazzaz, *Forgotten Saints: History, Power and Politics in the Making of Modern Morocco* and Emilio Spadola, *The Calls of Islam: Sufis, Islamists, and Mass Mediation in Urban Morocco*.

[58] For the context surrounding al-Kattānī's career and ignominious death at the hand of the Moroccan Sultan ʿAbd al-Hafīẓ (rl. 1908–12) see Sahar Bazzaz, *Forgotten Saints*. For al-Kattānī's views of the importance of renewing the proper practice of Islam and Sufism in order to strengthen the Muslim community in the face of European aggression, see the writings collected in Muḥammad al-Kattānī, *Min rasāʾil al-Imām Muḥammad b. ʿAbd al-Kabīr al-Kattānī fī-l-adāb wa-l-sulūk*, especially 48, 51–53, 66, 103–05.

[59] See the writings collected in ʿAllāl al-Fāsī, *Aḥādīth fī l-falsafa wa-l-tārīkh wa-l-ijtimāʿ*, especially the article printed in 1934 entitled "Africa's Youth: Their Current State and

inherently Islamic nature of Sufism from Orientalist allegations of its Christian origins, al-Fāsī was also at pains to stress the importance of a true Sufism, and like other Islamic modernist thinkers of his time, he looked back to the eighth/fourteenth century iconoclastic thinker Ibn Taymiyya (d. 728/1328) to do so.[60] The intellectual history of Morocco during the centuries preceding the French colonial presence played a marginal role in his thinking, where the needs of the Moroccan nation would be best met by recovering and practicing a pure form of Islam in conjunction with acquiring the modern sciences. For al-Fāsī, as for Muslim scholars in the Middle East in the late nineteenth and early twentieth century, the social, intellectual, and national benefits of the modern sciences were clear.

Scholars of the history of education in Morocco during the twentieth century have in this context argued persuasively that what French and Moroccan administrators considered to be traditional Moroccan education during the protectorate and post-independence eras was decisively influenced by the French colonial presence.[61] Faced with the overlapping imperatives of undermining local and foreign Salafi intellectuals, needing to build up the religious legitimacy of the Moroccan regime that they claimed to be helping develop a modern government, and wishing to control the creation of a class of modern Moroccan intellectuals, French colonial officials played an important role in crafting a vision of traditional Moroccan Islamic education. Following their occupation of Morocco in 1912 under Lyautey (d. 1934), French administrators worked both to support and revive a form of traditional Islamic education in the Qarawiyyīn in Fez a process that involved them carefully curating and determining what that tradition entailed and pruning it of the mathematical and natural sciences as well as history.[62] This curriculum, closely tied with their curation of Fez's architectural monuments to preserve what French scholars and administrators saw as pure Moroccan heritage, ran under the name of "Traditional Education System" between 1933 and 1959 and determined the content of what the Moroccan government restarted in 1988.[63] As with other aspects of Islamic intellectual production – Islamic

What They Should Become" (*al-shabāb al-ifrīqī: ḥālatuhu wa kayf yajib an yakūn*), 165–82, at 171, 177–79.

[60] See ʿAllāl al-Fāsī, *Al-Taṣṣawuf al-Islāmī fī-l-Maghrib*, 6–10.

[61] I have drawn here on Geoffrey Porter, "At the Pillar's Base: Islam, Morocco, and Education in the Qarawiyyīn Mosque, 1912–2000." Porter lays out his overall argument in brief in his conclusion (357–79). For a discussion of education in Morocco in general during the protectorate period, see Spencer D. Segalla, *The Moroccan Soul: French Education, Colonial Ethnology, and Muslim Resistance, 1912–1956*.

[62] Porter, "At the Pillar's Base," 114–16. [63] Ibid., chapter 2.

jurisprudence being the example par excellence – colonial and postcolonial Muslim understandings of Islamic intellectual history were decisively influenced by colonial interventions.[64] For our purposes, what is most significant regarding Porter's analysis of the French influence on Islamic education in colonial and postcolonial Morocco is that it shows how this influence hardened differentiations between religious and nonreligious sciences and projected a modern understanding of "traditional Islam" into the premodern period.

SCIENCE IN THE MIDDLE EAST IN THE NINETEENTH CENTURY

Within the broader discussions of Islamic reform in the nineteenth century, science has played an important role, as it was a tool for Protestant missionaries in Lebanon who used it to attract both Maronite Christian and Muslim students to the schools they opened in Beirut.[65] Using "the Gospel of Science" at the Syrian Protestant College (founded in 1866, it was renamed the American University of Beirut in 1920), American professors attempted, with various degrees of enthusiasm, to use the promise of Western technological advances to impart a narrative of spiritual superiority.[66] The success in conversion was limited, and more ironically, as seen most dramatically in the College's leadership to fire one of its professors, Edwin Lewis, for lecturing on the evolutionary theory of Charles Darwin in 1882, the year of the latter's death, it struggled with balancing its own theological and scientific priorities.[67] This struggle and its outcome disillusioned two of its local instructors, Yaʿqub Sarruf (d. 1927) and Faris Nimr (d. 1951), who had frequently discussed Darwin's ideas in *al-Muqtaṭaf*, an Arabic journal that they had founded in 1876 that was dedicated to popularizing trends and discoveries in the natural sciences. Dismayed by the outcome of the "Lewis affair," Sarruf and Nimr moved to Cairo, taking their journal with them. This incident, along with the broader mixed Arab Christian and Muslim reception of Darwinian thought in the late nineteenth and early twentieth centuries,

[64] For the example of French and British influence on determining the content and practice of Islamic jurisprudence see the already cited study of David Powers, "Orientalism, Colonialism, and Legal History."

[65] For the following see Adel A. Ziadat, *Western Science in the Arab World: The Impact of Darwinism, 1860–1930* and Marwa Elshakry, *Reading Darwin in Arabic*. Elshakry's book covers much the same ground as Ziadat, albeit in much more detail and with a greater awareness of contemporary debates on the globalization of science.

[66] Elshakry, *Reading Darwin in Arabic*, chapter 1. [67] Ibid., 65–72.

emphasizes the ways in which Middle Eastern scholars experienced and contributed to modernization through the discourse of science.[68] A number of these authors stressed the compatibility of Islam correctly understood with recent European scientific discoveries, and in a fashion not dissimilar to Muslim reformers such as al-Afghani and ʿAbduh (who themselves had stressed how as a religion Islam was inherently rational) linked their openness to scientific developments to the intellectual achievements of a past age of Islamic intellectual glory.[69] Talking about modern science was to talk about reform, European knowledge production, colonialism, and, in some fashion, about modernity.

Modern science played a similar role as an index of social and intellectual attitudes in Istanbul, at the heart of the Ottoman Empire, during the last decades of the nineteenth century when debates regarding political and intellectual reform were intensifying.[70] Here too, the narratives of reformers and Orientalists overlapped and drew upon each other, as the modernist aspirations of the reform movement harkened back to a Golden Age of Islamic scientific achievement as support for their own modernization agenda.[71] Within a community of Ottoman intellectuals struggling with defining their own identity along religious and ethnic lines – Islamic Empire or Turkish Nation – the attitude toward a legacy of Arabic science was more complicated than in Egypt.[72] In both cases, however, science remained a symbol of modernity, the danger of European political and intellectual superiority, and the memory of past glory in need of recovery. The relay-race narrative of the history of science remained intact as did the implicit intellectual sterility of the premodern Muslim world.

*

[68] For a diverse number of contemporaneous Muslim responses to Darwinism, see Ziadat, *Western Science*, 82–122.

[69] For one particularly striking case, consider the writings of the Lebanese Shiʿa scholar Hussein al-Jisr, who argued for the compatibility of Islam and Science while also leveling a harsh critique against materialism. Al-Jisr was praised by al-Afghani and was one of Riḍa's teachers, although the two would later disagree bitterly over the proper role of the ʿulama. See Ziadat, *Western Science*, 91–95; Elshakry, *Reading Darwin in Arabic*, 132–41 and 158–59, and compare with M. Alper Yalcınkaya, *Learned Patriots: Debating Science, State, and Society in the Nineteenth-Century Ottoman Empire*, 155–56.

[70] See Yalcınkaya, *Learned Patriots: Debating Science, State, and Society in the Nineteenth-Century Ottoman Empire*.

[71] See ibid., 15, 17–18. It is not surprising that Yalcınkaya drew productively on Steven Shapin's discussion of the importance of social status in seventeenth century England for the production of scientific truth (see Shapin, *A Social History of Truth*).

[72] See Yalcınkaya, *Learned Patriots*, chapter 5.

But this book is about a different story, one that can be told more easily once the fault lines in the narrative of decline and decadence have been uncovered. It is one of the significances of the natural sciences in one corner of the premodern Islamic world in the long seventeenth century. This is a story of marginal importance within the broader historiographical landscape, and yet it speaks to the intellectual dynamism of the scholarship of the age it examines and aims to add to our understanding of the complex ways in which different bodies of knowledge were related to each other. The narrative here draws on the work of a generation of scholars fascinated by science in all its guises and by the ways in which modernity has distorted our understandings of our pasts. To grasp what it does offer the reader, we need to set aside the narratives presented above and turn to what we currently know regarding the intellectual history of the premodern Muslim Middle East during the very centuries that Europe experienced the developments commonly glossed as the Scientific Revolution and the Enlightenment and often equated in the European context with Early Modernity.

THE INTELLECTUAL LANDSCAPE OF THE MUSLIM MIDDLE EAST IN AN AGE OF EMPIRES

It is curious, and a testimony to the power of the historical narratives sketched out above, that the intellectual history of the Muslim world in the centuries immediately prior to European colonialism would have been comparatively neglected for so long. This was, after all, a period in which much of the Muslim world enjoyed great political unity, demographic growth, and economic productivity – all due in part to the stability offered by three great empires, Ottoman, Safavid, and Mughal, that united the Muslim world from Algeria to Bangladesh between the sixteenth and eighteenth centuries and beyond.[73] To be sure, beginning in the seventeenth century the first of these three empires would suffer a series of military defeats in Europe, and by the eighteenth century the latter two would collapse.[74] But more pertinent to the question of the period's

[73] For a survey of the diverse economic strengths of these three empires during the sixteenth century, see Stephen Dale, *The Muslim Empires of the Ottomans, Safavids, and Mughals*, chapter 4, and compare with Baki Tezcan, *The Second Ottoman Empire* and Nelly Hanna, *Ottoman Egypt and the Emergence of the Modern World 1500–1800*.

[74] The seventeenth century, which witnessed the decline of the Safavid Empire, was a difficult one globally. See Rudi Matthee, *Persia in Crisis: Safavid Decline and the Fall of Isfahan* and compare with Geoffrey Parker, *Global Crisis: War, Climate Change, & Catastrophe in the Seventeenth Century*.

intellectual history in European scholarship is that its production did not register among European scholars: Renaissance translations of Arabic texts focused almost entirely on the known medieval authors, and when European visitors did visit these empires, they seldom interacted with indigenous scholarly networks and when they did, they played down the achievements of local scholars in order to accentuate their own.[75] Furthermore, the areas in which the vast majority of the local scholars of these empires were interested – for the most part the religious sciences – were of little interest to European travelers, who had their own religious agendas.[76]

The inability to parse internal developments within Islamic jurisprudence, theology and Sufism, not to mention literature, facilitated arguments that played up the negative effects of religious movements such as the Ottoman Kadizadelis, whose attack in the first half of the seventeenth century on what they considered to be innovations in spiritual practice gave some the impression of a static Islamic orthodoxy that rejected new developments. Recent scholarship has both disputed this understanding of the Kadizadelis, and has offered a fuller account of their intellectual context, as well as problematizing the use of "orthodoxy" as an explanatory concept.[77] But the most important revisions of this period's intellectual history have been offered by three recent projects that contain sufficient detail and historiographical sophistication to genuinely revolutionize the field's understanding of the intellectual production of this period. *Revealed Sciences* draws on all three to varying degrees. The first is found in a series of articles and two monographs by Khaled El-Rouayheb, in which he first explored the vibrant nature of Arabic logic as a field into the nineteenth century, and subsequently took up new

[75] See the list of translated authors in Hasse, *Success and Suppression: Arabic Sciences and Philosophy in the Renaissance*, 318–19, all of whom were from the Early or High Middle Ages; on European travelers to the Ottoman and Safavid Empires, see the studies gathered in Sonja Brentjes, *Travellers from Europe in the Ottoman and Safavid Empires, 16th–17th Centuries: Seeking, Transforming, Discarding Knowledge*, especially "Early Modern Western European Travellers in the Middle East and their Reports about the Sciences" [2004], and "Pride and Prejudice: The Invention of a 'Historiography of Science' in the Ottoman and Safavid Empires by European Travellers and Writers in the Sixteenth and Seventeenth Centuries" [2005].

[76] See Brentjes, "Early Modern Western European Travellers in the Middle East."

[77] Compare Madeline Zilfi, "The Kadizadelis: Discordant Revivalism in Seventeenth-Century Istanbul," with the valuable overview of the field of Ottoman intellectual history between the sixteenth and eighteenth centuries in Marlene Kurz, *Ways to Heaven, Gates to Hell: Fażlīzāde ʿAlī's Struggle with the Diversity of Ottoman Islam*, 9–17, and Khaled El-Rouayheb, *Islamic Intellectual History in the Seventeenth Century*, 14–18 and 190–92.

developments in dialectics, scholarly methodology, Islamic theology and Sufism in the long seventeenth century.[78] El-Rouayheb demonstrated that during the seventeenth century the central Ottoman Empire and its Arab territories experienced an intellectual revival through the influx of three groups previously marginal to it: Kurdish scholars drawing on the broader intellectual legacy of the Timurids in the rational sciences who promoted the practice of "independent logical demonstration" (*taḥqīq*), North African scholars who excelled in logic and who built on the Algerian theologian al-Sanūsī (d. 895/1490)'s strident opposition to blind imitation (*taqlīd*) in matters of belief, and South Asian Sufis who championed Ibn ʿArabī's (d. 638/1240) often controversial theory of the unity of existence (*waḥdat al-wujūd*).[79] This revival and the intellectual richness that accompanied it, was qualitatively different from previously studied eighteenth-century reform efforts of scholars as disparate as al-Zabīdī (d. 1205/1791) and al-Shawkānī (d. 1250/1834), which were centered on reviving the study of Prophetic tradition and which were critical of the rationalist sciences that form the heart of El-Rouayheb's study.[80] Further, El-Rouayheb argued for the rise during this period of "deep reading" (*ādāb al-muṭālaʿa*), in which learning happened directly through books, and independently of the teacher-student relationship that had been at the heart of education in the Islamic world both before and after the emergence of the *madrasa* in the eleventh century.[81] This is a significant insight into postclassical trends in the transmission of knowledge in Muslim societies, although its overall applicability will need to be confirmed by future research.

[78] See Khaled El-Rouayheb, *Relational Syllogisms and the History of Arabic Logic, 900–1900* and ibid., *Islamic Intellectual History.* Compare with Asad Ahmed, "The *Sullam al-ʿulūm* of Muḥibballāh al-Bihārī: A Milestone in Arabo-Islamic Logic."

[79] This summary is insufficient. For a fuller summary of the book and its importance, see my review of *Islamic Intellectual History* in the *Journal of the American Oriental Society* 137 (2017), 437–40.

[80] Note, however, El-Rouayheb's point that studies of Prophetic Tradition in the Ḥijāz were revived already in the seventeenth century (*Islamic Intellectual History,* 164).

[81] El-Rouayheb explores this argument at greater length in "The Rise of 'Deep Reading' in Early Modern Ottoman Scholarly Culture." On the nature of instruction in and outside of the madrasa, see the discussion and cited literature in Michael Chamberlain, *Knowledge and Social Practice in Medieval Damascus, 1190–1350*, especially chapter 2, and compare with Sonja Brentjes, *Teaching and Learning the Sciences in Islamicate Societies (800–1700).* In this context, the example of the Damascene scholar and Sufi ʿAbd al-Ghānī al-Nābulusī (d. 1731) is striking for both being known for his studies in isolation and for having written a treatise praising the practice (Samer Akkach, *'Abd al-Ghani al-Nabulusi: Islam and the Enlightenment*, 34–39). But see also Houari Touati, *Entre Dieu et les Homees: Lettrés, saints et sorciers au Maghreb (17ᵉ siècle)*, 34–38, for a more restrained view of the independence of the book in the seventeenth century.

Overall, El-Rouayheb's work established criteria by which progress and developments in the fields he examined could be measured and the master narratives he offered do an enormous amount to push back against prevalent narratives of decline. As he notes in both the introduction and conclusion of his survey of Islamic intellectual history in the seventeenth century, the narratives of decline that he has attempted to dispel are shared by both past generations of Orientalists and Middle East scholars, although one suspects that the intellectual work that is needed to create more nuanced intellectual histories will differ in distinct cultural and social contexts.[82] Unlike El-Rouayheb's recent book, *Revealed Sciences* offers less a series of master narratives or debates that can be used to chart the principal developments in specific sciences during the seventeenth century than a close reading of the ways in which the natural sciences were interwoven, institutionally and intellectually, into the scholarly landscapes of Morocco during this period.

The second scholar, whose work parallels my attempt in *Revealed Sciences* to recover neglected aspects of the premodern intellectual history of the Islamic world, is Matthew Melvin-Koushki, whose argument for the centrality of esoteric thought in postclassical thought in the Mughal, Safavid, and Ottoman Empires challenges many of the field's core assumptions. Melvin-Koushki's work was initially situated around the figure and writings of the occult philosopher Ṣāʾin al-Dīn Ibn Turka (d. 835/1432), but his prodigious output consists of a frontal assault on the intellectual history of the Islamic post-formative period, in which he argues that the esoteric and occult occupied a much more important position than previously acknowledged, one that was much closer to the one it occupied in Europe.[83] In doing so, he argues for a decolonialization of premodern Islamic intellectual history that is no longer filtered through the disenchanted criteria of a modernist historiography that mines the premodern for its precursors.[84] While acknowledging opponents of occult disciplines such as lettrism – a prime example here would be his decisive critique of

[82] See El-Rouayheb, *Islamic Intellectual History*, 1–3 and 360–61.

[83] Melvin-Koushki has two forthcoming monographs with Brill, *The Lettrist Treatises of Ibn Turka: Reading and Writing the Cosmos in the Timurid Renaissance* and *The Occult Science of Empire in Aqquyunlu-Safavid Iran: Two Shirazi Lettrists*, a third volume on the life of Ibn Turka and over ten articles or book chapters dealing with related matters. As such, my summary of his scholarly intervention is necessarily abbreviated and partial.

[84] He makes this point most clearly in two review essays, where he takes up his attack on what he calls the "science-religion-magic" triad: "(De)colonizing Early Modern Occult Philosophy"; ibid., "*Taḥqīq* vs. *Taqlīd* in the Renaissances of Western Early Modernity." Melvin-Koushki's refusal to separate the intellectual history of Christian Europe and the

the famed North African historian Ibn Khaldūn's (d. 808/1406) attack on esotericism – he argues that their importance and representativeness have been exaggerated, significantly distorting our understanding of the premodern Islamic world.[85] The occult is not central to this book, and despite its reputation in the premodern period as a land of magicians, North Africa is at the margins of Melvin-Koushki's work, which is principally engaged with the scholars of the Ottoman, Safavid, and Mughal Empires that claimed a Timurid legacy.[86] Yet, like Melvin-Koushki, my work is based on the assumption that our understanding of the intellectual life of the premodern Islamic period – and specifically of the nature and significance of the natural sciences therein – has been profoundly distorted by the reform-oriented modernizing discourses of the nineteenth century.

The third intellectual project that has influenced my thinking profoundly in writing the book is that of Sonja Brentjes in her work on the teaching and studying of the natural and mathematical sciences in Muslim societies, work that has recently culminated in her impressive survey *Teaching and Learning the Sciences in Islamicate Societies (800–1700)*. Brentjes' wide-ranging historical and historiographical writings, largely focused on Egypt, Iran and the Levant, have shown that the natural and mathematical sciences were part of institutional learning in Muslim societies in the post-formative period of Islamic intellectual history. In doing so, she has provided a powerful and productive counter to the influential observations of George Makdisi in his foundational work on the *madrasa* or college that the philosophical sciences were not taught in institutional settings.[87] Yet, she has also drawn attention to the importance of tracing the diffusion of the sciences in Muslim societies through genres including introductory manuals, mnemonic poems, and the rich commentary literature. In this regard, her work has contributed significantly to a framework for a social as opposed to a purely intellectual history of the sciences.

Muslim Middle East is especially evident in "Afterword: Conjuncting Astrology and Lettrism, Islam and Judaism."

85 See Melvin-Koushki, "In Defense of Geomancy: Sharaf al-Dīn Yazdī Rebuts ibn Khaldūn's Critique of the Occult Sciences."

86 A useful survey of the importance of Timur (d. 1405) for the subsequent Ottoman, Safavid, and Mughal Empires is given by Stephen Dale in chapter 2 of his *The Muslim Empires of the Ottomans, Safavids, and Mughals*, but compare with the introduction and second chapter of A. Afzar Moin, *The Millennial Sovereign: Sacred Kinship & Sainthood in Islam*, whose argument that Timur provided a valuable millennial precedent for later rulers such as the Mughal ruler Akbar (d. 1605) places astrology at the heart of both the political and intellectual life of the time.

87 See George Makdisi, *The Rise of Colleges: Institutions of Learning in Islam and the West*, 10, 75–76.

Revealed Sciences draws on Brentjes' contributions to investigate both the broader institutional settings of the sciences in seventeenth-century Morocco as well as their diffusion through a broad number of genres.

REVEALED SCIENCES

The title of this book comes from a key passage in a work on epistemology by the seventeenth-century Moroccan scholar al-Yūsī (d. 1102/1691), in which he disputes the taxonomy of knowledge of the fourteenth-century Granadan jurist Ibn al-Juzayy (d. 741/1340).[88] The passage is found in a section that is fascinating primarily for its elucidation of al-Yūsī's blurring of what was otherwise, even in his writings, a clear differentiation between those sciences that had a religious origin and those that did not. Al-Yūsī's initial point engages with Ibn al-Juzayy's breakdown of knowledge into three types: religiously sanctioned sciences, those sciences that are subsidiary to these sciences and facilitate them, and those sciences that are not religiously sanctioned. Al-Yūsī refutes the validity of this breakdown by challenging the validity of the distinction between the second and third categories. Ibn al-Juzayy's second category was to contain sciences that are not explicitly mentioned in revealed texts, but which are integral to religious sciences such as jurisprudence: these, in Ibn al-Juzayy's view, should also be considered religiously sanctioned or revealed (*sharʿī*) by virtue of their supplementary role. Before parsing the best translation for this last Arabic term, let us turn to al-Yūsī's own words, where he moves beyond the starting point to consider a more functional description of all knowledge that benefits the Muslim community having been revealed. He begins by raising the question of the scope of the term *sharʿī:*

> It can then be objected to Ibn al-Juzayy that that it is not right for one to intend with the use of religiously sanctioned in this context (*annahu lā yaṣiḥḥ an yurīd bi-l-sharʿī fī hādhā al-bāb*) only what has acquired this name from Revelation, according to the technical meaning of this phrase for the legal theorists. There remains nothing but to intend by it either what has been explicitly legally permitted by Revelation – as one says of a proper sale that it was religiously sanctioned – with everything else not being so. Or, instead, it refers to what is known in this community.

Here al-Yūsī focuses on Ibn al-Juzayy's limitation of religiously sanctioned sciences to those which aid in the pursuit of those sciences that deal with

[88] See al-Yūsī, *al-Qānūn fī aḥkām al-ʿilm wa aḥkām al-ʿālim wa aḥkām al-mutaʿallim*, 294–95. I discuss the work at greater length in Chapter 2.

religious matters proper, and opens the possibility that all knowledge available to the Muslim community should be considered *sharʿī*, or revealed. He then moves to unpack the implications of such a move:

If I intend the first, then what is intended is what is permitted concerning the science as a whole, or in its entirety and in parts, in a fashion that does not include anything outside of it. If I intend the first, then we hold that all of the sciences that contain in them something that has been permitted are religiously sanctioned (*kullu-hā sharʿiyya*) due to their overall inclusion of religious and temporal benefits. If I intend the second, it implies it is necessary that nothing but the book itself is considered revealed (*sharʿī*) – concerning which "Falsehood cannot come from before it or behind it" [Q41:41] – and following it, the established example of the Prophet (*al-sunna al-thābita*).

The passage is somewhat confusing due to al-Yūsī's use of "the first" and "the second" in the reverse order that one would expect, but it is clear that he is setting up an opposition between the term *sharʿī* referring to solely the Qur'an and Prophetic Tradition on the one hand, and to all knowledge that benefits the Muslim community in this world or the next on the other. Before developing his argument further, al-Yūsī offers the following aside on how while any given body of knowledge contains elements that were not reliable, this does not invalidate it as a whole:

In every science there are necessarily false matters that proceed from ignorance, mistakes, or delusion, which are not permitted in a given science, nor in listening to it, or in the teaching of it. It is not for you to consider correcting accomplished jurists (*mujtahids*), for the sciences also contain attested (*qad ishtamilat bi-l-mushāhida*) delusions and mistakes, as we have said, and not only authoritative scholarly opinions (*ʿalā siwā al-ijtihādāt*).

The point here is that all knowledge is fallible and the result of human effort. Every science contains aspects that are false because of this fallibility, but this does not change the nature of a science that is religiously sanctioned. Al-Yūsī's emphasis on contingency may surprise, but he is taking a basic interpretive principle from Islamic jurisprudence – that authorities in the field can err, even when trying their best – and applying it to all sciences.[89] The fact that a given science contains erroneous knowledge does not detract from its overall permissibility or that it has substantial benefits in this world and the next.

[89] The *locus classicus* on this topic is Aron Zysow, "The Economy of Certainty: An Introduction to the Typology of Islamic Legal Theory."

If I mean by revealed what is known in the community in its entirety (*fa-in urīdu ʿalā iṭlāqihi*), then this includes what the community knows of medicine, arithmetic, logic and so on. If I mean what is known only in it, then the science of rational theology (*kalam*) is not religiously sanctioned, for it is a metaphysical science that has been trimmed and pruned (*li-annahu huwa al-ilāhī hudhiba wa nuqiha*), as the other sciences have been trimmed.[90] It is false and not religiously sanctioned. Yet how is this possible when it is the foundation of the revealed injuctions (*al-sharʿiyyāt*) and their head, and it has been counted as one of them?

The syntax of this passage is convoluted in the Arabic, but the overall point is clear: revealed knowledge is defined by what is useful and present within the Muslim community, not only what is contained in the Qur'an and Prophetic Tradition. If it were limited to the latter, then much of what Muslims considered the religious sciences would in fact not be revealed, and this would in fact be true of Islamic law itself:

If I consider religiously sanctioned what is contained in the revealed rulings (*al-aḥkām al-mashrūʿa*), and what is specifically sought after (*fa-in urīdu al-maṭlūba bi-l-dhāt*), then, save the quarter that deals with ritual obligations, jurisprudence would necessarily not be counted as religiously sanctioned – the rest of it dealing with what is permitted in its essence and which may be sought after. Such is the case with medicine, despite those whom God has placed at the station of understanding causes being legally enjoined to provide treatment. It is similar to seeking sustenance.

Al-Yūsī makes here what is clearly a *reductio ad absurdum* argument that if revelation were limited to the textual basis of the Qur'an and Prophetic Tradition, even Islamic law could not be considered revealed. In this regard, Islamic law was similar to the science of medicine. Both were based on general injunctions, Islamic law dealing with ritual obligations and the regulation of permitted matters, while medicine met the requirement of providing treatment for body's ills. Al-Yūsī concludes by taking up what he clearly sees as the logical implication of the preceding passages — the study of all sciences, because they are revealed, is permissible:

If I have a broader understanding of the term, then generally speaking, none of the sciences contains anything that infringes on their permissibility. This is the correct legal judgment (*wa hiya ḥukm sharʿī ʿalā l-ṣaḥīḥ).*

It should be clear why I have spent this much time on al-Yūsī's reasoning for including all sciences – including the natural ones – within the bounds of

[90] In my translation *al-ʿilāhī* as metaphysical, I have drawn on the discussion in al-Urmawī's (d. 672/1274) *Risāla fī farq bayna nawʿay al-ʿilm al-ilahī wa-l-kalām*, 71–72. My thanks to Zakaria El-Houbba for the reference.

the religiously sanctioned and the revealed, the two translations overlapping here. In doing so, he blurs categories and shows how for this scholar in seventeenth-century Morocco there was no firm border between religious and natural sciences or between religious and secular knowledge. It is worth dwelling on this point for a moment.

In a recent article, Rushain Abbasi has made an insightful argument for this type of blurring of epistemological boundaries having had a solid precedent in none other than al-Ghazālī (d. 505/1111).[91] In his nuanced survey of the prevalence and the productive nature of the contrasting categories of *dīn*, religion, and *dunyā*, temporal world, Abbasi shows how for some scholars, including al-Ghazālī, this differentiation broke down when one considered how worldly activities such as travel or even acquiring worldly goods contributed on a secondary level to religious imperatives such as acquiring knowledge of God.[92] Abbasi uses his careful parsing of the overlapping realms of religious and worldly to argue for a new understanding of "the secular" as a lens to understand the premodern Islamic intellectual history. For us here, the value of Abbasi's analysis is that it shows how what al-Yūsī does in his canon by including the natural sciences within revealed knowledge had conceptual precedent in Islamic scholarship, although no other scholar before him, to my knowledge, had framed this inclusion as broadly. The title, *Revealed Sciences*, speaks therefore to al-Yūsī's innovative work, but more pertinently to its relevance as a description of seventeenth-century Moroccan attitudes to the natural sciences more broadly and as an opportunity for us to rethink the categorizations of the sciences we bring to the study of Islamic scholarship.

INTERTWINED INTELLECTUAL WORLDS: GENRE AND DISCIPLINE IN ISLAMIC SCHOLARSHIP

For some generations of scholarship, now, we have known that while scholars of the premodern Muslim world often wrote works specializing in a specific religious or rational science, their own intellectual lives were not as compartmentalized as our curricula today make them out to be. While scholarly giants such as Ibn Sīna (d. 428/1037), al-Ghazālī (d. 505/1111), Ibn Rushd (d. 595/1198), Ibn ʿArabī (d. 638/1240), Ibn Taymiyya (d. 728/1328), and al-Suyūṭī (d. 911/1505), with all the many differences

[91] Rushain Abbasi, "Did Premodern Muslims Distinguish Between the Religious and the Secular? The *Dīn-Dunyā* Binary in Medieval Islamic Thought."
[92] See ibid., 26–30.

between them, may have been exceptional in the depth of their engagement with a large number of religious and rational sciences, they were representative of a rich and dynamic world of scholarship in which scholars could draw on jurisprudence, medicine, theology, logic, Sufism, astrology, grammar, and mathematics, to mention only a few sciences. Although there is heuristic value in discussing developments in a specific legal or theological school – or in devoting books or classes to specific subjects in individual sciences – such a focus does scant justice to the broader intellectual worlds inhabited by individual scholars and often prevents us from comprehending their own projects. This insight, simple though it is, cannot be realized without the ability to treat multiple fields of knowledge as dynamic, each with their own debates that can impact developments in other fields. Thus, some of the best recent histories of the natural sciences in the premodern Islamic world – Nahyan Fancy on Ibn al-Nafīs (d. 687/1288), Robert Morrison on Nīẓām al-Dīn al-Nīsabūrī (d. c. 730/1330), or Miquel Forcada on Ibn Bājjah (d. 533/1138) – have shown how the respective scholars' insights into medicine or astronomy were contextualized by their theological or philosophical convictions.[93] It is no accident that these histories focused on a period of Islamic thought whose contours are reasonably well understood. Our generally insufficient understanding of post-formative developments in all areas of Islamic intellectual history – due in part to a scholarly focus on the early and modern periods to the detriment of the so-called Middle period and equally if not more so to the fact that the vast majority of the intellectual production of that same Middle period is still in manuscript – has impeded the field's ability to produce nuanced intellectual histories of this period.[94]

Revealed Sciences takes up this challenge with regard to Morocco in the seventeenth century, a corner of the Muslim world outside of the great empires that controlled much of the remainder of the Mediterranean, Middle Eastern, and South Asian worlds. This was a tumultuous century for Morocco, beginning with the dissolution of the Saʿdī dynasty and ending with the ʿAlawites' consolidation of power – the dynasty that continues to rule Morocco until today. Chapter 1 will lay out the political and social context for the remainder of the book, focusing especially on the

[93] Nahyan Fancy, *Science and Religion in Mamluk Egypt*; Robert G. Morrison, *Islam and Science*; Miquel Forcada, *Ética e ideología de la Ciencia*. The same could, of course, be said of the work of Khaled El-Rouayheb on logic and Matthew Melvin-Koushki on lettrism.

[94] Exceptions here include many of the works that I rely upon for this study, including those by Ḥajjī, Berque, Touati, and Warscheid – all of which rely extensively on sources still in manuscript.

importance of the Sufi lodges (*zāwiyah*, pl. *zawāyā*) that emerged in North Africa in the fifteenth century as sites of knowledge production and transmission. This contextualization is especially needed for an English-reading audience, who have not been able to access the few surveys of Moroccan intellectual and social history during this period that have been written in Arabic and French.

This book intends primarily to trace the broader significance of the natural and rational sciences and the natural world in the intellectual landscape of seventeenth-century Morocco, and not to establish a parallel with developments taking place in Europe.[95] In the belief that more such local studies are needed before broader generalizations can convincingly be made, it reads deeply across the scholarship produced by Muslims in Morocco's long seventeenth century. This scholarship was largely not concerned with European expansion or intellectual developments in Europe, for although these were not ignored, neither was seen as central to the main intellectual pursuits of the day.[96] The nature and status of what could be known and studied is therefore the subject of Chapter 2. Drawing on scholarly autobiographies (*fahrasa,* pl. *fahāris*), biographical dictionaries, and classificatory discussions of the sciences, this chapter shows that the natural sciences were an important if minority pursuit in Morocco during the seventeenth century and that their study and transmission were closely interwoven in both urban and rural contexts with Sufism, jurisprudence and theology.

Chapter 3 turns from the place of the natural sciences in the transmission of knowledge to the role they played in jurisprudence. While Islamic jurisprudence has in recent years been rightly critiqued for all too often being used as a synecdoche for Islam itself, it is difficult to dispute its importance as one discourse among several through which Muslims understood their relation to the divine. The genre of legal opinion (*fatwā*, pl. *fatāwā*) has in recent decades been used to write the social history of North Africa, and as a source it also offers valuable insights onto how and when the authority of the natural sciences manifested itself in law. As a case study, this chapter pays special attention to the great tobacco debate that echoed throughout the Muslim Mediterranean and beyond in the seventeenth century, and to the role the body and its states of consciousness played in it.

95 For the latter approach, see Akkach, *'Abd al-Ghani al-Nabulusi*, as well as the above discussion of nineteenth-century reform efforts in Egypt.

96 Here my approach is at odds with the teleological line taken by Nabil Matar, when he argues that an absence of engagement with European thought and refusal to acknowledge cultural decline on the part of seventeenth century Arab scholars was proof of their intellectual stagnation ("Confronting Decline in Early Modern Arabic Thought").

In Chapter 4, the book turns to a series of close readings of works of natural science produced in Morocco in the seventeenth century, including discussion of works of astrology and alchemy alongside works of the more traditionally privileged disciplines of astronomy and medicine. It takes up the example of the best-known Moroccan astronomer of this century, Muḥammad b. Sulaymān al-Rūdānī (d. 1094/1683), whose works, while written in the Ottoman East, returned west and were preserved, among other places in the Ḥamziyya-ʿAyyāshiyya lodge in the High Atlas. Al-Rūdānī's writings are then compared with those of the lesser known al-Mirghitī al-Sūsī's (d. 1089/1678) astronomical and alchemical work. Turning from astronomy to medicine, the chapter examines Aḥmad b. Ṣāliḥ b. Ibrāhīm al-Darʿī's (d. 1144/1731) writings on *materia medica*, and contextualizes them in relation to other medical works read and written in the Islamic West at this time.

Between these chapters I have placed four short excursuses addressing in broad terms some of the central methodological suppositions that have determined my approach in writing *Revealed Sciences*. These are intended to be general in tone and to help the reader think through what may seem self-evident propositions that nonetheless have in the past led to partial understandings of the history of the natural sciences in the Muslim world. As such, they take up themes introduced elsewhere in the book and consider them with greater focus. The first, "The Poverty of Intellectual History as a Series of Great Men," explores the self-evident proposition that summarizing the achievements of major scholars results in an insufficient understanding of a field of study. The second, "The Horizons of Causality or How to Think about Causes, Nature, and Ghosts of Scientific Methods," traces how modern preconceptions regarding causality have shaped depictions of premodern and non-Western ways of thinking about the natural world. The third, "Kuhn and the History of Science in Islamicate Societies," argues that the historian of science Thomas Kuhn's work offers productive tools for us to think beyond the category of progress that has limited many previous debates regarding the natural sciences in post-formative Muslim societies. The fourth, "Sufism and the Spiritual Life or Balancing the Exoteric and Esoteric Sciences," offers reflections on the vexed category of Sufism in the writings of Muslim reformers and Western scholars of Islam, noting how it is often drawn into the older conflict narrative of religion opposing the natural sciences.

The Conclusion of *Revealed Sciences* returns to the broader historiographical concerns of this Introduction and poses the question of what a history of the natural sciences of the early modern Muslim world would look like if not measured by the teleological standards of modern science. As this book argues, it would first and most simply entail addressing and recognizing the amount of energy and activity Muslim scholars in the premodern world put into scientific pursuits that today might seem useless or not even scientific. Here, the critique of a traditional history of science that the book advances is also tied up with the recognition that its own interest is a very modern one – it is not free from the object of its critique and this is a tension that it is not able to resolve. Despite this, exploring the benefits of considering the importance of the natural sciences in the broader worldview of Muslim scholars of this period opens up new ways to appreciate the pervasive importance of their contemplation of the natural world during the long seventeenth century and more importantly it offers us the possibility of more fully reconstructing their scholarly worlds.

I

A Landscape of Learning in the Far West

During his [Abū ʿAbdallāh Muḥammad al-Ḥājj] days and those of his father [Maḥammad b. Abī Bakr] the lodge enjoyed a wonderful reputation. The exchange of the sciences took place there: they were studied fervently and read night and day so that there emerged from it a group of the best and most distinguished scholars such as the Shaykh al-Yūsī and those like him. This was so much the case that the student of knowledge in Morocco went no further than it, nor did the one desiring knowledge aspire to any other place.

Al-Nāṣirī, *An Examination of the Accounts of the Dynasties of Morocco*[1]

During my student days in the south, I acquired a bit of Arabic. It then happened that I moved to the region of Marrakesh, and this was during the rule of the Sultan Muḥammad al-Shaykh. I acquired other arts, such as legal theory, logic, and theology and left off from the study of Arabic. When I later came to the farthest part of the Sūs, I met with my teacher Abū Fāris ʿAbd al-ʿAzīz Aḥmad al-Rasmūkī, God have mercy on him. I found the people of this region busying themselves with the conjugation of verbs, and discussing them in the context of texts from the *Epitome* and similar texts. I spent time with them and found that verses of the *Epitome* no longer came to mind due to the long period of time I had not studied them. When I saw this, I desired to renew my acquaintance with them. I said to the students, "Let come whoever wishes to study the *Epitome*." So, we began its study, sitting together after the evening prayer for an hour or more or spending the entire night together. I transmitted in writing everything that was in the

[1] Al-Nāṣirī, *Kitāb al-istiqsā li-akhbār duwal al-maghrib al-aqsā*, vol. 5, 303.

commentary of al-Murādī with the utmost care and we completed our study in about a month and ten days.

Al-Yūsī, *al-Muḥāḍarāt*[2]

1.1 INTRODUCTION

In the politically and socially fractured landscape of seventeenth-century Morocco, several generations of scholars addressed the natural sciences in a variety of genres and discourses from within urban and rural intellectual institutions. These scholars were Arab and Amazigh – although they wrote almost entirely in Arabic – and were themselves the product of an educational landscape that had gradually come into being over the previous millennium. This chapter begins by sketching the political history of Morocco in order to draw attention to when legal, theological, and spiritual schools of thought and institutions appeared in the region and to provide the reader with the social and historical context for subsequent chapters. While this focus on the Moroccan political and institutional context may seem a tangential fashion to begin a book on the natural sciences, this regional focus speaks to the argument laid out in the introduction that the best way to move forward with the writing of the history of the sciences in the Muslim world is rigorously contextual.

It then turns to those aspects of Morocco's seventeenth century that have attracted the most attention from previous historians – political uncertainty, pirates, Jewish diplomats, holy warriors, and Sufi lodges – to situate scholars' engagements with natural science within the prior historiography on the period. The third and final part of the chapter returns to the intellectual currents referenced in the first part to sketch the institutional landscape of learning in seventeenth-century Morocco in greater depth and to consider how best to frame it historiographically.

1.2 BEFORE THE SEVENTEENTH CENTURY

The history of Morocco, "the furthest part of the West (*al-maghrib al-aqsā*)" as Muslim historians tended to describe it, has often been told in terms of dynasties. Considering their convenient packaging of time into units, this is not surprising or untypical of history in general and draws attention to certain continuities and changes while eliding others. It remains a useful way to

[2] Al-Yūsī, *al-Muḥāḍarāt*, 391–92.

introduce the origins and development of the main intellectual trends relevant to the long seventeenth century, while its artificial nature cautions us against ignoring continuities with regard to institutions and intellectual currents.

The initial Arab Muslim conquest of North Africa in the late seventh century CE led both to the conversion and revolt of the indigenous peoples there and to a series of Kharijite polities viewed as heretical by the supporters of the Umayyad (661–750) and ʿAbbasid (750–1258) dynasties based, respectively, in Damascus and Baghdad.[3] The first dynasty to establish itself in Morocco were the Idrīsids in the late eighth century. They were descendants of the Prophet from the Hijaz who, after settling in the former Roman settlement of Walīla (Volubilis) in the Middle Atlas, founded Fez nearby as their capital.[4] It is difficult to tell precisely what understanding of Islam the Idrīsids possessed – their founder Idrīs I (d. 175/791) was a follower of the Zaydi branch of Shi'ism and their chief political opponents at the beginning of their career were Kharijites whose political influence stretched over what is now southern Algeria and Morocco – but what evidence we have suggests that they maintained an adherence to Zaydi Shi'ism, a fact that would be elided in most later histories.[5] Due to successive waves of Arab immigration from Qayrawan and al-Andalus in the early ninth century, the city of Fez became characterized by an Arab population, in contrast to the Amazigh countryside surrounding it and the Kharijite imamates further east and south in Tāhart

[3] In the following, I will use the term Amazigh to refer to the indigenous people of North Africa, in large part because the conventional term "Berber" is a construction of Arabic Muslim historiography. See Ramzi Rouighi, "The Andalusi Origin of the Berbers?," and now, ibid., *Inventing the Berbers: History and Ideology in the Maghrib*. The Kharijites were Muslims who had comparatively idealistic standards for Muslim leadership and who developed a distinct identity in the first/seventh–second/eighth centuries prior to the solidification of Sunni and Shiʿa communities.

[4] For a quick overview of the history of North Africa (organized largely by dynasties) see Jamil M. Abun-Nasr, *A History of the Maghrib in the Islamic Period* but compare with the nineteenth century Moroccan historian al-Nāṣirī, whose *Kitāb al-istiqsā* was the first Moroccan history to organize its narrative around the country instead of individual dynasties (internally it was still organized by dynasties). On al-Nāṣirī, see Eric Calderwood, "The Beginning (or end) of Moroccan History: Historiography, Translation, and Modernity in Ahmad b. Khalid al-Nasiri and Clemente Cerdeira," and on the early history of the Idrīsids, see Najam Haider, "The Community Divided: A Textual Analysis of the Murders of Idrīs b. ʿAbd Allāh (d. 175/791)." The best single volume survey of Moroccan history remains Abdallah Laroui, *L'histoire du Maghreb: Un Essai de synthèse. 2ème ed.*

[5] See Laroui, *L'histoire du Maghreb*, 106–8; Wilfred Madelung, "Notes on Non-Ismāʿīlī Shiʿism in the Maghrib"; Ruggero Vimercati Sanseverino, *Fès et sainteté, de la foundation à l'avènement du Protectorat (808–1912)*, 122–25. The clearest analysis of the historiography of the early Idrīsids has been done by Chafik Benchekroun. See "Les Idrissides: L'histoire contre son histoire," and "Écriture et réécriture de l'histoire des Idrissides."

and Sijilmāsah.[6] Following the death of Idrīs' son Idrīs b. Idrīs (d. 213/828), born of an Amazigh concubine after his father's assassination by an agent of the ʿAbbasid caliph Hārūn al-Rashīd (d. 193/809), the territory under his rule was divided into principalities ruled over by his descendants. While we know little of the intellectual legacy of the Idrīsids, they remain important in Moroccan history for their founding of Fez and its grand mosque al-Qarawiyyīn and their establishing the precedent in Morocco of Prophetic descent as an important criterium for legitimate rule.[7]

The rise of the Ismāʿīlī Fatimid caliphate in North Africa in the tenth century and its initial struggle with the newly announced Umayyad caliphate of al-Andalus reshaped the political and intellectual landscape, and brought about the end of the Idrīsids as well as the Kharijite imamates. Even after the center of Fatimid activity moved east to Egypt in the second half of the tenth century, Morocco remained politically divided into small principalities. It was, however, during this century that we have first evidence of the arrival of Malikism, the Sunni legal school that would become dominant there as it had done a century previously in al-Andalus.[8] It was not until the reformist Amazigh movement of the Almoravids (al-Murābitūn) united West Africa and al-Andalus in the eleventh century that this political division ended. The Almoravids emerged out of the Ṣanhāja tribal confederation located in present-day Mauritania, and while the origins of their reformist ardor are lost in legend, their policies included calling for the adherence to the Sunni Mālikī school of jurisprudence and the ceasing of reprehensible innovations.[9] For

[6] For the complicated politics of North Africa in the eighth–ninth centuries and the relationship of the Kharijite states of North Africa with both Umayyad Iberia and the ʿAbbasid caliphate, see Adam Gaiser, "Slaves and Silver across the Strait of Gibraltar: Politics and Trade between Umayyad Iberia and Khārijite North Africa"; Paul Love, "The Sufris of Sijilmasa: Toward a History of the Midrarids," and D. Jacques-Meunie, *Le Maroc Saharien des origins a 1670,* vol. 1, 190–206.

[7] The Idrīsids and their Sherifian origins played an important role in the political imagination of Moroccan ruling dynasties from at least the Merinids onwards. See Mercedes García-Arenal and Eduardo Manzano Moreno, "Idrīssisme et villes idrīssides." Chefik Benchekroun has argued persuasively for the Qarawiyyīn having been founded by Dāwud b. Idrīs, a son of Idrīs II, in 263/877, and not by Fāṭima al-Fihriyya (Benchekroun, "Les Idrissides," 184–88).

[8] For the introduction of the Mālikī school of jurisprudence into Morocco see the biography of Abū Maymūna Darrās b. Ismāʿīl al-Fāsī (d. 357/967) in Muḥammad Makhlūf, *Shajrat al-nūr al-zakiyya fī ṭabaqāt al-mālikiyya*, vol. 1, 153–54; Muḥammad al-Kattānī, *Salwat al-anfās*, vol. 2, 197–200. Compare with the reviews of the evidence given in ʿUmar al-Jīdī, *Mabāḥith fī al-madhhab al-mālikī*, 15–21; Vincent Cornell, *The Realm of the Saint*, 33–35, and Ruggero Vimercati Sanseverino, *Fès et sainteté*, 150–52. For Malikism in al-Andalus see Maribel Fierro, "Proto-Malikis, Malikis, and Reformed Malikis in al-Andalus."

[9] Amira Bennison has suggested that the meeting of Yaḥyā b. Ibrāhīm, leader of the Guddala tribe (one of the three main tribes of the Ṣanhāja confederation) with the great Mālikī scholar Abū ʿImrān al-Fāsī (d. 1038–39) in Qayrawan during which Abū ʿImrān chose to

roughly three generations – a time the North African historian Ibn Khaldūn (d. 808/1406) would later generalize as the period needed for a new dynasty to lose its social solidarity – the Almoravids ruled over a vast empire uniting al-Andalus, North and West Africa from their capital in Marrakesh, a city they founded at the foot of the northern slopes of the High Atlas.[10] During this time, the Mālikī school became widespread in the areas they controlled and scholars introduced the Ashʿarī school of theology into the Islamic West.[11] It was also at this time that Sufi orders made substantial inroads throughout Morocco.[12] By the middle of the twelfth century, however, the empire the Almoravids had assembled fell to another Amazigh reformist movement, that of the Almohads (al-Muwaḥḥidūn), which expanded to cover much of the same territory as its predecessor.

The Almohads claimed their legitimacy from the messianic claims of their spiritual founder Ibn Tūmart (d. 524/1130), who advanced far-reaching social and intellectual reforms, including a broad expansion of literacy. Where the premodern Arabic historiographical tradition had placed the origins of the Almoravid movement in an encounter with the great Mālikī scholar of Qayrawan, Abū ʿImrān al-Fāsī (d. 430/1039), the Almohads were defined by Ibn Tūmart having reportedly met the great Sufi and theologian al-Ghazālī (d. 505/1111) in Baghdad.[13] In this rather tidy narrative, the former dynasty's support for Mālikism and their burning of al-Ghazālī's masterpiece, the *Revival of Religious Sciences* (*Iḥyā ʿulūm al-dīn*) for its heterodox nature, is contrasted with the Almohads' emphasis on theology and their support of the philosophical projects of Ibn Ṭufayl and Ibn Rushd (d. 595/1198).[14] By the time the Almohad dynasty splintered into client states in the thirteenth

send him to Wajjāj b. Zalwī in the Sūs of Morocco was legendary (*The Almoravid and Almohad Empires*, 27). For Almoravid control as far south as the oasis of Azagui, in what is Mauritania today, see Erin Pettigrew, "The History of Islam in Mauritania."

[10] On the length of dynasties' rule in Ibn Khaldūn's thought, see the remarks of Aziz al-Azmeh, *Ibn Khaldūn: An Essay in Reinterpretation*, 36–40.

[11] On the development of Ashʿarism in the Islamic West, see Delfina Serrano Ruano, "Later Ash'arism in the Islamic West"; Yūsuf Iḥnāna, *Taṭawwur al-madhhab al-Ashʿarī fī-l-gharb al-Islāmī*. Ibrāhīm Ḥarakāt suggests that Ashʿarism was not widely spread in Morocco before the rule of the Almohads. See his *Madkhal ta'rīkh al-ʿulūm bi-l-maghrib al-muslim ḥattā al-qarn 9/15h*, vol. 2, 378–81.

[12] Cornell, *The Realm of the Saint*, 44.

[13] On Ibn Tūmart not actually having met al-Ghazālī, see Frank Griffel, "Ibn Tūmart's Rational Proof for God's Existence and Unity, and His Connection to the Niẓāmiyya *Madrasa* in Baghdad."

[14] For a more nuanced and interesting version of this story, see the work on Ashʿarism by Delfina Serrano Ruano (inter alia, "Por qué llamaron los almohades antropomorfistas a los almorávides?") and that on the Almohads of Maribel Fierro (inter alia, "The Legal Policies of the Almohad Caliphs and Ibn Rushd's *Bidāyat al-Mujtahid*").

century, the Moroccan intellectual landscape was characterized by the Mālikism and Sufi orders that had taken root with the Almoravids, and the Ashʿarism that had been forwarded under the Almohads.

The Banū Marīn or Merinids (1258–1465), an Amazigh tribal confederation that had served the Almohads, rebelled against their former rulers in the first half of the thirteenth century, and for the next two centuries established themselves over an area corresponding roughly to today's Moroccan state.[15] It was the Merinid rulers who embarked on the extensive project of building religious colleges (*madrasa*, pl. *madāris*) throughout the cities of Morocco, including eleven in their capital Fez alone.[16] In expanding institutional education beyond the mosque, they continued the islamization of Morocco the previous two Amazigh dynasties had undertaken, and gave new support to the Mālikī school of law that had at times suffered repression under the Almohads. It was also in the time of the Merinids that Sufi lodges (*zāwiya*, pl. *zawāyā*) began to be built in greater numbers throughout North Africa, though their social and spiritual functions varied widely.[17] Finally, it was under the Merinids that we find extensive historiographical attention being paid to the Idrīsids, both as founders of Fez as well as the Sherifian precedents for legitimate Muslim rule over Morocco.[18] This attention went hand in hand with their granting a representative of the Prophet's descendants (*naqīb al-shurafāʾ*) an important public role, especially in the commemoration of the Prophet's birthday (*mawlid al-nabī*).[19] It is possible that while both their building of colleges as well as their interest in reviving the importance of the Idrīsids was to strengthen their own religious legitimacy in the wake of a dynasty that had begun with messianic pretensions, that their interest in Fez's putative Sherifian origins contributed to their own downfall.

In 1438 the bones of Idrīs II were miraculously discovered in Fez by the representative of the Prophet's descendants, leading both to the Merinids

[15] For the dissolution of the Almohad empire into the Merinid, ʿAbd al-Wādid (1236–1555), and Ḥafṣid (1229–1574) emirates, see first the nuanced historiographical comments in Ramzi Rouighi, *The Making of a Mediterranean Emirate: Ifrīqiyā and Its Andalusis, 1200–1400*, 1–18. On the Merinids, we have the detailed monographs of Ahmed Khaneboubi: *Les premiers Sultans Mérinids 1269–1331: Histoire politique et sociale*, and *Les Institutions gouvernementales sous les Mérinids (1258–1465)*, as well as the articles of Maya Shatzmiller collected in *The Berbers and the Islamic State: The Marīnid Experience in Pre-Protectorate Morocco*.

[16] Khaneboubi, *Les Institutions gouvernementales*, 247. Seven of these have survived until today. For their role in Moroccan education in the first half of the twentieth century, see Porter, "At the Pillar's Base," chapter 2.

[17] Ḥarakāt, *Madkhal ta'rīkh al-ʿulūm*, vol. 3, 40–49.

[18] See García-Arenal and Manzano Moreno, "Idrīssisme et villes idrīssides."

[19] Kugle, *Sufis and Saints Bodies*, 63–64.

constructing a shrine to the eighth century founder of Morocco's first Sherifian dynasty and the growing social authority of the Prophet's descendants in general.[20] The Merinid emirate was already in political decline, and while the last years of their rule over a diminished polity were characterized by a complicated array of events, it is telling that when their last ruler was killed in 1465 it was during a revolt by the city's *shurafā'*.[21] Their rule did not last long and for much of the next century northern Morocco was ruled by the Banū Waṭṭāṣ, an Amazigh tribal confederation related to the Merinids who had served as regional administrators under them. But the political dynamics within Morocco had shifted. When the next dynasty that would unite all of the territories of the Furthest West arose at the beginning of the sixteenth century, it drew on the renewed salience of Prophetic decent and the increasingly well-organized Sufi orders and their growing network of lodges.

Earlier French colonial historiography framed the rise of the Sherifian Saʿdī dynasty (1565–1603) as the result of a nativist *jihād* movement energized by Spanish and Portuguese encroachment along the Moroccan coast that went hand in hand with the rising power of rural Sufi saints, or marabouts.[22] More recent scholarship has argued convincingly for the beginnings of the Saʿdī dynasty having an internal dynamic that preceded Christian incursion into Morocco, and which combined the increased salience of Sharifism with the rising political legitimacy of Sufi orders – particularly that of the Shādhilī al-Jazūlī (d. 870/1465) – and messianic expectation.[23] The shift from the earlier Amazigh dynasties to Arab dynasties from the south of the Atlas – the Draʿ valley in the case of the Saʿdīs, Sijilmāsah in that of the ʿAlawites – with their claims to Sherifian authority through their descent from Muḥammad Nafs al-Zakiyya (d. 145/762) was decisive in determining the subsequent political and intellectual landscape of Morocco in the precolonial period.[24] Saʿdī control over a unified Morocco lasted only half a century, but on the political front these years

[20] See Herman Beck, *L'image d'Idrīs II, ses descendants de Fās et la politique sharīfienne des sultans marīnides (656–869/1258–1465)* and Kugle, *Sufis and Saints Bodies*, 60–79.

[21] For the ways in which contemporary and later sources described this revolt and the varying depictions of Fez's Jewish population in it, see García-Arenal, "The Revolution of Fās in 869/1465 and the Death of Sultan ʿAbd al-Ḥaqq al-Marīnī."

[22] Hence the term, "maraboutic crisis." For a critical survey of the role of the French colonial historian Alfred Bel (d. 1945) in promoting maraboutism as the cause of the decline of Islam in Morocco, see Cornell, *The Realm of the Saint*, xxv–xxix.

[23] For the origins of the Saʿdīs, see the two articles of García-Arenal, "*Mahdī, Murābit, Sharīf*: l'avènement de la dynastie Saʿdienne," and "Sainteté et pouvoir dynastique au Maroc," and on al-Jazūlī and the Jazūliyya see Cornell, *The Realm of the Saint*, chapters 6–8.

[24] On Saʿdī and ʿAlwaite claims to be, like the Idrīsids, descended from the Alid rebel Muḥammad Nafs al-Zakiyya (d. 145/762), see al-Nāṣirī, *Kitāb al-istiqṣā*, vol. 5, 11.

witnessed Aḥmad al-Manṣūr (d. 1012/1603) coming to power after his brother's death and his victory over the Portuguese and his own rival for the throne at the Battle of the Three Kings in 1578. Al-Manṣūr's own imperial ambitions led him to order the Moroccan expansion across the Sahara in the campaign of 1591.[25] The latter conquest brought the West African cities of Timbuktu and Gao under Moroccan control, albeit a control that only lasted for a few decades in any meaningful sense.[26] By strengthening the political ties between West Africa and Morocco's urban centers north of the Atlas, the trade route in the south of Morocco gained in importance and was realigned from Sijilmāsah in the Tifilalt region further west to the Draʿ valley, where the Nāṣirī order and lodge was founded in the seventeenth century at Tamgrout, south of Zagora. In sum, then, the political and economic changes that took place under the Saadians established what was for the premodern period a strong Moroccan state with imperial ambitions and important economic and intellectual ties with the Sahara and West Africa.[27] When Aḥmad al-Manṣūr died of the plague in 1603 after twenty-five years of rule, this state shuddered and split, fracturing the Moroccan political landscape for much of the seventeenth century and opening up space for new political and intellectual centers to emerge.

1.3 POLITICAL TURMOIL, HOLY WARRIORS, PIRATES, AND THE NATURE OF OUR SOURCES

If one were to focus solely on dynastic leaders and followed the last members of the Saʿdī dynasty, one could find them ruling parts of Morocco almost until the beginnings of the nascent ʿAlawite dynasty in the 1660s and the intervening decades of political chaos could be glossed over. This would make for tidy historiography, yet it would elide the limits of caliphal power even during the periods of the Sherifian state's greatest strength and would fail to explain how quickly the territory under Saʿdī control slipped away from them following al-Manṣūr's death.

[25] The eventful reign of Aḥmad al-Manṣūr is addressed in two recent monographs: Mercedes García-Arenal, *Ahmad al-Mansur: The Beginnings of Modern Morocco*, and Stephen Cory, *Reviving the Islamic Caliphate in Early Modern Morocco*. On the ways in which the dramatic events of August 4, 1578 were remembered on both sides of the straits, see Lucette Valensi, *Fables de la Mémoire: La glorieuse bataille des Trois Rois (1578)*.

[26] See Elias Saad, *Social History of Timbuktu: The Role of Muslim Scholars and Notables 1400–1900*, 174–96.

[27] For an important reminder of just how weak premodern states were, see Patricia Crone, *Pre-Industrial Societies*, chapter 3 ("The State").

Many of the recent historical accounts of Morocco during the seventeenth century have in fact been interested precisely in the individuals and groups who flourished in the wake of the Saʿdī loss of control over Morocco. This was a period when North Africa was closely connected with Spain, in part due to Spain's expulsion of roughly 300,000 Moriscos between 1609 and 1614 – a substantial minority of whom settled in the Moroccan port of Rabat-Salé and to the proliferation of pirate enclaves in the Maghreb that carried out frequent raids on European ships and coasts, in the process enslaving significant numbers of Christians.

Moroccan political history during this period is complicated and eventful, but it can be broadly summarized as Morocco breaking initially into three centers, each of which exerted shifting amounts of influence on the surrounding areas.[28] The grandson of al-Manṣūr, ʿAbdallāh b. Muḥammad al-Shaykh ruled over Fez until his death in 1627 while his father Muḥammad (d. 1022/1613) briefly collaborated with the Spanish in fruitless attempts to seize power over Morocco, helping them gain control of the port al-ʿAra'ish (Larache). Further south, al-Manṣūr's son Zaydān ruled over the imperial capital in Marrakesh until 1627, though his rule was interrupted by the messianic rebel Ibn Abī Maḥallī's conquest of Marrakesh in 1610 and his subsequent three-year rule before he was killed by Zaydān's ally from the High Atlas, Yaḥyā al-Ḥāḥī (d. 1035/1626). Between these two urban centers with their shaky control over their hinterlands was the pirate republic of Rabat-Salé, where the Hornacheros and other Moriscos had collectively settled in 1609 following their expulsion from Spain.[29] For a few decades this republic was largely autonomous before it was drawn into the influence of the Dilā' lodge on the one hand and the holy warrior Muḥammad al-ʿAyyāshī (d. 1051/1641) on the other.

Located in the Middle Atlas near the modern city of Khenifra, the Sufi lodge of the Dilā'iyyah played not only an important intellectual role during this period – more on which in Section 1.3.1 – but came to represent the single most powerful political and military force in Morocco before the rise of the ʿAlawites in the 1660s.[30] It was founded around 974/1566 by Abū Bakr b. Muḥammad b. Saʿīd al-Dilā'ī (d. 1021/1612), an

28 For a brisk narrative account of the political events of these years, see Jamil M. Abun-Nasr, *A History of the Maghrib in the Islamic Period*, 219–27.

29 Leïla Maziane, *Salé et ses corsaires (1666–1727): Un port de course marocain au XVIIe siècle*, 50–51. The Hornacheros were distinct from other Moriscos in having maintained Arabic and open adherence to Islam, as well as having resettled in Morocco as an intact community. See Roger Coindreau, *Les Corsaires de Salé*, 42–44.

30 The locus classicus on the Dilā' *zāwiyah* remains Muḥammad Ḥajjī, *al-Zāwiyah al-dilā'iyyah wa dawru-hā al-dīnī wa l-ʿilmī wa l-siyāsī*.

Amazigh scholar of the Majāṭ branch of the Ṣanhājah tribe, with the aim of spreading the Shādhilī order of Sufism and of feeding the poor and travelers. Under the Saʿdīs, it enjoyed good relations with the ruling dynasty in Marrakesh and attained local fame due to Abū Bakr's generosity. After Aḥmad al-Manṣūr's death in 1603 and the subsequent political chaos, the Amazigh population of the Middle Atlas looked increasingly to the lodge to provide stability and the political role of the Dilā'iyyah increased under Abū Bakr's son Maḥammad (d. 1046/1636), despite the latter's preference for spiritual pursuits.[31] Maḥammad organized the Amazigh tribes who were loyal to the lodge militarily and in some cases, such as with his support for the holy warrior al-ʿAyyāshī's efforts to push back the Spanish encroachment on the Moroccan coast, sent them beyond the Middle Atlas.[32] The apex of the lodge's political power, however, took place under Maḥammad's son Muḥammad al-Ḥājj, who oversaw the lodge's military ventures from the time of his father's death in 1046/1636 until its destruction by the ʿAlawites in 1079/1668 and was arguably the most important political figure in Morocco during the second third of the seventeenth century. Muḥammad al-Ḥājj founded a new lodge near the old one, settled five Amazigh tribes of the Ṣanhājah near it, and began actively seeking to unite northern Morocco under his rule.[33] Over the next two decades he carried out campaigns against the remnants of the Saʿdīs, the Spanish, al-ʿAyyāshī, and the incipient ʿAlawite dynasty in Sijilmāsah, and exercised control over Fez from 1051–61/1641–51, without ever taking Marrakesh, which was under the influence of the lodge at Ilīgh, in the Sūs.[34] The Dilā'iyyah's star began to fade in the 1650s when they lost control of Fez to an Arab tribe, al-Khadr al-Ghaylān, which briefly exercised control over both Fez and Rabat-Salé in the 1660s before the rise of the first ʿAlawite ruler Moulay Rashīd (d. 1082/1672) in Sijilmāsah. It was under Rashīd and then more importantly his half brother and successor Moulay Ismāʿīl (d. 1139/1727), who ruled for half a century and firmly established the ʿAlawite dynasty, that the political turmoil Morocco had slipped into following the death of al-Manṣūr finally ended. Moulay Ismāʿīl's consolidation of power during the first decade of his rule involved

[31] Ibid., 141–42. For the lodge becoming famous during Maḥammad's time for its celebration of the Prophet's birthday which included the public recitation of al-Buṣīrī's *Burdah*, see ibid., 49.

[32] Ibid., 154–61. [33] Ibid., 167–77.

[34] Ibid., 227. Significantly, the Dilāʾiyyah also may have struck coinage of their own, square in shape as with the earlier Almohad coins (ibid., 232–33). On the lodge at Ilīgh and its political importance, see Muḥammad al-Mukhtār al-Sūsī, *Īlīgh qadīman wa ḥadīthan*, and specifically on the relations of Īlīgh with Dilā' see ibid., 131–50.

not only considerable violence, but the creation of a standing army of black slaves, a controversial innovation that played a central role in his establishing control over the country.[35]

Within this fractious political context, much of the scholarship (and especially that in English) on Morocco's seventeenth century has focused on the tumultuous political events of the time and their social context. Historians have been especially interested in the activities of the pirates of Rabat-Salé, the enigmatic figure of the messianic rebel Ibn Abī Maḥallī, the fascinating story of the prominent Jewish family of the Pallaches, and the remarkable career of the Morisco Aḥmad al-Ḥajarī (d. after 1640).[36] The last of these figures is particularly striking as an example of the continued porous border between Iberia and North Africa, for after leaving Spain before the official expulsion of the Moriscos, he returned to Spain and then northern Europe as ambassador for Moulay Zaydān.[37] In this regard, a new generation of scholarship has substantially revised Andrew Hess' 1978 argument that in a shift from preceding centuries of interconnectivity the Western Mediterranean was increasingly compartmentalized into discrete cultural and social zones in the sixteenth century.[38] Recent studies, some of them excellent, on the role of piracy and slavery in the Mediterranean have continued this critique, drawing attention to the myriad and complex ways in which polities on both littorals of the Mediterranean were extensively economically and socially connected.[39]

[35] Chouki El Hamel, *Black Morocco: A History of Slavery, Race, and Islam*, chapter 4: "Racializing Slavery: The Controversy of Mawlay Isma'il's Project."

[36] On Samuel Pallache and his family, see Mercedes García-Arenal and Gerard Wiegers (trans. Martin Beagles), *A Man of Three Worlds: Samuel Pallache, a Moroccan Jew in Catholic and Protestant Europe*. For an overview of the life and career of Ibn Abī Maḥallī and an edition of his own *fahrasa*, see ʿAbd al-Majīd al-Qaddūrī, *Ibn Abī Mahallī al-faqīh al-thā'ir wa riḥlatu-hu al-iṣlīt al-khirrīt*, which should be compared with Mercedes García-Arenal, *Messianism and Puritanical Reform: Mahdis of the Muslim West*, chapter 12: "Ibn Abī Maḥallī and his Adversaries."

[37] For an overview of al-Ḥajarī, see Gerald Wiegers, "A Life Between Europe and the Maghreb: The Writings and Travels of Aḥmad b. Qâsim ibn Aḥmad ibn al-faqîh Qâsim ibn al-shaykh al-Ḥajarî al-Andalusî (born c. 977/1569–70)"; for al-Ḥajarī's work itself, see Aḥmad b. Qāsim al-Ḥajarī (d. after 1640), eds. P. S. van Koningsfeld, Q. al-Samarrai, and G. A. Wiegers, *Kitāb nāṣir al-dīn ʿalā 'l-qawm al-kāfirīn* ("The Supporter of Religion against the Infidels").

[38] Andrew Hess, *The Forgotten Frontier: A History of the Sixteenth-Century Ibero-African Frontier.* See the 2013 special issue of *Medieval Encounters* devoted to revising Hess' argument: "*Spanning the Strait: Studies in Unity in the Western Mediterranean.*"

[39] For one example, see the work of Daniel Hershenzon, *The Captive Sea: Slavery, Communication, and Commerce in Early Modern Spain and the Mediterranean*; ibid., "Traveling Libraries: The Arabic Manuscripts of Muley Zidan and the Escorial Library."

Less attention has been paid in English language scholarship to the internal intellectual life of North Africa during this period, where the majority of Moroccan scholars at the time paid scant attention to connections with European powers and focused on debates and issues that were closely related to the Islamic and rational sciences taught in the region's Sufi lodges and urban colleges (*madrasa*, pl. *madāris*). Here, scholarship in French and Arabic of the past century has been considerably more productive, in the former case due doubtlessly in part to the long French colonial presence in Algeria and Morocco, which produced several generations of scholars with deep knowledge of the region, of whom Jacques Berque (1910–95) is perhaps the most famous. Strikingly productive and learned, three of Berque's many works touched on the intellectual history of seventeenth-century Morocco, focusing in particular on the prominent polymath al-Ḥasan al-Yūsī.[40] Their work built on the earlier work of Évariste Lévi-Provencal (1894–1956), who was the first to offer an overview of the historiography in a European language of the Saʿdī and ʿAlawite dynasties in his 1922 *Les historiens des chorfas*. In Arabic, Morocco's intellectual history for this period is deeply indebted to the scholar and editor Muḥammad Ḥajjī (1923–2003), whose books on the Dilā' lodge and the intellectual life of the Saʿdī period should be read in counterpoint with those of Berque as the distinct approaches of the two scholars supplement each other well.[41] Where Ḥajjī's work offers a comprehensive overview of the ways in which Morocco's intellectual landscape changed and developed under the Saʿdīs and shaped subsequent generations, Berque provided a series of extended analytical meditations on individual scholars or works or specific subjects. The two differed somewhat in their overall evaluation of the intellectual production of the period, with Ḥajjī arguing that the Saʿdīs contributed to an institutional and social rebirth of intellectual life in Morocco after a century of comparative stupor and Berque seeing the sixteenth–seventeenth centuries

[40] Jacques Berque, *Al-Yousi: Problèmes de la Culture Marocaine au XVIIème Siècle*; ibid., *L'intérieur du Maghreb: XVe–XIXe siècle*; ibid., *Ulémas, fondateurs, insurgés du Maghreb: XVIIe siècle*. Impressive in their erudition and for his use of a wide array of unpublished manuscripts, these books are a necessity for anyone working on the period, though they should now be read against the more recent work of Houari Touati and Ismail Warscheid on the scholars of Algeria and the oasis of Touat, which during the seventeenth century owed allegiance to Morocco. Houari Touati, *Entre Dieu et les Hommes*; Ismail Warscheid, *Droit musulman et société au Sahara prémoderne: La justice islamique dans les oasis du Grand Touat (Algérie) aux XVIIe–XIXe siècles*.

[41] Muḥammad Ḥajjī, *al-Ḥaraka al-fikriyya bi-l-maghrib fī l-ʿahd al-saʿadiyyīn*. This work appeared originally in French in 1976 as the author's doctoral dissertation, completed at the Sorbonne under the direction of Charles Pellat.

as ones of general intellectual decline in which figures such as al-Ḥasan al-Yūsī (d. 1102/1691) were the exception that proved the rule.[42] This difference in evaluation did not impede both scholars characterizing the scholarly production of the time as largely one of commentaries and mnemonic poems in contrast to an earlier period of more original works. Both were also writing before the critical revaluation of the genres of summary (*mukhtaṣar*) and commentary (*sharḥ*, *ḥāshiya*) that has taken place in the field of Islamic intellectual history over the past two decades and the increasingly accepted argument that many examples of these genres contained creative and independent insights.[43] As this chapter turns from this overview of Morocco's political history to sketch its intellectual landscape during the seventeenth century, it also considers the nature of our sources for this period and how they relate to what was taught and written by scholars of the time. Doing so provides the needed context for the argument I advance in Chapters 2 and 3 that the natural sciences played an important role in religious discourses during this period.

1.3.1 Ulamalogy and Educational Institutions

It has been said, with some exaggeration, that all social and intellectual history of the premodern Islamic Middle East is the history of its scholars or *ʿulamāʾ* (sing. *ʿālim*).[44] While this group can be difficult to define in that it included figures from all social classes, conceptually it was straightforward in that it referred to those who had studied a body of knowledge or science (*ʿilm*) with someone who had mastered it to at least a nominal level. What differentiates scholars in Islamic contexts from some other intellectual traditions is that from comparatively early on, they compiled biographical dictionaries to keep track of a scholar's death dates, his or her – women were marginal but present in these works – teachers, subjects

[42] Ḥajjī, *al-Ḥaraka al-fikriyya*, vol. 1, 54–55; Berque, *Al-Yousi*, 10–11.

[43] See Robert Wisnovsky, "The Nature and Scope of Arabic Philosophical Commentary in Post-Classical (ca. 1100–1900 AD) Islamic Intellectual History: Some Preliminary Observations," and the articles in the 2013 issue of *Oriens* edited by Asad Q. Ahmed and Margaret Larkin, especially their introductory overview: "The *Ḥāshiya* and Islamic Intellectual History."

[44] See Stephen Humphries, *Islamic History: A Framework for Inquiry. Revised Edition*, 187–208 (chapter 8: "A Cultural Elite: The Role and Status of the *ʿUlamāʾ* in Islamic Society"), where he is drawing on the work of Roy Mottahedeh. Such a generalization, while insightful, contains a textualist prejudice and ignores the considerable contributions archeology and art history have made to our understanding of the social history of Middle Eastern Islamicate societies.

studied, and works authored. While these dictionaries could be organized in a variety of ways with various levels of comprehensiveness, they frequently took the form of generations of scholars *(ṭabaqāt)*, describing who in each generation transmitted knowledge to the next and what their role was in that process. A prominent and early example of the genre is Ibn Saʿd's (d. 230/845) massive *Large Book of the Generations* (*Kitāb al-Ṭabaqāt al-Kabīr*), which laid out the scholars who had lived during and following the Prophet's time generation by generation. The science that Ibn Saʿd was interested in was Prophetic Tradition and his encyclopedic work was designed to aid his colleagues in verifying the transmission of this tradition and in ascertaining which words or actions attributed to the Prophet Muḥammad were accurate. In this way, the act of recording the lives of scholars was not simply a recording of the past, it could also represent a vision of what kind of knowledge was valid and the identity of its representatives. This enterprise erred toward the comprehensive and recorded the names and brief biographies of numerous minor scholars who never wrote a work of their own, in this way offering historians today the possibility of writing the social history of entire communities through the lens of what we know of its literate members.[45] It is in this sense no accident that Ibn Saʿd was writing during the third/nineth century when the importance of verifying scholarly lineages was heightened due to the intense theological debates taking place in ʿAbbasid society and the solidifying of both sectarian and juridical identities that was reflected in emerging legal and theological schools. As Muslim scholars pursued sciences beyond that of Prophetic tradition, biographical dictionaries came to address these as well, though they often remained formally focused on schools of jurisprudence.[46] Prominent examples of biographical dictionaries focusing on scholars specializing in the natural sciences include Ibn Abī Uṣaybiʿa's (d. 668/1270) *ʿUyūn al-anbāʾ fī ṭabaqāt al-aṭṭibāʾ*, Ibn al-Qiftī's (d. 646/1248) *Taʾrīkh al-ḥukamāʾ*, and Shams al-Dīn al-Shahrazūrī's (fl. seventh/thirteenth century) lesser-known *Tawārīkh al-ḥukamāʾ wa l-falāsifa*. As with those biographical dictionaries devoted to jurists, these were not simple lists of scholars, but reflected the personal and intellectual predilections of their authors, blending the prescriptive with the descriptive and at times omitting mention of scholars of whom the author

[45] The locus classicus for such as study being Richard Bulliet, *The Patricians of Nishapur: A Study in Medieval Islamic Social History.*

[46] For a nuanced and careful reading of a ninth/fifteenth century biographical dictionary of Shāfiʿī jurists that has shaped my understanding of the genre, see R. Kevin Jaques, *Authority, Conflict, and the Transmission of Diversity in Medieval Islamic Law.*

disapproved. Biographical dictionaries remained important in the Arab Middle East down into the twentieth century with reference works such as ʿUmar al-Kaḥḥāla's (d. 1987)'s *Encyclopedia of Authors* (*Muʿjam al-muʾallifīn*) and Khayr al-Dīn al-Ziriklī's (d. (1976) *Prominent Scholars* (*al-Aʿlām*) remaining critical references for scholars today.

When Berque, Ḥajjī and Khaled El-Rouayheb describe the intellectual history of Morocco in the sixteenth–seventeenth centuries, their narratives take as their primary material works of jurisprudence, law, theology, and logic. Yet the context for their readings and narratives, which situates the scholars and scholarship that they discuss, is supplied by the numerous biographical dictionaries describing the scholars of Morocco of the period, each of which has its own geographical and chronological organization. None of these dictionaries focuses explicitly on scholars known for their devotion to the natural sciences, though they do record those who did study them and when taken together with autobiographical accounts of scholars' studies, offer an invaluable window onto the place of the natural sciences within the broader field of learning. I will take up these dictionaries more closely in Chapter 2 along with the autobiographical accounts of a given scholar's studies (*fahrasa,* pl. *fahāris*) that were written during this period and which represent a particularly popular genre in the Islamic West. Here, building on the political context given in the first section of this chapter, I wish to present an overview of the institutions of learning in Morocco in the seventeenth century and of the main types of scholarship they produced – an overview that itself draws substantially on Ḥajjī's reading of the sources of the period.[47]

The consolidation of Saʿdī power over Morocco in the middle of the sixteenth century brought with it not only increased political stability, it also witnessed the rise of new institutions of learning in both cities and rural areas, and a concerted effort on the part of the ruling dynasty to forward the study and transmission of religious and philosophical sciences. This new wave of intellectual activity, especially with regard to the rational sciences, has been connected by El-Rouayheb to the arrival in Morocco of Tunisian scholars who resurrected the study of logic, a subject previously neglected that went on to be widely studied in Morocco in the seventeenth century.[48] Ḥajjī, for his part, situates the origins of Morocco's intellectual revival in the sixteenth–seventeenth centuries in the origins of the Saʿdīs

[47] Ḥajjī relies on chronicles, travel literature, biographical dictionaries, and *fahāris* as the primary sources for his valuable survey of institutions of learning (Ḥajjī, *al-Ḥaraka al-fikriyya*, vol. 1, 18–33).

[48] El-Rouayheb, *Islamic Intellectual History*, 148–49. El-Rouayheb is thinking specifically of the Moroccan scholars Ahmad al-Manjūr (d. 1587) and Muhammad al-Qaṣṣār al-Gharnāṭī

themselves, noting that before attaining political power the founder Muḥammad al-Qā'im bi-Amr Allāh (d. 923/1517–18) had been a jurist who taught in the southern Darʿa region, his son Muḥammad al-Mahdī al-Shaykh (d. 964/1557) memorized the Diwan of the celebrated poet al-Mutanabbī, and all of his sons and grandsons, including Aḥmad al-Manṣūr and Moulay Zaydān were poets and students of literature.[49] This personal devotion to learning was matched by the dynasty's investing in educational institutions. In Fez, the celebrated mosque and university al-Qarawiyyīn was restored and expanded, the al-Ḥazzābīn lodge and the Grand library were added, and the colleges founded by the Merinids were restored. Marrakesh regained its status as a center of learning with the restoration and expansion of the Ibn Yūsuf college and the founding of a number of mosques, libraries, and hospitals – Ḥajjī lists seven principal ones – that forwarded the teaching and transmission of the sciences. South of the Atlas, in the Sūs, Muḥammadiyya (today known as Taroudant), named after the Saʿdī ruler Muḥammad al-Mahdī al-Shaykh, become a regional intellectual hub, with its grand mosque expanded to provide additional space for teaching.[50] In these cities, students received financial support from the religious endowments (*ḥabs*, pl. *aḥbās*, in the Mashriq *waqf*, pl. *awqāf*) funding the institutions at which they studied, that consisted of guaranteed housing for ten years, a daily ration of bread and at times a small stipend.[51] In rural areas, where numerous Sufi lodges, many of which were founded in the sixteenth–seventeenth centuries, functioned as regional centers of learning, students were supported by the tribes living in the area, with the tribes of the Sūs region being particularly known for their generosity.[52] To the extent that the economic basis of the rural lodges

(d. 1603), who studied with the Tunisian scholar Muḥammad Kharrūf al-Anṣārī (d. 1558) who had settled in Fez.

49 Ḥajjī, *al-Ḥaraka al-fikriyya*, vol. 1, 55. Compare with ʿAbdallāh al-Targhī's emphasis on the importance in Fez of Ibn Ghāzī's (d. 910/1513) students as a bridge between the Waṭṭāṣid and Saʿdī periods (*Fahāris ʿulamāʾ al-maghrib mundhu al-nashaʾ ilā nihāyat al-qarn al-thānī ʿashr li-l-hijra*, 25).

50 Ḥajjī, *al-Ḥaraka al-fikriyya*, vol. 1, 55, 82.

51 Ibid., vol. 1, 127. On the theory and practice of pious endowments in the Mālikī school, see David Powers, "The Maliki Family Endowment: Legal Norms and Social Practices."

52 Ibid., vol. 1, 128–29. The lodge in Wazzān was founded at the beginning of the eleventh/seventeenth century (ibid., vol. 2, 472); the Ayt ʿAyyāsh was founded in 1044/1634–35 and then due to the great number of students it attracted, expanded in 1066/1655–56 after its founder Muḥammad b. Abī Bakr al-ʿAyyāshī (d. 1067/1651) studied at both the Nāṣirī and Dilāʾ lodges (ibid., vol. 2, 508 and Ḥajjī, *al-Zāwiya al-Dilāʾiyya*, 68–69); in Tamgrout both the Nāṣirī and the Sīdī ʿAlī lodges were founded at the beginning of the tenth/sixteenth century (*al-Ḥaraka al-fikriyya*, vol. 2, 545, 549), as was the nearby Sayyid al-Nās lodge (ibid., vol. 2, 542); the lodge at Ilīgh in the Anti-Atlas was founded in 1021/1612

was subject to crop failures and adverse weather conditions, the stability of the educational support respective tribes could supply was uncertain. It is significant, in this regard, that two of the most significant lodges in the seventeenth century, the Dilā' in the Middle Atlas and the Nāṣirī in Tamgrout on the edge of the Sahara, were somewhat protected from these seasonal pressures, the first due to the dense forests of the Middle Atlas and their flocks, the latter due to the income it received from the Saharan trade.[53] These two lodges also distinguish themselves from the majority of rural lodges by their offering their students advanced studies in specific subjects, an ability generally only found in the above mentioned urban areas, and one certainly linked to the economic affluence of these lodges, which allowed the Dilā' lodge, for example, to support hundreds of students at a time.[54]

A student's initial education began when he, or she, although the sources suggest that girls were more frequently instructed within the household, was taught literacy through the recitation of the Qur'an. A late fifteenth century Algerian manual on the instruction of boys, Aḥmad b. Abī Jumaʿa al-Maghrāwī's (d. 920/1514) *The Great Collected Summary and Explanation of what Is Presented to Teachers and the Parents of Young Boys* (*Jāmiʿ jawāmiʿ al-ikhtiṣār wa al-tibyān fī-mā yuʿraḍ li-l-muʿallimīn wa ābā' al-ṣibyān*), gives us a window onto the kind of education students could expect before they arrived at a college or lodge.[55] Al-Maghrāwī was a student of the influential theologian al-Sanūsī, and as such would likely have spent much of his time occupied with the theological issues of his day; yet, if we take the rhetorical trope seriously, he responded willingly to his colleague's request for him to write a treatise on the subject of primary education. Much of the treatise deals with the issues of whether it was permissible or not to pay a teacher for teaching the Qur'an – the founder of the Mālikī school, Mālik b. Anas (d. 179/795) had said it was, while the founder of the Ḥanafī school, Abū Ḥanīfa (d. 150/767) had said it was not – and how to administer corporal

(ibid., vol. 2, 605); the Fāsī lodge was founded by Abū al-Maḥāsin Yūsuf ibn Muḥammad al-Fāsī (d. 1013/1604), in 1004/1595 (ibid., vol. 2, 364–65).

53 Ibid., vol. 1, 129–30. Note that Ḥajjī describes the students at the Nāṣirī lodge as less protected than those in Dilā'. For more background on the social and intellectual context of the Nāṣirī lodge, see also Muḥammad al-Mannūnī, *Ḥaḍarat wādī Daraʿ min khilāl al-nuṣūṣ wa-l-athār.*

54 Ibid., vol. 1, 125, 127. See also below for both of these lodges boasting sizable libraries.

55 Al-Maghrāwī believed that girls and boys should be instructed in separate classes after they reached the age of puberty in order to avoid corruption (al-Maghrāwī, *Jāmiʿ jawāmiʿ al-ikhtiṣār wa al-tibyān*, 43).

punishment.[56] But there is also some discussion of what children should be taught beyond Qur'anic recitation itself, with the Sufi authority al-Jazūlī being quoted as to students needing to being instructed not just in matters of ritual, jurisprudence, and theology, but also in the names of both Arabic (lunar) and foreign (solar) months as well as the basics of mathematics and inheritance law.[57] The number of hours that schoolchildren spent on study was distinctly less than what they would experience if they continued their studies, with the teaching concentrated in the periods between the morning prayer and the appearance of sunlight and then between the noon and afternoon prayers. In rural areas, the author notes, most Qur'anic recitation took place in the evenings.[58] Matters changed substantially for those students who continued their studies beyond this initial stage and who sought out instruction in a college or lodge.

Within these institutions, there was no defined curriculum, and each student could, depending on the teachers available, choose his subject of study, including the natural and mathematical sciences. Correspondingly, there was no institutional diploma of graduation – instead, a teacher would grant a student a certificate (*ijāza*) testifying to his mastery of everything the teacher had taught him in a specific discipline or to his having read a particular work with that teacher.[59] The manner of instruction during the Saʿdī period was generally based on the teaching of individual works and began with the advanced students sitting in a semicircle around the teacher who would choose one of them as reader, with the other students sitting behind them. The reader would read the text and any commentaries on it that the teacher asked of him. The reader would throughout the class be the one the teacher would come back to in order to read any additional texts. This was the first stage of instruction and along with a basic explanation of the text at hand, may well have been the only type of instruction in many rural institutions. With more advanced students, however, teachers would pursue a variety of more sophisticated methods: the exploration of the text's difficulties (*ḥakk al-masāʾil*); the teacher lecturing on the text, which seems to have involved a comprehensive exploration of a text's context, content, and transmission, especially in the case of works of

[56] Ibid., 27–29 and 41–45. The subject of corporal punishment for young boys recurs in al-Tamanartī's seventeenth century autobiography, where we find similar deliberations on how to adjust the punishment according to the constitution of the child (al-Tamanārtī, *Fawāʾid al-jamma fī isnād ʿulūm al-umma*, 503–12).

[57] Al-Maghrāwī, *Jāmiʿ jawāmiʿ*, 47–49. [58] Ibid., 51.

[59] Ḥajjī, *al-Ḥaraka al-fikriyya*, vol. 1, 85, 101–05. Compare here with the overviews given in Makdisi, *The Rise of Colleges*, 80–98 and Ahmed El Shamsy, "The Social Construction of Orthodoxy," *The Cambridge Companion to Classical Islamic Theology*.

Prophetic Tradition; and finally, through debate, which was limited to a small number of accomplished scholars and advanced students.[60] Classes were generally taught from November through May, according to the solar calendar, and during this time began an hour after the morning prayer, before sunrise, and continued – with the interruption of the daily prayers – until an hour after sunset. Still, some teachers taught students throughout the year, and travel was such a standard part of many scholars' lives due to their carrying out the pilgrimage, taking part in military ventures against the Spanish and Portuguese, and traveling to different parts of the kingdom to take part in public celebrations, that studying with a teacher while traveling was commonplace.[61] Advanced and ambitious students would seek out renowned teachers throughout the region, and while the urban centers of Fez and Marrakesh continued to attract a large number of students, the career of one of the most celebrated scholars of the time, al-Ḥasan al-Yūsī (d. 1102/1691) is notable in that the two main locations at which he studied, taught and wrote were the rural Dilā' and Nāṣirī lodges.

This overview of the education landscape would ideally be supplemented with the daily routine of an individual student to help us better understand the circumstances in which specific sciences were acquired. We are fortunate enough to have such an account in the writings of Abū Sālim Ibrāhīm b. ʿAbd al-Raḥmān b. ʿIsā, known as al-Kalālī (d. 1047/1637) that details his studies in Fez during the reign of al-Manṣūr, soon after he first came to Fez as a teenager in 994/1586 and took up residence in the Miṣbāḥiyya college.[62] Here is a brief passage from this account that paints a vivid picture of one particular class:[63]

> At that time, I attended daily after the morning prayer the session of our teacher Sīdī Muḥammad al-Sharīf al-Mrī al-Tilimsānī[64] [d. 1018/1609–10] on al-Shaykh's *Treatise*[65] at the chair at the back of the allotted space (*al-khaṣṣa*) of the aforementioned communal mosque [i.e., al-Qarawiyyīn]. The following students were attending this session at that time: our blessed teacher, the jurist, grammarian,

[60] Ibid., vol. 1, 94–99. [61] Ibid., vol. 1, 111–14.

[62] Al-Mannūnī, "Faṣla taṣif al-dirāsah bi-l-qarawiyyīn ayyām al-manṣūr al-saʿdī." The Miṣbāḥiyya college is one of the many colleges in Fez founded by the Merinids, built in the 740s/1340s.

[63] Ibid., vol. 3, 1135–36.

[64] On this scholar, including the spelling of his *nisba*, see the entry in the biographical dictionary of al-Qādirī, *Nashr al-mathānī li-l-ahl al-qarn al-ḥādī ʿashr wa-l-thānī*, included in Muḥammad Ḥajjī (ed.), *Mawsūʿat aʿlām al-maghrib*, vol. 3, 1174.

[65] This is the famed treatise in Mālikī law of the tenth century jurist Ibn Abī Zayd al-Qayrawānī (d. 386/996).

prosodist, litterateur, Abū ʿAlī Sīdī al-Ḥasan b. Mahdī al-Zayyātī, who was staying then at the al-Ḥalfāwiyyīn college; our brother the jurist Abū ʿAbdallāh Sīdī Muḥammad al-Baṭīwī Makhshān al-Shafshāwnī; and the jurist Abū ʿAbdallāh Sīdī Muḥammad al-Baṭīwī. All of them were staying at the aforementioned college with a group of others of whom we knew nothing then nor do we now. Another person who attended the mentioned session was the teacher Abū ʿAbdallāh Sīdī Muḥammad al-Sharīf al-Jarāī al-Qaṣarī, who was based at al-ʿAṭṭārīn college. The following students of the college of Sīdī Miṣbāḥ were there attending: the teacher and verifier Sīdī Abū Qāsim al-Filālī, the teacher and polymath Sīdī Aḥmad al-Farkalī and a group of devoted students from Fes. One of those present was the student Sīdī ʿAbd al-ʿAzīz al-Zamrānī. The Imām al-Qalshānī was also present. The Shaykh did not look at him, instead relying in what he said on Sīdī Yūsuf and Ibn Nājī[66], and if anyone mentioned to him the speech of al-Qalshānī,[67] he said to that person: What does al-Qalshānī have to do with me?

When he finished with the *Treatise*, he would descend to the ground and teach al-Sanūsī's *Short Creed.* Some students would leave and others would remain sitting. The knowledgeable and insightful jurist Abū Ḥafṣ Sīdī ʿUmar b. ʿAbd al-ʿAzīz al-Khaṭṭāb[68] attended this session and spent much time in intense discussion with the mentioned Shaykh, to the point that the students would become bored of it. Sometimes he taught Ibn Mālik's *Alfiyya,* relying on Sīdī al-Makūdī.[69] At other times he taught al-Sanūsī's *Long Creed.* All of this was at odds with the desires of his students regarding what he offered, after finishing what was his contractual teaching, to teach of his own accord.

We see here students and teachers from a number of colleges attending a class together and we get a sense of what the dynamics of class discussion may have looked like to a young, recent arrival to Fez, albeit one who tells us that he had memorized prior to his arrival, along with the Qur'an, Ibn Abī Zayd's *Treatise* as well as Ibn Mālik's *Alfiyya*.[70] His description reveals that students not only noted which commentaries were being referred to during their classes, but were also sensitive to the academic politics underlying their instructor's choices in this matter: here, al-Kalālī stresses his teacher's rejection of the pressure of referring to a commentary written by

[66] The authors of two commentaries on Ibn Abī Zayd's treatise: Sīdī Yūsuf b. ʿUmar al-Anfāsī (d. 761/1359) and Abū al-Faḍl b. ʿĪsā b. Nājī (d. 837/1433 or 838/1434).

[67] This seems to be a reference to the Tunisian scholar Aḥmad al-Qalshānī (d. 863/1459), who wrote a commentary on Ibn Abī Zayd's treatise entitled *Taḥrīr al-maqāla fī sharḥ al-risāla*.

[68] A jurist and grammarian who was killed in 1002/1594 at the age of thirty. See Ibn al-Qāḍī, *Durrat al-ḥijāl fī asmāʾ al-rijāl*, vol. 3, 206.

[69] Abū Zayd ʿAbd al-Raḥmān al-Makūdī (d. 807/1404–05) wrote a commentary on Ibn Mālik's (d. 674/1274) *Alfiyya*, a seventh/thirteenth-century introductory work on Arabic grammar.

[70] Al-Mannūnī, "Faṣla taṣif," vol. 3, 1132.

an ancestor of someone in attendance. Yet he also describes how Sīdī Muḥammad al-Sharīf, who seems to have been paid to teach Ibn Abī Zayd's *Treatise*, a legal text, would once he finished with the first for the day, switch to introductory theological and grammatical texts.[71] From a student's perspective, in other words, there were many places to seek out instruction that were not set down in the texts contractually taught in accordance with either endowed chairs or the directions of those who managed the endowments of the respective mosques and colleges. The question of what a student could study was, however, also linked to the availability of books.

The geographic spread of studying and writing beyond the traditional urban centers was facilitated by the founding of numerous public and private libraries – most often affiliated with mosques, lodges, or colleges – that went hand in hand with a greater demand for book copying.[72] While we have mention of private libraries in Morocco as far back as the Idrīsids and the Almoravids, public libraries appeared under the Almohads and Merinids and expanded in size and number under the Saʿdīs. Aside from the Saadians own interest in learning, this increase was in part due to the books brought by Muslims arriving from Spain during the sixteenth-seventeenth centuries, as well as books being imported from the Mashriq and West Africa. Thus, we know of public libraries in Fez, Marrakesh, Taroudant, Meknes, Rabat, and Asfi but also at Sufi lodges in the rural areas of Figuig, Dilā', and Tuwat.[73] This was in additional to hundreds of private libraries, some of them substantial, in urban areas, as well as those of the Habtī family in the Rif, the ʿAyyāshī family at their lodge in the High Atlas, the Nāṣirī family at their lodge at Tamgrout in the Draʿ valley in the south, and the Samlālī family at their lodge at Ilīgh in the anti-Atlas.[74] This proliferation of books throughout Morocco deep into rural areas reinforces the impression that the knowledge production of these areas during this

[71] On the relationship between al-Sanūsī's various creeds, see Caitlyn Olson, "Beyond the Avicennian Turn: The Hierarchy of Creeds of Muḥammad b. Yūsuf al-Sanūsī."

[72] Ḥajjī, *al-Ḥaraka al-fikriyya*, vol. 1, 134–35. For a broader understanding of the significance of libraries in the premodern Middle East, the work on Konrad Hirschler is essential. See now his *A Monument to Medieval Syrian Book Culture – The Library of Ibn 'Abd al-Hādī*. For the donation of individual works to libraries, with a striking example of Ibn Khaldūn donating a copy of his history to the Qarawiyyīn library, see Aḥmad Shawqī Binebine, "Ẓāhirat waqf al-kutub fī tārīkh al-khizāna al-maghribiyya."

[73] For an overview of the rise of privately endowed libraries in the late Medieval period, see Hirschler, *The Written Word in the Medieval Arabic Lands*, chapter 4: "Local Endowed Libraries and Their Readers."

[74] Ḥajjī, *al-Ḥaraka al-fikriyya*, vol. 1, 182–94.

period, if not marginalizing urban centers, relativized their importance considerably.[75] The results of this period can still be seen today, as any visitor to the Ḥamziyya-ʿAyyāshiyya or Nāṣiriyya lodge can attest, for both still possess substantial libraries dating back to the seventeenth century, which in the former case is of this writing not yet entirely under the control of the centralizing state's Ministry of Pious Endowments.[76]

The process of studying and learning continued after a student finished his studies at a lodge, mosque, or college, and he could continue to collect certificates throughout his life. Yet unless he was independently wealthy or otherwise employed – we know of several examples of scholars holding a full-time job in trade or manual labor – he would need to seek employment or patronage. Here there were the official state offices of judge (*qāḍī*) or jurisconsult (*muftī*) that he could be appointed to by the ruler, both of which seem to have guaranteed comfortable salaries. Or they could apply to be official witnesses, imams, or preachers, offices affiliated with specific mosques or *madāris* in the case of the first, only with mosques in the latter cases. These offices were paid from the endowments associated with the institution, and in the case of the imams and preachers, included free housing adjacent to the mosque or madrasa.[77] What of those students who chose to follow their interests in the natural sciences for their professions? Here the two most likely options were to become a timekeeper (*muwaqqit*) affiliated with a mosque, or to become a doctor.[78] The little evidence that we have regarding the professional formation of these offices in the seventeenth century suggests that timekeepers were also paid from the endowments of religious institutions, and that during the reign of al-Manṣūr at least, doctors were tested before being allowed to practice.[79]

[75] For an eloquent argument for the importance of reframing rural areas in North Africa – in this case the oases of Touat – as centers of knowledge production and not merely as recipients of urban knowledge, see Warscheid, *Droit musulman et société au Sahara prémoderne*, 48–57.

[76] For a general narrative of the fate of Morocco's libraries and manuscripts during the Protectorate and their restructuring following independence, see Ahmed Chouqui Binebine, *Tāʾrīkh khazāʾin al-kutub bi l-maghrib*, 147–58. This is the Arabic translation of the French original, which I was unable to consult.

[77] Ḥajjī, *al-Ḥaraka al-fikriyya*, vol. 1, 116–24.

[78] On the office of timekeeper see David King, "On the Role of the Muezzin and Muwaqqit in Medieval Islamic Societies," and specifically on Morocco at 644–45.

[79] For a reference to the licensing of doctors in Saadian Morocco, and their having a leader – during the reign of al-Manṣūr it was al-Wazīr al-Ghassānī (fl. 1012/1603) – who oversaw their affairs and tested each one of them, see Ḥajjī, *al-Ḥaraka al-fikriyya*, vol. 1, 160, where he is relying on ʿAbd al-Ghanī al-Zammūrī's (d. 1030/1621) *al-Qānūn al-mufīd fī ʿalāj al-ḥaṣā bi-qawl sadīd*, a work that remains in manuscript. I consulted the copy in the ʿAllāl al-Fāsī Library.

Taken together, all of these opportunities were limited by the number of available institutions, which while substantially greater than before the Saʿdīs, would not have been able to provide employment for all students, many of whom would have likely been absorbed into the court and the regional administrations of the Saʿdī state.[80] That certain disciplines, jurisprudence chief among them, offered a greater number of employment possibilities than others, must have shaped students choices – much as it does today. Yet, social prestige must have also played a role – it is striking that endowed chairs at *madāris* were mostly made in the fields of Prophetic tradition, jurisprudence, or Qur'anic exegesis, disciplines the endowing of which would have counted as a charitable act for the public good, much like endowing a college or public fountain (*sabīl*).[81] Yet, there is also the example of Ibn Ghāzī, who held teaching chairs in mathematics and inheritance law, which shows that chairs were not solely in the transmitted religious sciences narrowly defined (see Chapter 2).

1.3.2 *Intellectual Trends, Schools of Thought, and the Problem of "Religion"*

Within this educational landscape, scholars pursued a variety of intellectual interests that fell within a series of established disciplines including grammar, rhetoric, Prophetic tradition, Qur'anic interpretation, jurisprudence, legal theory, theology, Sufism, logic, mathematics, geometry, medicine, astronomy, astrology, and timekeeping. Students started with the linguistic sciences before they began their studies at a lodge or college, mastering, as we saw with al-Kalālī, basic works of grammar and memorizing the Qur'an before specializing further. In Chapter 2 I will take up divisions of the sciences from this period in detail to elucidate the overall place of the natural sciences in contemporary epistemologies. Here, I wish to stress that the overall intellectual landscape of the time was one in which jurisprudence, theology, and Sufism formed an inextricable backdrop for all intellectual activities. Thus, while a given scholar may not have specialized in jurisprudence, theology, or Sufi texts, his understanding of these disciplines was profoundly shaped by the Māliki legal, Ashʿarī theological,

80 This point is speculative.

81 See Ḥajjī, *al-Ḥaraka al-fikriyya*, vol. 1, 120, where, drawing on a number of *fahāris*, he notes that such chairs were often named after the first scholar to hold them instead of the endower, and that they were held for life. Compare with Gabriel Baer, "The Waqf as a Prop for the Social System (Sixteenth–Twentieth Centuries)," especially at 285–87.

and Shādhilī Sufi traditions, which represented the majority of the scholarship produced in Morocco during this period. It may be obvious to say so, then, but an astronomer, physician, alchemist, or philosopher of the period would have begun his studies in those fields after having completed at least elementary studies of the Qur'an and Prophetic tradition and, more likely, would be studying jurisprudence or theology in conjunction with his studies in the natural sciences. This was, after all, a period when specialization in a single science was still comparatively rare, and most acclaimed scholars were well-known for their mastery of multiple fields. One example of this is the case of the astronomer Muḥammad al-Rūdānī (d. 1094/1683), who attained fame through both his astronomical works and his writings on Prophetic tradition.[82]

In practical terms, the acquisition of knowledge in any science began with the mastery of a series of texts, a canon of inherited authority. I have combined here Ḥajjī's brief survey of the most popular works in a variety of sciences with the curriculum laid out by al-Yūsī in his canon (*al-Qānūn*).[83] The two are distinct in that Ḥajjī has collated works from the intellectual autobiographies that formed the basis for his study – it is a descriptive list – whereas al-Yūsī laid out an overview of the most relevant works studied in each of the four schools of jurisprudence, organized by whether they were summaries, of medium length, or extensive in nature. His is, in other words, a normative list that does not address the natural sciences. This survey should be read alongside those works mentioned in Chapter 2 where I take up individual accounts of scholars' studies (*fahāris*):[84]

Qur'anic exegesis:

Summaries: Ibn al-Jawzī's (d. 597/1200) *Zād al-masīr*; al-Wāḥidī's (d. 468/1076) *al-Wajīz*; Ibn Juzayy's (d. 741/1340)'s *Tashīl*.

[82] On Rūdānī see El-Rouayheb, *Islamic Intellectual History*, 160–70, and the discussion in Chapter 4.

[83] See Ḥajjī, *al-Ḥaraka al-fikriyya*, vol. 1, 136–37, and al-Yūsī, *al-Qānūn*, 363–72. I have slightly changed the arrangement of the works listed to distinguish the sciences involved, and added al-Sanūsī, the North African theologian whose importance Ḥajjī stresses elsewhere and whose creeds were commented upon repeatedly during this period. Compare with those authors and works listed by the sixteenth/tenth century Algerian scholar Muḥammad ibn Maryam al-Tilimsānī (d. 1014/1605) in his biographical dictionary of the scholars of Tilimsān, *al-Bustān* (Osama Abi-Mershed, "The Transmission of Knowledge and the Education of the 'Ulama in Late Sixteenth-Century Maghrib: A Study of the Biographical Dictionary of Muhammad ibn Maryam," at 23).

[84] It is of interest to compare the works listed here with those studied in the Qarawiyyīn in the last decades of the nineteenth century, as there is a great deal of overlap (see G. Delphin, *Fas, son université et l'enseignement supérieur musluman*, 31–41).

Medium works: al-Baghawī's (d. 516/1122) *Tafsīr*; al-Kawāshī's (d. 680/1281)'s *Tafsīr*; al-Māturīdī's (d. 333/944) *Tafsīr*; al-Wāḥidī's *al-Wasīṭ*; al-Zamakhsharī's (d. 538/1144) *al-Kashshāf*; Abū Bakr Ibn al-ʿArabī's (d. 543/1148) *Aḥkām al-Qur'an*.

Extensive works: Fakhr al-Dīn al-Rāzī's (d. 606/1209) *al-Tafsīr al-kabīr*; Ibn ʿAṭiyya al-Gharnāṭī's (d. 542/1147) *al-Muḥarrir al-wajīz*.

Prophetic Tradition:

Collections: Mālik's *Muwaṭṭa'*; the *Ṣaḥīḥayn* of al-Bukhārī and Muslim, the *Sunan* of Ibn Mājah, Abū Dāwūd, al-Tirmidhī, al-Nasāʾī, and al-Dāraquṭnī (d. 385/995); the *Musannadāt* of Ibn Ḥanbal (d. 241/855), Ibn Abī Shayba (d. 235/849), and al-Bazzār (d. 292/905).[85]

Criticism: al-Nawawī's (d. 676/1277) *Taqrīb al-Taysīr*; al-Nīsābūrī's (d. 405/1015–16) *Ma'rifat 'ulūm al-ḥadīth*; al-Khaṭīb al-Baghdādī's (d. 463/1071) *Kifāya*; Ibn al-Ṣalāḥ's (d. 643/1245) *Muqaddima*.

Prophetic Biography: Qāḍī ʿIyāḍ's *Shifā'*, Imām al-Suhaylī's (d. 581/1185) *Rawḍ al-unuf*.

Creed and theology:

Summaries: al-Ghazālī's (d. 505/1111) *Qawā'id al-ʿaqā'id*;[86] the Creeds of al-Sanūsī (d. 895/1490); al-Nasafī's (d. 537/1142) *ʿAqā'id al-burhān*.

Medium works: Fakhr al-Dīn al-Rāzī's *Muḥaṣṣal* and *Maʿālim*; al-Juwaynī's (d. 478/1085) *Irshād*; al-Urmawī's (d. 693/1294) *Lubāb al-arbaʿīn*.

Extensive works: Fakhr al-Dīn al-Rāzī's *Nihāyat al-ʿuqūl*; al-Samarqandī's (d. 575/1179) *Ṣaḥā'if*.

Sufism:

al-Qushayrī's (d. 465/1072) *Risālah*, al-Ghazālī's (d. 505/1111) *Iḥyāʾ ʿulūm al-dīn*

Legal theory:

Summaries: al-Juwaynī's *al-Waraqāt*; Ibn al-Sāʿātī's (d. 394/1004) *al-Qawā'id*; Ibn Bājī's (d. 473/1081) *Kitāb al-ḥudūd*.

Middle works: Ibn al-Ḥājib's (d. 646/1248) *Mukhtaṣar*; al-Bayḍāwī's (d. 685/1286) *Minhāj*; al-Urmawī's *al-Tanqīḥ* and *al-Taḥṣīl*.

Extensive works: al-Ghazālī's *al-Mustaṣfā*; Fakhr al-Dīn al-Rāzī's *Maḥṣūl*; al-Āmidī's (d. 631/1233) *Iḥkām*.

[85] On the canonical books of Prophetic Tradition having been increased from six to ten in the Maghrib under the Almohads (albeit a slightly different ten than found here), see Goldziher, *Muhammedanischer Studien*, vol. 2, 265.

[86] As Ḥamānī notes in a footnote, this refers to a section of the *Iḥyā'* (al-Yūsī, *al-Qānūn*, 369).

(Mālikī) Jurisprudence:[87]

Summaries: Ibn Abī Zayd al-Qayrawānī's (d. 386/996) *Risāla*; al-Qādī Abū Muḥammad al-Baghdādī's (d. 422/1031) *Talqīn*; Qādī ʿIyāḍ's *al-Qawāʿid*; Ibn Jallāb's (d. 378/988) *al-Tafrīʿ*.

Medium works: al-Qādī Abū Muḥammad al-Baghdādī's *al-Muʿawwanna*; Ibn Shāsh's (d. 610/1213)'s *al-Jawāhir*; Ibn al-Ḥājib's *Jāmiʿ al-Ummahāt*; Ibn Rushd's(d. 520/1126) *al-Muqaddimāt*; al-Khalīl's (d. 726/1326)'s *Mukhtaṣar*.

Extensive works: Mālik's *Muwaṭṭaʾ*, Saḥnūn's (d. 240/854) *Mudawwana*; al-Qarāfī's (d. 684/1285) *Dakhīra*; Ibn ʿArafa's (d. 803/1401) *Mukhtaṣar*; Ibn Rushd al-Jadd's *al-Bayān wa l-taḥṣīl*

Poetry: The Hanging Odes, the pre-Islamic and early Islamic poets al-Shanfarā, Kaʿb b. Zuhayr and Ḥassān b. Thābit, the anthology of Abū ʿAlī al-Qālī (d. 356/967).

Grammar:

Summaries: Ibn al-Ḥājib's *al-Shāfiyya*; Ibn Mu'ṭī's (d. 672/ 1231) *Alfiyya*; Ibn Mālik's *Alfiyya* (d. 672/1274); Ibn ʿAjurrūm's (d. 723/1323) *Muqaddima*.

Medium works: al-Zamakhsharī's *al-Mufaṣṣal*; Ibn Mālik's *al-Tashīl*.

Extensive works: Sībawayh's (d. ca. 180/796) *Kitāb*; al-Suyūṭī's (d. 911/1505), *Jāmiʿ al-jawāmiʿ*.

Books on Language: the works of Quṭrub (d. 206/821), Thaʿlab (d. 291/904), al-Zajjājī (d. 337/948 or 339–40/949–50), al-Jawharī's (d. between 393/1003 and 400/1010) *Ṣiḥāḥ*, al-Zabīdī al-Andalusī (d. 379/989)

Geometry: Euclid's *Elements*

Mathematics: Ibn al-Yāsmīn's (d. 601/1204) works.[88]

Medicine: Ibn Sīnā's (d. 428/1037) *Qānūn*

This list is less an indication of the most important works of the period as it is an impression of what works a scholar would have thought of as canonical for a given science. As such can be supplemented with an overview of the works in astronomy, astrology, and medicine written in Morocco during the Saʿdī period, which also shows that canonical works were not often the ones on which commentaries were written:[89]

[87] Al-Yūsī also gives the respective books for the other three Sunni law schools, which I have omitted here (ibid., 365–66).

[88] For an insightful survey of the teaching of mathematics in the premodern Middle East, see Abdelajouad, "Issues in the History of Mathematics Teaching in Arab Countries."

[89] The source is again Ḥajjī's survey of the intellectual life of the period (*al-Ḥaraka al-fikriyya*, vol. 1, 158–61). These works are not necessarily extant and this list should be compared with the manuscript list in Appendix 2, which contains only extant works.

Astronomy and timekeeping:

- Numerous commentaries on Ibn al-Banna' al-Marrakushī 's (d. 721/1321) *al-Yasārah fī ta'dīl al-sayyārah*, including those of Aḥmad b. Ḥumaydah al-Muṭarrifī (d. 1001/1592), Aḥmad b. Ma'yūb al-Andalusī (d. 1020/1610–11), and 'Abd al-Raḥmān al-Sūsī al-Bu'aqīlī (d. 1019/1609–10).
- Many commentaries on al-Jādirī's (d. 839/1435) work on timekeeping, *Rawḍat al-azhār fī 'ilm waqt al-layl wa l-nahār*, including those of al-Muṭarrifī, Ibn al-Qāḍī (d. 1025/1616), and al-Bu'aqīlī.
- al-Bu'aqīlī's versification of Muḥammad Sibṭ al-Māridīnī (d. 912/1506)'s *al-Risāla al-fatḥiyya fi al-a'māl al-jaybiyya.*
- al-Muṭarrifī's *Jam' al-muhimmāt al-muḥtāj ilayhā fī 'ilm al-mīqāt*, in which he also talks about the use of the astrolabe; al-Muṭarrifī's *al-Muqrib fī waṣf al-mujayyab*, which details the use of the sine quadrant.
- al-Zammūrī's (d. 1030/1621) *Urjūza fī waṣf al-manāzil.*
- Ibn 'Āshur's (d. 1040/1631) *Manẓūma al-rub' al-mujayyab.*
- Muḥammad b. 'Ali al-Shāṭibī (d. 936/1529)'s *Nubdha falakiyya.*

Medicine:

- Abū Qāsim b. Muḥammad b. Ibrāhīm al-Ghassānī *(d. 1019/1611)'s Ḥadīqat al-azhār; Ikhtiṣār ḥadīqat al-azhār; Tafsīr ba'ḍ al-a'shāb wa l-'aqāqīr*; *al-Rawḍ al-maknūn fī sharḥ rajaz Ibn 'Azrūn* (written in 999/1586, the poem on fevers and tumors, a supplement to Ibn Sīnā's poem); *Mughnī al-labīb 'an kutub a'dā' al-ḥabīb* (a translation of European works on medicine).[90]
- 'Abd al-Ghanī b. Mas'ūd al-Zammūrī's *al-Qānūn al-mufīd fī 'ilāj al-ḥaṣā bi-qawl sadīd*, dedicated to the ruler Aḥmad al-Manṣūr; *Khawāṣṣ al-nabātāt*, an explanation of the Greek, Syriac, Persian, and foreign names of medicines.
- Abū Qāsim al-Ghūl al-Fishtālī (d. 1059/1649) wrote in 1038/1628 a 1410-line poem entitled *Ḥāfiẓ al-mizāj* as well as a poem on the plague.[91]
- Muhammad b. Ahmad al-Bu'aqīlī's (fl. 17th century) *Majmū'at al-manāfi' fi 'ilm al-ṭibb al-nāf'i.*[92]
- 'Abd al-'Azīz al-Rasmūkī's (d. 1065/1665) *Kashf al-rumūz*, in which the author explained medical issues, herbs and sicknesses in Shilha (Amazigh)

[90] On this last work, part of which may be extant, see Muḥammed al-Mannūnī, "Ẓāhira ta'ribiyya fī l-maghrib ayyām al-sa'diyyin," 332–33.

[91] On his poem on the plague, which does not seem to be extant, see al-Ifrānī, *Safwa man intashara*, 248.

[92] Also known as *al-Ṭibb al-Bu'aqīlī*, of which the Moroccan National Library has numerous copies (see Appendix 2). Note that this is a different al-Bu'aqīlī than the one who lived at the beginning of the seventeenth century and whose works were cited above.

- The many works of ʿAlī b. Ibrāhīm al-Andalusī (fl. seventeenth century) who dedicated a number of medical poems to Moulay Zaydān when he ruled Marrakesh: *Urjūzat al-fawākih al-ṣayfiyya wa l-kharīfiyya, Urjūzat ʿilāj al-ʿuyūn, Urjūzah fī al-aʿshāb wa khawāṣṣihā fī shifāʾ al-amrāḍ, Manzūmah fī l-nikāḥ.*

Even a quick survey of these lists confirms Berque and Ḥajjī's impression that many of the works written under the Saʿdīs were commentaries or mnemonic poems, although scholars also wrote monographic works or long commentaries on poems they themselves had written.[93] These commentaries and summaries were not principally on the canonical texts listed above as they were based on later texts by respected regional authors such as the eighth/fourteenth-century Ibn al-Banna' al-Marrakushī. These were teaching texts and the profusion of commentaries on them indicates a scholarly culture that placed a great amount of emphasis on instruction, a focus that also goes a certain way toward explaining the large number of poems written during this period. The spread of educational institutions sketched above in rural areas and lesser urban centers in Morocco created a demand for summaries and commentaries, and even without drawing on the revisionist opinion that these genres were often quite innovative in their content, this expansion of literacy and education should be seen as a sign of intellectual vibrancy instead of decline. It is worth comparing this situation with Sonja Brentjes' observations about the growth of commentaries in the field of mathematics in the late Medieval period in the Mamluk Empire:

> The increase of elementary texts as well as commentaries, super-commentaries, and glosses and the decreased importance of texts by ancient and classical authors in the educational cultures of post-classical societies are features that came out of the integration of the mathematical disciplines into the madrasa. These phenomena observable in the mathematical literature of the period reflect their function as introductory and survey for a student population that aimed primarily to acquire legal and religious knowledge and studied mathematical texts as an auxiliary means for their later activities as judges and other legal officials or as members of the fiscal administration. Thus, the emergence of a huge quantity of basic, elementary texts on mathematical topics should no longer be interpreted as signs of decline, but recognized as an expression of a profoundly altered social and cultural function of mathematical education.[94]

[93] See Appendix 2, however, for an attempt to establish a list of astronomical, medical, astrological, and lettrist texts composed in Morocco between the sixteenth and eighteenth centuries.

[94] Brentjes, "Teaching the Mathematical Sciences in Islamic Societies Eighth–Seventeenth Centuries," 104. See now also Brentjes, *Teaching and Learning the Sciences.*

The educational system and scholarly environment described here was a religious one. That this was the case seems clear, especially when one considers the importance of Sufi lodges and that the colleges were maintained by pious endowments that were frequently closely associated with mosques. Yet it is far less clear what the category of religion means, and how the student of the period is best advised to avoid importing into her understanding of this period the nineteenth-century baggage of the term alluded to in the Introduction. The predicament of how to accurately parse the significance of the term is especially pressing when we turn to the natural sciences, considering how closely modern science and secularism have been intertwined in our collective narrative of modernity. The more triumphalist apologists for science today depict it as having superseded religion, and less strident defenders are content with arguing that science and religion operate in different spheres that coincide, if at all, only in their appreciative contemplation of creation. What, then, does it mean to discuss the study and writing of the natural sciences at Sufi lodges and Islamic colleges in Morocco in the seventeenth century?

In the Introduction, I alluded to the debate between Cunningham and Grant regarding how to frame the nature of Newton's seminal seventeenth-century work the *Philosophiae naturalis principia mathematica.*[95] For Cunningham, since Newton had stated that the work was about God's creation and titled it as he had, it was a work of natural philosophy and reflected a religious worldview in which a rigorous investigation of the laws of physics went hand in hand with a belief in God's presence in the natural world. There is no division between religion and science as these two terms have not yet acquired their modern meaning. For Grant, on the other hand, Newton's work is one of science, his talk of God is mere rhetorical window dressing and we can find a separation of science and religion as far back as the division between the faculties of Arts and Theology in the Medieval European University.[96] In a narrow sense, the discipline of the history of science is of course more closely aligned with Grant's view, for if there is no justifiable teleology through which to look for antecedents to contemporary understandings of the world, then what justification is there for the discipline itself? The history of science would necessarily become absorbed into intellectual history writ large. Yet, even if not all would accept Cunningham's approach, many if not most

[95] Their choice of this work as an anchor for their debate was hardly accidental. On the ways in which Newton and his work came to symbolize the Enlightenment tout court, see J. B. Shank, *The Newton Wars and the Beginning of the French Enlightenment.*

[96] Edward Grant, *God and Reason in the Middle Ages.*

historians of science in the European context have in the past generations become more skeptical of their discipline's central teleology and have adopted a more contextualized and contingent understanding of their remit.[97] For reasons described in the Introduction, the same has not been the case for many historians of science working on the Islamic world, resulting in part in a scripturalist understanding of religion and Islam in particular that has privileged the self-understandings of jurists and theologians over those of other Muslims. More simply: when discussing the place of the natural sciences in Islamic societies and the critiques occasionally leveled at their practitioners, historians have often taken at face value the objections to these sciences leveled by self-appointed guardians of orthodoxy, and this has in turn shaped their understanding of the category of religion itself.[98] At its most reductive, this can lead to the definition of Islam as primarily a religion of law or Scripture, a curious characterization that could be credited to nineteenth-century Protestant critiques of religions of law, Muslim reform movements from the eighteenth century onwards, or Colonial attempts to codify Islamic law for administrative use.[99] Few students of Islam or the Middle East continue to profess such a narrow understanding of the Islamic tradition, but the legacy of such descriptions continues to influence the presuppositions of much scholarship in the field.

A first step in moving beyond an understanding of Islam as law has been to reintegrate Sufism and the social organization of Sufi orders into the intellectual and social history of the Muslim world as a means of questioning the right of the disciplines of law and theology to define what is the essence of the tradition and what is an accretion.[100] The scholarship of the

[97] Shank's *The Newton Wars* is an excellent example of this trend.

[98] For an example and critique of this trend, see Sonja Brentjes, "On Four Sciences and Their Audiences in Ayyubid and Mamluk Societies."

[99] See Bauer, *Die Kultur der Ambiguität*, chapter 6: "Die Islamisierung des Islams," and Ahmed, *What Is Islam?*, chapter 2: "Islam as Law." The Christian characterization of Judaism as a religion of law and therefore lacking goes back to the writings of Paul in the New Testament and was taken up by later figures as diverse as Augustine and Martin Luther. For the history of the role of Judaism in Christian self-definition, see David Nirenberg, *Anti-Judaism: The Western Tradition*. What was new to the nineteenth century was the force with which these critiques were applied to Islam.

[100] A corollary to the Protestant critique of Judaism and Islam as religions of law and therefore lacking a pure spiritual core would be the tendency of such Catholic scholars as Miguel Asín Palacios to depict Sufism as the result of Christian mysticism. In both cases, the Islamic tradition is defined in a way to be lacking when compared with Christianity. For a brief but insightful look at the scholarship of Asín Palacios, see Patrick Henriet, "Miguel Asín Palacios (1871–1944), le christianisme et l'islam: theses, idéologie, reception."

past generation has largely accomplished this goal, culminating in the larger historiographical efforts of Shahab Ahmed and Thomas Bauer mentioned in the Introduction. In the case of Morocco, the work of scholars such as Vincent Cornell and Ruggero Vimercati Sanseverino has demonstrated conclusively the centrality of Sufism and Sufi order to the intellectual milieu from at the latest the sixth/eleventh century onwards. Instead of an accretion or a supplement to a legal orthodoxy, we have come to approach Sufism properly as a yet another facet of Muslims' social and intellectual experience of religion, one that was co-constitutive of legal and theological understandings of the Islamic tradition. Like these other understandings, in the way it has reached us, it was an elite and largely urban phenomenon, albeit one that in certain contexts such as the Moroccan one during our period, was equally present in rural areas.[101] This acceptance of Sufism as quintessentially Muslim has gone hand in hand with recognizing that not all behavior or belief that had once been glossed as ascetic, mystical, or esoteric had anything to do with Sufism at all.[102] In this recent scholarship, Sufism emerges as a hierarchical social organization or order based around the veneration of saints, past and present, that uses supererogatory rituals and gatherings to heighten its followers' awareness of the divine and their moral responsibility to their fellow believers. For our purposes in this chapter, it is the social aspect that is worth stressing for, as we have seen, Sufi lodges played an important role in the production and dissemination of knowledge during this period.[103]

But recognizing the central importance of Sufism in all its diversity in Islamic intellectual history is only a first step to addressing the ways in which that history has been distorted by the historiographic framings of the nineteenth and early twentieth centuries. This has been demonstrated, for example, by a body of recent scholarship that argues that the study of the occult played a far greater role than previously acknowledged in the

[101] On the urban, elite nature of early Sufism see Ahmet T. Karamustafa, *Sufism: The Formative Period*, 19–26.

[102] For a nuanced discussion of scholarly piety in thirteenth–fifteenth century Egypt, see Megan Reid, *Law and Piety in Medieval Islam*. On the nature and importance of *ʿirfān* in the Safavid context, see Ata Anzali, *"Mysticism" in Iran: The Safavid Roots of a Modern Concept*.

[103] For the local history of Sufism in Morocco in the centuries before our period, see Cornell, *The Realm of the Saint*, but also and especially Scott Kugle, *Rebel Between Spirit and Law: Ahmad Zarruq, Sainthood, and Authority in Islam*. Zarruq's (d. 899/1494) example was especially important for many of the Sufis of Morocco in the seventeenth century, as can be seen by their frequent citation of him. For one example, see al-Yūsī, *The Discourses*, vol. 1, 193, 301, 411.

centuries following the dissolution of the ʿAbbasid caliphate, one that was far more similar with its importance in Europe.[104] One could envision a scenario in which disciplines or intellectual and social projects are added individually to what is considered "properly Islamic." It is more accurate and productive to revisit our general use of the concept of religion. At first, this may seem a curious move at the end of a chapter on the intellectual landscape of Morocco in the seventeenth century. Conversely, it may be the only way to properly understand what we are seeing when we look back over where we have been.

Students of religion have struggled a good deal with clarifying their object of inquiry and have ritually noted the diversity of the many definitions their predecessors have given the concept. I am less interested here in definitions than in pulling back from the Moroccan Islamic context to reconsider however briefly this study's broader framing before returning to the Moroccan case study. The discipline that shaped my intellectual formation – it could be variously defined as Islamic Studies, Middle East History or Orientalism – has a long history of textual bias in its privileging the voices of those who wrote: a small, almost entirely male literate elite, many if not most of whom were closely linked to urban centers of power. In this my discipline and its object of study are hardly unique, indeed our sources suffer from a bias common to most if not all fields of premodern history. This bias affects our understanding of the concept of religion itself, for we can catch only glimpses from our sources of how the vast majority of Muslims in the premodern period understood their faith.[105] I am not referring here to what was once called "popular religion" – a concept scholars have now largely discarded – although the recent turn to exploring the embodied experience of Muslim religious practice is relevant here in that it challenges the centrality of the textual to our understanding of the Islamic tradition.[106] Nor am I wholly comfortable with the distinction Marshall Hodgson made between religion and civilization in his truly remarkable introduction to *The Venture of Islam*, or with its related

[104] See my comments in the Introduction on the work of scholars such as Matthew Melvin-Koushki and Noah Gardiner on this topic. Note the importance in this body of work in stressing the distinction between Sufism and the occult.

[105] Yet see now the impressive work of Jack Tannous, *The Making of the Medieval Middle East: Religion, Society, and Simple Believers*.

[106] This trend has been evident in both legal and Sufi studies, in different ways. See Leor Halevi, *Muhammad's Grave: Death Rites and the Making of Islamic Society*; Kugle, *Sufis & Saints Bodies*; Kugle (ed.), *Sufi Meditation and Contemplation: Timeless Wisdom from Mughal India*; Shahzad Bashir, *Sufi Bodies: Religion and Society in Medieval Islam*; Reid, *Law and Piety in Medieval Islam*.

differentiation between the categories of religion and culture.[107] Hodgson's definition of the religious as a person's "ultimate cosmic orientation and commitments and the ways in which he pays attention to them, privately or with others" on the other hand is broad enough to get at some of what I am interested in.[108] Finally, while deeply appreciative of Shahab Ahmed's recent magisterial *What is Islam? The Importance of being Islamic* – and of his own critique of Hodgson – I would like to direct attention to the gap between Ahmed's definition of Islam as a specific religious tradition and broader definitions of religion as a category. Ahmed has defined Islam as "*meaning-making for the self in terms of hermeneutical engagement with Revelation to Muḥammad as Pre-Text, Text, and Con-Text.*"[109] Compare this definition with Thomas Tweed's recent definition of religion: "*Religions are confluences of organic-cultural flows that intensify joy and confront suffering by drawing on human and suprahuman forces to make homes and cross boundaries.*"[110] Tweed's definition, and to some degree Hodgson's, help direct our attention more readily than Ahmed's to the ways in which the study of the natural sciences was a religious endeavor for seventeenth-century Moroccan scholars, something that proceeded from and not despite their self-understanding as Muslims. The study of the natural sciences proceeded from both what Tweed calls "dwelling," a desire to understand and shape one's lived and social environment in light of an awareness of the divine, what Hodgson termed a cosmic orientation.[111] It is true that this took place through an engagement with what Ahmed has called revelation's Pre-Text, or the philosophical heritage of Antiquity, yet for some of the scholars in question, most notably al-Ḥasan al-Yūsī, these sciences and all beneficial knowledge in general was part of God's Revelation, part of the religious and scientific inquiry into the workings of God's Creation. They did not reside in the past to be venerated or preserved, but in the present, to be practiced.[112]

[107] Marshall Hodgson, *The Venture of Islam*, vol. 1, 85–90. Hodgson's aim in this section of his "Introduction" was admittedly distinct from my concern here, as he was invested in creating space for his concept of Islamdom, a cultural or civilizational rubric that could encompass the writings and actions of non-Muslims living under Muslim rule.

[108] Ibid., vol. 1, 88. [109] Ahmed, *What is Islam?*, 405.

[110] Tweed, *Crossing and Dwelling: A Theory of Religion*, 54. As with Ahmed's, Tweed's definition is less useful by itself than as the basis for an extended meditation and exploration of the respective subjects.

[111] Hodgson, *Venture of Islam*, vol. 1, 88.

[112] For al-Yūsī's thinking on this issue, see the "Introduction" regarding his comments in his *al-Qānūn*.

1.4 CONCLUSION

This chapter has attempted to give a brief overview of Morocco's political and intellectual history up until the seventeenth century, alongside a similarly curtailed survey of the educational milieu that came into being in Morocco in the sixteenth century and which defined Moroccan institutions of learning throughout much of the century after the Aḥmad al-Manṣūr's death in 1603. It has also taken up some of the principal historiographical and conceptual challenges of writing an intellectual history of this period: the narrative of Moroccan intellectual decline during this period that has characterized much of the work of such influential scholars as Jacques Berque and Muḥammad Ḥajjī, on whom I have drawn extensively here, but also the categories of Sufism and religion. In doing so, it has begun a longer theoretical engagement with these terms I that will continue in the chapters to come.

This chapter began with two epigraphs concerning the Dilā' lodge and its most famous scholar, al-Ḥasan al-Yūsī, and in concluding it is worth returning to them to remind ourselves of some of the central aspects of the intellectual life of the period. During this time, rural Sufi lodges supplemented when they did not supplant the urban centers of Fez and Marrakesh, in the process speaking to widespread literacy and engagement with received scholarly traditions throughout Morocco and specifically at such lodges as the Bakrī, Nāṣirī, and Samlālī lodges at Dilā', Tamgrout, and Ilīgh, respectively, that hosted sufficient scholars and boasted enough economic means to support advanced studies in a wide variety of sciences. How scholars conceived of these sciences precisely, is the subject of Chapter 2, which takes up the classification of knowledge in seventeenth-century Morocco in greater detail in order to clarify the meaning of the term "natural sciences."

EXCURSUS

The Poverty of Intellectual History as a Series of Great Men

It is a perpetual temptation when not a necessity for the historian to extrapolate from the available evidence to shape a narrative that elides or papers over what is unknown or does not fit. Without such a narrative one is left only with a cluster of facts, individual points of data lacking the requisite generalization to make them legible. The temptation of narrative presents itself not only when evidence is lacking, as it is for much of the premodern period, but also when the painstaking work of previous generations of scholarship has made it comparatively abundant. The perils of the historian's narrative work and the familiar tropes that are available – the rejection of narrative for impressionistic sketches is itself a familiar narrative ploy – is poignant in its inevitability: like a political dynasty, each school of thought has a beginning, a flowering, and a legacy, however varied. And as the historian seeking to understand the nature and significance of a political dynasty will look to establish a list of rulers, whose success she measures in terms of their length of rule, their administrative, military, economic, and cultural successes and failures, so the intellectual historian studying a school of thought or an idea will look for the founder, prominent students and exponents, relations with opposing schools, the influence on it of previous and of it on subsequent schools of thought. After establishing an initial narrative of progression and development, the historian nuances first impressions, questions self-representations of significant actors more deeply, and uses case studies to support or question the established narrative, a narrative that rests in large part on the names and reputations of great men.

When it comes to the intellectual history of the Islamicate world, the above comments, banal in themselves, are especially relevant only to the extent that the field (at least its Western academic recension) as a whole is undeveloped when compared to its European counterpart. For while tremendous advances have been made during the last generation in our understanding of the formative period of Islamic thought from the seventh to eleventh centuries, our understanding of Islamic thought in the post-formative period is still undeveloped, and nowhere more so than with regard to the natural sciences. At issue here is not questioning the notion of post-formative intellectual decline, but exploring the related implications of this history consisting of the legacy of a progressive series of achievements of extraordinary individuals instead of a geographically and institutionally diverse body of engagements with the Islamic reception of Late Antique thought.

Thus, the past generation of scholarship has convincingly shown us that the early history of the four Sunni law schools that would ultimately become prevalent not only needs to proceed on the assumption that in some cases the eponymous founders had limited influence on the early formation of the schools that bore their names, but also that individual schools possessed a great deal of internal diversity and dynamism. Similarly, the range of such diversity and dynamism in the first three centuries with regard to theology has been convincingly demonstrated by the past fifty years of scholarship. This same generation of scholarship has productively directed our attention beyond the names of prominent scholars to the institutional settings in which knowledge was transmitted, and this focus on institutions has in recent scholarship on mathematics and the natural sciences expanded our understanding of the role of these sciences in institutional settings between the eleventh century when the madrasa spread through Iran and the Eastern Mediterranean and the rise of the Gunpowder Empires in the sixteenth century. And yet both this expansive revisiting of the narrative of intellectual history being a largely progressive series of great men and this welcome turn to institutional curricula and transmission has not been applied to the post-formative period, where our grasp of intellectual history is decidedly weaker. Unsurprisingly, there is a direct correlation between our lack of detailed knowledge regarding the intellectual production of a period and our tendency to fall back on the views of the most prominent chronologically proximate scholar whose works are known and this often without acknowledging possible regional diversity. At its most egregious, this tendency can contribute to such claims as al-Ghazālī having put an end to the study of philosophy in the Muslim world. It is worth thinking through the implications of such a claim, for not only does it rest on the understanding that a single scholar was able to articulate a critique of a field that made that body of knowledge anathema to subsequent scholars from the Indian subcontinent to West Africa but also the more pernicious assumption that in charting the trajectory of an idea or claim it is sufficient to focus on the figure of a certain area and period who would at some later date become the most influential. It would instead be more reasonable to assume – this being the case in other intellectual traditions and in the modern period as well – that influence varies across time and place and that a more accurate way of ascertaining the place of an idea or discipline would be to read broadly among the writings of scholars who had varying levels of influence on subsequent generations. This much seems common sense, and where modern scholarship has taken the time to contextualize particular scholars or ideas, they have been remarkably less successful in producing nuanced intellectual histories that reveal complex debates and occasionally bitter disagreements on central claims (the exceptions proving the rule). Yet, while this particular role attributed to al-Ghazālī, for example, has now been put firmly to rest within the fields of Islamic studies and Middle East History (no matter what influence it may still have more generally), other influential examples remain, including such figures as Ibn al-ʿArabī and Ibn Taymiyya whose contested legacies in the realms of Sufism and theology have been productively explored in the last decades. The situation is less clear in the natural sciences.

As with many other branches of Islamicate intellectual history, the broad contours of fields of study are generally located in the third/ninth century. It was in the context of the much-discussed translations in Baghdad of primarily Greek, Syriac, and Sanskrit texts that the foundations were laid for subsequent generations of scholars living in the Muslim world who worked on astronomy, medicine, alchemy, and other natural and mathematical sciences. Our understanding of the subsequent role and nature of these sciences has rested, first, on these translations and then on the work of those scholars who subsequently developed syntheses in Arabic of these fields. In astronomy, for example, we look at al-Bīrūnī (d. after 442/1050) and then at later scholars such as Naṣīr al-Dīn al-Tusī (d. 672/1274) or Shams al-Dīn al-Khafrī (d. after 1525) for an understanding of how these Muslim scholars appropriated, developed, and critiqued a Ptolemaic understanding of the universe. In medicine, the synthetic efforts and lasting influence of Ibn Sīnā are widely acknowledged, while the later efforts of Ibn al-Nafīs (d. 687/1288) have attracted welcome attention. Less so, unfortunately, Dāwūd al-Anṭākī (d. 1008/1599), who, while famous, has had few studies devoted to him. As for alchemy, a field that has admittedly enjoyed less attention than the first two, we have the central figures of Jābir b. Ḥayyān (third/ninth century) and al-Jildakī (d. ca 743/1342).

The dates and locations of these figures alone speak to both to the importance in this narrative of the formative Islamic period (eighth–twelfth centuries) and the central Middle East, principally the region between Egypt and Iran. As a field, we have far less understanding of how these sciences were studied and transmitted outside of the urban centers in these areas, much less in distinct geographical areas (West and sub-Saharan Africa, the subcontinent), although matters have much improved in the past two decades. Still, where we do not have the benefit of a detailed study of the fate of a text's study in both institutional and noninstitutional settings, commentaries on the text, or notes of a text's reception (either original or anecdotally in other works), our evaluation of the study and practice of a specific field is drawn, and understandably so, back to the circulation of prominent texts by prominent figures. The commentaries, glosses, and presumably minor works by presumably minor figures, almost always still in manuscript, are read, when read at all, in search of evidence of progress or development, not to chart the broader presence or popularity of a given science.

As a critique, mine is a sympathetic one, and one to which I do not consider myself an exception. We have simply not had access to the necessary sources to allow us to write the kinds of nuanced histories of the natural sciences that our colleagues in the European history of science have been able to provide in the past generation. Our field is, after all, smaller and the necessary philological skills, in addition to the detailed knowledge of a given science, have set a high bar for the needed qualifications for scholars to produce innovative scholarship. In addition, our historiographical narratives have been significantly impacted by the pressures outlined in the Preface, where we see that the pressure to discover "influence" or antecedents to European scholarship has significantly impacted research programs in the field. Still, when one casts even a cursory eye over recent students of the legal history of specific Ottoman communities in the Early Modern period, or of theology and philosophy in early modern Iran, it is hard not to yearn for a comparable history of astronomy in seventeenth-century Mauritania, or of alchemy in eighteenth-century Tunis. The practical challenges faced by our field should not keep us from aspiring to continuing the work on the importance of institutions in transmitting the natural sciences that has been done for the late Medieval period and to bring it in into the Early Modern world.

Attempting to avoid a focus on the achievements of great men in the study of the natural sciences in the Muslim world leads naturally to the social lives of texts, to their reception, study, and transmission – and ideally beyond the elite and literate worlds that produced and valued these texts to popular understandings of these sciences and the apprehension of the natural world and its functions by the illiterate, who represented the majority of the population. Yet since such a sociological focus on the broader social life and influence of the natural sciences is challenging, if not quixotic due to the limitations of our sources, it is likely our narratives of the natural sciences (and especially the so-called occult sciences, for which we have even fewer sources) will predictably return to the works and names of a number of great men. Much of this current book strives to push back against this temptation by painting the intellectual landscape of Morocco with a broad brush and including multiple creative and dynamic voices. Yet our narrative hunger for the individual genius may lead to the reader to dwell too much in the following on figures such as al-Yūsī or al-Mirghitī at the expense of the institutions and communities from which they emerged. This would be a mistake, and one that would simply expand the canon of great men instead of refocusing on the pervasiveness of the study and transmission of scientific knowledge.

2

Constructing Knowledge in Morocco between the Sixteenth and Eighteenth Centuries

Abū ʿAbdallāh said, "I found that great rancor had developed between the jurist Abū ʿAbbās and the students of religious knowledge in Egypt. The cause was a curious coincidence – namely, that he had one day gone to the book market at a time when he was still unknown. He came upon a rare volume of a Qurʾānic exegesis and opened it to the exegesis of Sūrat al-Nūr. This passage addressed a curious juridical matter and described a disagreement in precise detail. He memorized all of this immediately, being a man with this ability. Soon after this, it happened that the scholars of the city came together for some occasion and he was present. When the gathering was underway, someone holding a card asked about precisely this issue. The card was given to the first of the great scholars attending the gathering. He looked at it, and it appeared that nothing occurred to him on the subject. He gave it to the person next to him. That person then gave it to the next, and so on, until it reached the aforementioned Abū ʿAbbās. When he had read it, he called for an inkwell and wrote out the answer he had memorized. They looked at him with amazement. When he was finished writing, they asked, 'Who wrote this?' and he said, 'So-and-so wrote it in his exegesis of Sūrat al-Nūr.' They asked for the exegesis and found it to be as he had said. Their egos became troubled on account of this. This was not the first time this has occurred and, save for the few whom Exalted God has preserved from the vice, they continue to be prejudiced against those whom they perceive to be superior to themselves, or against those who compete with them for rank or distinction.

Like the fellow wives of a beautiful woman who say of her face;
Out of envy and hatred, 'Truly, it is hideous.'

> Because of this type of thing, a certain jurist issued a legal opinion that scholars should not testify against each other. No doubt this is not generally the case, but the prejudice is widespread. Other examples of this are what famously happened to the exemplar Sībawayhi with the people of Kufa. Or to Sayf al-Dīn al-Āmidī with the people of Egypt – when he demonstrated his superiority over them in the sciences, they disavowed him, accused him of heresy, and wrote an official decree to that effect. They passed this decree among themselves, so that each would sign it and bear witness. They did this until it reached one whom God had granted success and preserved from error, and he wrote underneath the testimonies:
>
> Unable to match him, the people envied the young man;
> they were his enemies and adversaries.
>
> Abū ʿAbbās refrained from anything similar while in Egypt. This is also what occurred to Ibn Tūmart, known as 'the Savior,' the leader of the Almohads. When he entered the city of Marrakesh, he was on his way back from the East and was advocating for the rational sciences (*al-ʿulūm al-ʿaqliyya*). The people of the countryside were ignorant of this and said, 'This fellow has introduced the sciences of the philosophers among us.' They denounced him to the Almoravid ruler, and matters took their course."
>
> Al-Yūsī, *The Discourses*[1]

Writing in his intellectual autobiography in the latter half of the seventeenth century, the celebrated Moroccan scholar, Sufi, poet, physician, and astronomer, Abū ʿAbdallāh Muḥammad b. Saʿīd al-Mirghitī (d. 1089/1678) described how to make a talisman to prevent pregnancy, and a few pages later offered his readers two invocations to protect against epidemics and plague, an occurrence of which would end his own life some years later.[2] Over a hundred pages later in the same work, after listing numerous medical remedies for a wide variety of ailments, he offered a corrective to his Moroccan predecessor, the timekeeper and mathematician Abū Zayd ʿAbd al-Rahman al-Jādirī (d. 818/1415) regarding an astronomical calculation he had made, citing his own observation using an astrolabe in Marrakesh in 1049/1639.[3] For al-Mirghitī, the use of both talismans and

[1] Al-Yūsī, *The Discourses*, 187–89.

[2] Al-Mirghitī, *al-ʿAwāʾid al-muzriyya bi-l-mawāʾid*, vol. 1, 317, 322. On his dying of the plague and being buried in Aghmāt, near Marrakesh, see ibid., 54–55. It is unclear precisely when he completed his *fahrasa*. It is worth noting that there is considerable variation in the spelling of his *nisba*, with al-Mirghītī and al-Mirghithī also being used.

[3] Ibid., 533. See below for more detail on these passages.

astrolabes was part of natural science, beneficial and worthy of being transmitted to future generations of scholars. For the modern reader, living in an age when the world has arguably become disenchanted and for whom there is likely a sharp division between magic and science, these two anecdotes may seem incongruous, the first evidence of superstition, the second of an early stage of now superseded rationality.[4] In this chapter, I seek to move past our contemporary taxonomies of knowledge and to review how Moroccan scholars in the long seventeenth century defined the sciences and how they situated the natural sciences within their own intellectual trajectories as well as within the biobibliographical literature that described their predecessors. The importance of establishing these definitions lies not only in working through yet again the influence of nineteenth- and twentieth-century scholarship on our understanding of the premodern period, but also, if not primarily, in clarifying the specificity of the Moroccan context. In doing so, I focus on the natural sciences broadly, but at times I include the mathematical sciences (especially astronomy) as they were studied in conjunction with or alongside the more narrowly defined category of natural sciences.

Muslim scholars had categorized the sciences since at least the third/ninth century – influential examples include the taxonomies of al-Fārābī (d. 339/950), Ibn al-Akfānī (d. 749/1348), Ibn Khaldūn (d. 808/1406), and Ṭāshkubrīzādah (d. 968/1561) – in a variety of ways and according to different criteria.[5] Running through many of these classificatory efforts is the binary of religious or transmitted sciences (*al-ʿulūm al-naqliyya*) and rational or philosophical sciences (*al-ʿulūm al-ʿaqliyya*). This differentiation does not imply an incompatibility between the religious and rational sciences – though some contemporary and modern commentators have understood it in this manner – as much as it does an awareness of their distinct epistemological bases.[6] Other authors used different categories

[4] On Weber's much cited disenchantment of the world not having been as successful as some believe, see Jason Ā. Josephson-Storm, *The Myth of Disenchantment: Magic, Modernity, and the Birth of the Modern Sciences*.

[5] For a fuller list, see Stearns, "Writing the History of the Natural Sciences," 930–31, and the literature cited there. The following secondary sources are of particular interest: Hans Hinrich Biesterfeldt, "Arabisch-islamische Enzyklopädien: Formen und Funktionen" and ibid., "Medieval Arabic Encyclopedias of Science and Philosophy," Biesterfeldt's overviews should be supplemented by the mentioned study of Heinrichs, Gerhard Endress, "The Cycle of Knowledge: Intellectual Traditions and Encyclopaedias of the Rational Sciences in Arabic Islamic Hellenism," and Osman Bakar, *Classification of Knowledge in Islam: A Study in Islamic Philosophies of Science*. Bakar's study looks closely at the work of al-Fārābī, al-Ghazālī, and Quṭb al-Dīn al-Shīrāzī (d. 710/1311).

[6] See Edward Grant, *Foundations of Modern Science in the Middle Ages*, 181.

altogether, such as Ibn al-Akfānī, who grouped religious and natural sciences together under the rubric of "principal theoretical sciences" (*al-ʿulūm al-aṣliyya al-naẓariyya*), though few went as far as al-Yūsī who defined all sciences that benefited the Muslim community as religiously sanctioned (*sharʿiyya*), and whose views we will turn to below.[7] The diversity of approaches taken by these scholars emphasizes the limitations in the generalizations employed by historians of the Muslim world in the nineteenth and twentieth centuries on the relationship between a singular Islam and a generic conception of science. The field of Islamic intellectual history has matured enough so that we can write more fine-grained local histories of the natural sciences in Muslim societies that accentuate geographic particularities and enable more effective comparisons both within the Muslim world and globally. It is in this context that this chapter takes up the specific place of the natural sciences within the Moroccan intellectual landscape in the sixteenth–eighteenth centuries.

It does so in three ways. First, it offers an overview of the presence of the natural sciences in the biobibliographical (*ṭabaqāt*) literature of the sixteenth–eighteenth centuries. It then turns to three extended attempts of scholars during these centuries to develop taxonomies of the sciences, and examines the place of the natural sciences in these divisions. Finally, it takes up the genre of the scholarly autobiography (*fahrasa*), which has a particular history in the Islamic West, and looks closely at the ways in which scholars framed the natural sciences in their own intellectual careers and surroundings. Taken together, these readings of different genres of scholarly self-examination build on the institutional and historical context established above even as they provide the necessary intellectual context for the following chapters.

2.1 GENEALOGIES OF KNOWLEDGE

As discussed in Chapter 1, the *ṭabaqāt* literature not only preserved biographical information about prominent scholars, but also, by its selection and arrangement of information, often functioned to present the compiler's vision of appropriate or desirable knowledge. The inclusion of the philosophical sciences in collections of the generations of scholars who studied and taught in the Maghrib between the sixteenth and eighteenth

[7] For Ibn al-Akfānī, see J. J. Witkam, *De egyptische arts Ibn al-Akfānī (gest. 749/1348) en zijn indeling van de wetenschappen*, 20–21 (English Introduction), 21 (Arabic text). For al-Yūsī, see al-Yūsī, *al-Qanūn*, 177, and the discussion in the Introduction.

centuries attests to broad support for the philosophical sciences, even as it also shows how the authority of these sciences was occasionally challenged by scholars interested in emphasizing the superiority of certain spiritual leaders or of reliance upon God.

For the period in question, Morocco did not boast any comprehensive examples of *ṭabaqāt* literature on the order of the Egyptian Mamluk scholar Abū 'l-Ṣafāʾ al-Albakī al-Ṣafadī's (d. 764/1363) massive *al-Wāfī bi-l-wafayāt* – individual examples are limited by place or period.[8] In establishing a rough overview of the number of scholars involved in the natural sciences during this period, I have relied on the following works, all of which, with the exception of the first, were compiled between the sixteenth and eighteenth centuries:[9]

1. Muḥammad al-Kattānī (d. 1345/1927), *Salwat al-anfās wa muḥādathat al-akyās bi-man uqbira min al-ʿulamāʾ wa'l-ṣulahāʾ bi-fās*
2. Ibn al-Qāḍī (d. 1025/1616), *Durrat al-ḥijāl fī asmāʾ al-rijāl*
3. al-Rasmūkī (eleventh/seventeenth century), *Wafayāt al-Rasmūkī*
4. ʿAbdallāh b. Muḥammad al-Fāsī al-Fihrī (d. 1131/1718), *al-Iʿlām bi-man ghabar min ahl al-qarn al-ḥādī ʿashar*
5. Muḥammad al-Ḥājj al-Ifrānī (d. 1150/1737), *Ṣafwa man intashara min akhbār ṣulahāʾ al-qarn al-ḥādī ʿashar*
6. Muḥammad al-Qādirī, *Nashr al-mathānī li-ahl al-qarn al-ḥādī ʿashar wa l-thānī* (d. 1187/1773)
7. Muḥammad al-Ḥuḍaygī (d. 1189/1775), *Ṭabaqāt al-Ḥuḍaygī*

The generalizations made from these works should be taken with a degree of caution: the criteria for each author's selection differ, and while they do borrow from their predecessors, where they overlap, their entries for the same scholar can differ. Still, taken together they offer a broadly representative depiction of the rational sciences in the biographical literature.

[8] See, however, Muḥammad Ḥajjī's collection of nine Moroccan *ṭabaqāt* works into the massive *Mawsūʿat aʿlām al-maghrib*.

[9] I have not included Ibn ʿAskar (d. 986/1578)'s *Dawḥat al-nāshir li-maḥāsin man kāna bi-l-maghrib min mashāyikh al-qarn al-ʿāshir* in this list, as, while formally a *ṭabaqāt* work, the author states in his introduction that it is a *fahrasa* and primarily consists of scholars he knew and with whom he studied (Ibn ʿAskar, *Dawḥat al-nāshir*, 1). Of the 153 biographies, only 4 are described as having studied the natural sciences: Muḥammad al-Shutabyī (d. 963/1556) studied alchemy (ibid., 16); al-Ḥusayn al-Maṣmūdī (d. 950s/1540s), lettrism and astrology (ibid., 45); ʿAbdallāh b. Muḥammd al-ʿAnnābī al-Darʿī, achemy (for more on him, see Chapter 4) (ibid., 91–92); Muḥammad al-Andalusī, astronomy, lettrism, alchemy, mathematics (d. 984/1576) (ibid., 103).

The *ṭabaqāt* work with the broadest scope for this period is that of Muḥammad al-Kattānī (d. 1345/1927), whose *Salwat al-anfās wa muḥādathat al-akyās bi-man uqbira min al-ʿulamāʾ wa'l-ṣulahāʾ bi-fās* (*Solace for the souls and the conversing with the astute regarding those scholars and saints who are buried in Fez*) records 1453 biographies of scholars who lived between the sixth/twelfth and thirteenth/nineteenth centuries. Of these, approximately 100 studied one or more of the rational sciences (principally logic, arithmetic, medicine, timekeeping, the calculation of inheritance shares, astrology, and astronomy); a handful of others are described as having combined transmitted and rational sciences.[10] This last category has fewer scholars than the category of those whom al-Kattānī describes as studying both rational and transmitted sciences, while the group who are explicitly described as philosophers is even smaller, numbering only two for the eighteenth century, and only two others from earlier centuries.[11] Although explicit opposition to the natural sciences is rare, students of the natural sciences are, admittedly, not always presented favorably in al-Kattānī's work, such as in the case of Muḥammad al-Akḥal (d. 1014/1606), who was ordered by his spiritual master Abū al-Maḥāsin to give up his study of alchemy and not to teach it to his children.[12]

The next collection is *The Pearl of the Bridal Canopy Containing Men's Names (Durrat al-ḥijāl fī asmāʾ al-rijāl)* of Ibn al-Qāḍī (d. 1025/1616), who lived through the reign of the Caliph Aḥmad al-Manṣūr, and who dedicated his works to the ruler after the latter ransomed him following his capture at sea in 944/1586 by Christian pirates.[13] *The Pearl* was intended by its author to be a continuation of Ibn Khallikān's (d. 681/1282) famous *Deaths of Prominent Men* (*Wāfayāt al-ayān*) and contains the biographies

[10] See the biographies of ʿAbd al-ʿAzīz b. Masʿūd al-Dabbāgh (d. 1131/1719) (al-Kattānī, *Salwat al-anfās,* 2: 227), Muḥammad b. al-Ḥasan al-Bannānī (d. 1194/1780) (ibid., 1: 176), ᶜAbd al-Qādir b. Aḥmad b. Shaqrūn (d. 1219/1804) (ibid., 1: 98). In addition, there is al-Sharqī b. Abī Bakr al-Dilāʾī (d. 1079/1668–69), who was "an imam in rational and traditional sciences" (ibid., 2: 105).

[11] See the biographies of ʿAbd al-Wahhāb b. Aḥmad Adarrāq (d. 1159/1746), a student of al-Yūsī's who is described as the "most knowledgeable of philosophers" (ibid., 2: 39–40) and Sulaymān b. Aḥmad al-Fishtālī (d. 1208/1793–94) (ibid., 3: 141–42). The two earlier examples are Muḥammad b. al-Baqqāl (d. 725/1325) (ibid., 2: 178) and the famous Ibn Bājja (d. 533/1139) (ibid., 3: 333–34).

[12] See ibid., 2: 368. Compare this episode with the account of how al-Ṭayyib b. Muḥammad al-Kattānī (d. 1253/1836–37) was able to miraculously heal the eye of a man whom no doctor had been able to heal (ibid., 2: 276). Such descriptions seem more designed to elevate the status of a given saint or to demonstrate his authority than to necessarily dispute the legitimacy of the natural sciences.

[13] See the article "Ibn al-Kādī" in EI2 by G. Deverdun.

of 1552 Moroccan and Andalusi scholars from the seventh/thirteenth to tenth/sixteenth centuries, many of whom were still alive when he composed it.[14] Of this number, fifty-five studied one of more of the rational and natural sciences, the most frequent of which were mathematics and inheritance law (eighteen studied medicine or astronomy).[15] This is a considerably lower percentage – 3 percent in absolute terms, or roughly a third – than the other collections examined here, in which ca. 8–9 percent of the scholars studied these sciences.

The third largest biobibliographical work to deal with this period is Muḥammad b. al-Ṭayyib al-Qādirī's (d. 1187/1773) *Unrolling the Scrolls Concerning the People of the Eleventh and Twelfth Centuries* (*Nashr al-mathānī li-l-ahl al-qarn al-ḥādī ʿashr wa-l-thānī*), which, as the title suggests, focuses on the scholars – largely but not only of Morocco – of the eleventh/seventeenth and twelfth/eighteenth centuries. Al-Qādirī's work was an important source for al-Kattānī's later collection, and al-Qādirī himself drew substantially on eleventh/seventeenth century sources, particularly al-Yūsī's *The Discourses* (*Muḥāḍarāt*) and al-ʿAyyāshī's massive *Riḥla*.[16] He organized the *Nashr al-mathānī* chronologically, giving the biographies of those who died in a given year followed by the momentous events of the year – generally a combination of accounts of natural disasters and political incidents. All in all, he gives biographies of some 843 scholars,

[14] I am grateful to Hind Ait Mout for her help in identifying relevant biographies in Ibn al-Qāḍī's work.

[15] This latter group of eighteen included the following from the tenth/sixteenth century: the Saʿdī caliph Aḥmad al-Manṣūr himself, who knowledgeable in mathematics, astronomy, and geometry and whom God had granted knowledge of Euclid (Ibn al-Qāḍī, *Durrat al-ḥijāl*, vol. 1, 107–08); Aḥmad al-Ghazzānī (d. 925/1514), taught inheritance law and mathematics, knew astronomy (ibid., 170); Aḥmad b. Muḥammad b. Maswīb al-Andalusī (alive at the time of writing), knew geometry, timekeeping (ibid., 173); Ibrāhīm al-Maṣmūdī (d. in 912/1506 or 913/1507), was the "Commander of the Faithful" in inheritance law and math (ibid., 199); Ibrāhīm b. al-Akḥal al-Suwaydī (alive at the time of writing), "unique in his time for his knowledge of astronomy and geometry" (ibid., 202); Muḥammad b. Muḥammad b. Abī al-Khayr (alive at the time of writing), studied geometry and astronomy with his father, knew Euclid well, excelled at the use of the sine quadrant (*al-rubal-mujayyab*) and other instruments of timekeeping; "I [Ibn al-Qāḍī] saw him in Egypt and studied geometry and a summary of the Almagest with him in 986 [1578–79]" (ibid., vol. 2, 104); Muḥammad b. Yūnis b. Riḍwān (b. 967/1560), skilled in medicine, math, and inheritance law (ibid., vol. 2, 211); ʿUmar b. ʿAlī b. Ṣaʿd al-Haṣīnī al-Lajāʾī (alive at the time of writing), knew math and astronomy (ibid., vol. 3, 193); ʿUmar b. ʿAbd al-Raḥmān (fl. 900/1495), knew math, inheritance law, wrote a poem on *ʿilm al-awfāq* (ibid., vol. 3, 204–05); Abū Qāsim b. Muḥammad b. Ibrāhīm al-Wazīr al-Ghassānī (alive at the time of writing), wrote several works on medicine, including a commentary on Ibn ʿAzrūn's book on fevers as well as his famous book on herbs (ibid., vol. 3, 289).

[16] See al-Qādirī, *Nashr al-mathānī*, vol. 1, 6, for al-Kattānī's indebtedness to al-Qādirī.

some of whom were from earlier periods or from the eastern Mediterranean Muslim world. Of these, fifty-eight studied the rational or natural sciences, the most frequently mentioned of which was logic, followed by timekeeping and medicine, with a few striking examples of scholars who were known for their mastery of the occult natural sciences such as the properties of letters, letter magic (*al-sīmīyā*), or magical squares (*al-jadwal*, pl. *al-jadāwil*).[17]

We find a similar pattern in the shorter and less illustrious *ṭabaqāt* works of the eleventh/seventeenth and twelfth/eighteenth centuries. In the biographical dictionary of al-Rasmūkī (fl. 1071–98/1662–87), focusing on scholars in the south of Morocco, we find mention of several scholars who studied mathematics and logic, and, we read of Aḥmad b. Muḥammad b. Yaʿzī b.ʿAbd al-Samīḥ al-Taghātīnī (d. 1080/1669), a friend of the author's, who is described as a philosopher (*faylasūf*), emphasizing the author's favorable attitude toward philosophy.[18] In the slightly later work of ʿAbdallāh b. Muḥammad al-Fāsī al-Fihrī (d. 1131/1718) devoted to scholars who died in the eleventh/seventeenth century, we find numerous scholars who are described as having mastered both rational and transmitted sciences, as well as others who were known for their expertise not only in those sciences that have a direct relationship to Islamic ritual practice, such as mathematics, timekeeping, and astronomy, but also in alchemy, letter magic, logic, and medicine.[19] This

[17] Examples of those scholars who know one or more occult sciences include most notably, al-Mirghitī (d. 1089/1678), for his writings on books on natural magic and the making of varied magic squares, and on whom see more below (al-Qādirī, *Nashr al-mathānī*, vol. 2, 245); ʿAbd al-Raḥman b. ʿAbd al-Qādir al-Fāsī (d. 1096/1685), wrote more than forty works on the terminology of different sciences, including mathematics, natural magic, magic squares, the secrets of the letters, geometry, timekeeping, astrolabes, and the occult sciences (ibid., vol. 2, 327); Ḥassan b. ʿAli al-ʿAjīmī (d. 1113/1701), worked with names, magic squares, the secrets of letters, invocations, things that the teachers try to keep hidden (ibid., vol. 3, 136).

[18] Al-Rasmūkī, *Wafayāt al-Rasmūkī*, 31. The scholars who studied mathematics were ʿAbd al-ʿAzīz b. Abī Bakr b. Aḥmad b. Yaʿqūb al-Rasmūkī al-Burjī (d. 1065/1655) (see ibid., 15), Dāwūd b. ʿAbdallāh b. Aḥmad al-Ḥāmidī (d. 1046/1637) (see ibid., 41), and Muḥammad b. Muḥammad b. ʿAlī al-Iyasī (d. 1055/1644–45) (see ibid., 45); while Muḥammad b. ʿAbd al-Wāsiʿ al-Rasmūkī al-Taghātīnī (no dates given) was known for his talent in logic (see ibid., 42).

[19] Al-Fihrī, *al-Iʿlām bi-man ghabara min ahl al-qarn al-ḥādī ʿashar*. See the following entries: Muḥammad al-Akḥal (d. 1014/1606), who was knowledgeable in alchemy (ibid., 70); Muḥammad b. Aḥmad al-Murrī (d. 1018/1609–10), who studied logic (ibid., 91); Abū Qāsim b. Muḥammad b. al-Qāḍī (d. 1022/1613–14), who studied mathematics and timekeeping (ibid., 104); Aḥmad al-Qāḍī (d. 1025/1616), who studied inheritance law and mathematics (a frequent pairing) (ibid., 124–25); Muḥammad b. Ismāʿīl al-Mūsāwī (d. 1064/1653–54), who studied letter magic (*sīmiyāʾ*) and alchemy (ibid., 160); Saʿīd Qaddūra (d. 1066/1655–56), who wrote many works on logic (ibid., 166); ʿAbd al-Qādir b. Muḥammad Bū Shaykh (d. 1069/1658–59), who excelled in mathematics (ibid., 171); Muḥammad al-Aḥmad al-Sabbāgh al-ʿAqīlī (d. 1076/1665–66), who wrote on both mathematics and astronomy (ibid., 209); ʿAbd al-Qādir b. ʿAlī al-Tulaytī (d. 1077/1666–67), who

trend is continued in the twelfth/eighteenth century *ṭabaqāt* work of Muḥammad al-Ḥuḍaygī (d. 1189/1775), who himself also summarized sections of al-Yūsī's taxonomic *al-Qānūn*, and who wrote on medicine and taught logic in the Afīlāl *zāwiya*.[20] Al-Ḥuḍaygī mentions scholars from the ninth/fifteenth to twelfth/eighteenth centuries who were known for their expertise in logic, medicine, astronomy, alchemy, mathematics, astrology, and the rational sciences in general. Of the 823 biographies in his compilation, some 56 studied one or more of the rational sciences.[21] Most notable here is his account of the astronomer and contemporary of al-Yūsī, Muḥammad b. Sulaymān al-Rūdānī (d. 1095/1685), who, in addition to his well-known work on the astrolabe, was famed for his skill in timekeeping, astronomy, and glassblowing.[22]

The examples drawn from the *ṭabaqāt* literature suggest that the philosophical sciences, while not the main focus of study for most Muslim scholars in North Africa in the sixteenth through eighteenth centuries, were seen as an accepted field of inquiry by the majority of scholars and were studied intensely by a small number. In making this claim, I am assuming that the authors of *ṭabaqāt* works were representative of the scholars of their time and that their inclusion of specific scholars reflected their judgment that these scholars were worthy of inclusion and that their intellectual pursuits were valuable and valid – assumptions I believe to be borne out by the literature on the *ṭabaqāt* genre. This survey does not, however, clarify how these sciences were defined and how the scholars of this period framed them. The next section turns to three attempts of seventeenth- and eighteenth-century Moroccan scholars to categorize

studied both logic and mathematics (ibid., 215); Muḥammad b. Aḥmad al-Fāsī (d. 1084/1673–74), who studied logic and other rational sciences (ibid., 237); Muḥammad b. Saʿīd al-Sūsī al-Mirghītī (d. 1088/1677–78), the authority on timekeeping and medicine who opened this chapter (ibid., 254–55); ʿAlī al-Marrākushī (d. 1090/1679–80), who studied mathematics (ibid., 263); ʿAbdallāh b. Muḥammad b. Abī Bakr al-ʿAyyāshī (d. 1090/1679), the famous traveler who studied logic (ibid., 264); Muḥammad Adarraq (d. 1090/1679–80), an excellent doctor (ibid., 273); Muḥammad al-Rūdānī (d. 1095/1684), the famous astronomer, discussed in greater detail in both al-Ḥudaygī and al-ʿAyyāshī (ibid., 304); ʿAlī b. Ibrāhīm al-Marrākushī al-Fulūs (d. 1096/1684–85), known for his medical knowledge (ibid., 307).

20 Al-Ḥuḍaygī, *Ṭabaqāt al-Ḥudaygī*, 1: 44–45, 74. The Afīlāl lodge was founded by Abū al-Qāsim al-Fīlālī of the Īsī tribe, to which al-Ḥuḍaygī belonged, and lies south-east of Tafraout, in the anti-Atlas (ibid., vol. 1, 34).

21 Some of these are, unsurprisingly, the same as found as in al-Fihrī's *al-Iʿlām*.

22 Al-Ḥuḍaygī, *Ṭabaqāt al-Ḥudaygī*, vol. 1, 307–09. On al-Rūdānī see especially al-ʿAyyāshī, *Riḥlat al-ʿAyyāshī 1661–1663*, vol. 2, 43–60. The treatise on the astrolabe he constructed was translated by Charles Pellat, "L'astrolabe sphérique d'ar-Rūdānī."

knowledge, focusing especially on the place of the natural sciences, and briefly considers their subsequent influence in Morocco.

2.1.1 *The Divisions of Knowledge of al-Ḥasan al-Yūsī, ʿAbd al-Raḥmān al-Fāsī, and Muḥammad b. Masʿūd al-Ṭurunbāṭī*

Al-Ḥasan b. Masʿūd al-Yūsī is a towering figure in the intellectual landscape of seventeenth-century Morocco. Known as a prominent scholar of Sufism, theology, and logic, and best known today for his long and admonishing letter to the ʿAlawite Sultan Moulay Ismāʿīl (r. 1672–1727), his significance in Moroccan history has been discussed by scholars of such varied interests and backgrounds as Jacques Berque, Clifford Geertz, ʿAbd al-Kabīr al-ʿAlawī al-Madgharī, Henry Munson Jr., and Kenneth Honerkamp.[23] Despite his prominence, al-Yūsī was something of an intellectual outsider in his day, never fully accepted by the scholars of Fez, and largely formed by his initial education in the south of Morocco and subsequent residence in the Dilā' *zāwiya* between 1653 and 1668. It was during his time at the *zāwiya* that he studied the use of the astrolabe with Muḥammad b. Saʿīd al-Mirghitī (d. 1089/1678), whose *fahrasa* we will turn to in Section 2.2, and with ʿĪsā al-Suktānī (d. 1062/1652), who had also studied with al-Mirghitī in Marrakesh. Despite the ambivalence of many Fāsī scholars toward al-Yūsī, in part the result of his having been favored by both Moulay Rashīd and Moulay Ismāʿīl (regardless of his famous criticisms of the latter), al-Yūsī attracted a large number of students, many of whom later settled in the Ottoman-controlled eastern Mediterranean, carrying his teachings and works with them.[24] His numerous writings include a commentary on the influential theological creed of al-Sanūsī (d. 895/1490), a long theological commentary on the *shahāda* entitled *The Source for the Common and Elect on the Words of Sincerity* (*Mashrab al-ʿāmm wa'l-khāṣṣ min kalimat al-ikhlāṣ*), an extensive

[23] See Jacques Berque, *al-Yousi*; Clifford Geertz, *Islam Observed*, 29–35; al-Madgharī, *al-Faqīh Abū ʿAlī al-Yūsī*; Henry Munson Jr., *Religion and Power in Morocco*, 1–34; Kenneth Honerkamp, "al-Ḥasan ibn Masʿūd al-Yūsī." Munson provides a close reading of al-Yūsī's letter to Moulay Ismāʿīl in *Religion and Power in Morocco*, 27–31. It is striking that when Paul Rabinow carried out his fieldwork in al-Yūsī's home village – Sidi Lahcen Lyussi – its inhabitants knew little of their ancestor. See Rabinow, *Reflections on Fieldwork in Morocco*, 132–33.

[24] See especially Khaled El-Rouayheb, "Sunni Muslim Scholars on the Status of Logic, 1500–1800," and ibid., "Was there a Revival of Logical Studies in Eighteenth-Century Egypt?"

discussion of the nature of knowledge entitled *The Definitive Collection* (*al-Qānūn*), the famous *adab* work *The Discourses* (*al-Muḥāḍarāt*), numerous letters, and his *Fahrasa* or scholarly autobiography. Throughout these works we find al-Yūsī expressing broad support for the study of the natural and rational sciences, a tendency displayed most clearly in the taxonomy of knowledge he presented in *al-Qānūn*, an encyclopedia on the nature of knowledge and the proper comportment of both teacher and student. This work, the title of which can be translated as *The Definitive Collection of Rulings regarding the Sciences and the Comportment of Teachers and Students*, was a late work of a wide-ranging scholar.[25] It repays a close reading as well as a comparison with other similar works of the same period, including ʿAbd al-Raḥmān al-Fāsī's (d. 1096/1685) *Kitāb al-uqnūm fī mabādi' al-ʿulūm,* written at approximately the same time as the *Qānūn* and Muḥammad b. Masʿūd al-Ṭurunbāṭī's (d. 1214/1799) *Bulūgh aqṣā al-murād fī sharaf al-ʿilm*, written approximately a hundred years later. All of these works draw significantly on the writings of the generations preceding them, perhaps most notably those of al-Ghazālī (d. 505/1111), among whose writings the *Iḥyā'* and the *Mustaṣfā* seemed to have enjoyed special popularity in early modern Morocco. Yet, while indebted to their predecessors, all three works, and this is especially the case with that of al-Yūsī, use the genre to advance their own intellectual projects. A final comment is that, typical of the Moroccan scholars of their day, all three authors followed the Mālikī law school, were Ashʿarites theologically, and were practicing Sufis, the significance of at least the second of which will become clear in the following.

Al-Yūsī's *al-Qānūn* is the most systematic of the three texts.[26] Broken up into three main sections dealing with knowledge, and the proper comportment of, respectively, teachers and students, it offers a comprehensive description for what can be known, what should be known, and how to go about learning it. In his first chapter al-Yūsī situates the reader by clarifying his own theological preconceptions, drawing largely on earlier works of Ashʿarism: God is the only true cause and the only entity to truly possess the qualities of will, power, knowledge, and life.[27] Four types of God's creation possess knowledge: Angels, Satan, Jinn, and humans, and all attain it through the power of reason:

[25] Berque had already grouped *al-Muḥāḍarāt* and *al-Qānūn* together in his comments on the latter (*al-Yousi*, 25–27).

[26] As Berque pointed out over sixty years ago, this systematic approach to structuring his writings is typical of al-Yūsī (Berque, *al-Yousi*, 26).

[27] Al-Yūsī, *al-Qānūn*, 109.

Know that God, when he chose these types to benefit from knowledge, he gave them His instrument, reason, which is the site of legal obligation. He did not give it to the created things, which have no obligation on them, including inanimate things and dumb beasts. Nor did he give it to children or the insane.[28]

In defining reason, al-Yūsī gives several definitions but seems to privilege al-Baqillānī's (d. 403/1013) gloss on al-Ashʿarī's definition: "Knowledge of the necessity of necessary things, and the impossibility of impossible things, and way in which habitual actions take place."[29] He continues with a longer discussion of the divisions of reason itself, drawing here repeatedly on the Persian scholar al-Taftazānī's (d. 792/1390) *Sharḥ al-Maqāṣid*, and then proceeds to summarize a general Ashʿarite phenomenology and epistemology, referring in the process to the writings of many prominent Ashʿarites, including al-Baqillānī (d. 403/1013), al-Juwaynī (d. 478/1085), al-Rāzī (d. 606/1209), and al-Ijī (d. 756/1355).[30]

These broader considerations take up the first nine chapters of the first section, and it is only in the tenth chapter that he takes up a more narrowly formal definition of the different types of knowledge, beginning with the different types of philosophical knowledge. Considering the efforts al-Yūsī will muster later in the text to refute certain views of earlier scholars such as al-Suyūṭī (d. 911/1505) and Ibn Juzayy (d. 741/1340) on the permissibility of studying philosophy, it is not surprising that he takes time to address philosophy in general before turning to its divisions:

We hold that concerning philosophy, there are things that the Muslim community accepts and those that it rejects. As for what is acceptable, there are matters that they cultivate, and those that they do not. Let us begin by defining the philosophical sciences along with the terms used in them, with an eye to what is acceptable and what not. We hold that knowledge is either desired in itself or for other than itself. The first kind is metaphysics (*al-falsafa al-ūlā*), the aim of which is the realization of the rational animal, and through exertion attaining the true meanings of things. It is either theoretical or practical. The first is either absolute and abstract (*mujarrad ʿan al-māddah muṭlaqan*) and it is metaphysics (*al-ʿilm al-ilāhī*), or it is only in the mind, and this is mathematical (*al-ʿilm al-riyāḍī*) or it is bound to matter, and this is natural science (*al-ʿilm al-ṭabiʿī*). The second is either related to the nature of a person's ego (*al-nafs*) and is called the rule of the ego or ethics, or it is related to the ego and its relevant desires in the matter of managing the house, or more generally exercising rule or power. The one who preserves control over the ego and who masters the external and internal precepts for governing it is a person whose identity has been indicated by the greater conjunctions (*al-qirānāt al-kibār*) and

[28] Ibid. [29] Ibid., 110. [30] See ibid., 112–45.

who has been distinguished from the rest of mankind by the virtue that he received from these immaterial entities (*quwā al-mujarradāt*). This person is the Prophet and this is the domain of prophecy. If a person were to master solely the external aspects of the ego and to be indicated solely by the middle conjunctions (*al-qirānāt al-mutawassiṭa*), the domain would be that of temporal rule (*al-salṭana*) and he would be a ruler (*al-sulṭān*) over a greater or lesser area.[31]

Al-Yūsī immediately follows this integration of the Prophet Muḥammad into the field of ethics by clarifying to his reader that planetary conjunctions themselves do not have any effect but that God can give his believers knowledge of those things that he creates in His Habit in relation to them. Any belief in planetary influence as well as the theory of emanation is false.[32] While he does not pause to expound on the implications of this statement, it contains a clear critique of those followers of astrology who believe that the stars have actual influence on the sublunar world, not to mention of Neoplatonism. It is, not, however, a rejection of astrology in general.

Al-Yūsī then notes that there are different paths to knowing God, with valid ones including that pursued by philosophers (*al-ḥukamā'*) through reason and that of ascetics – whom he calls "the Sufis of our community" – through self-discipline and self-denial.[33] Then there are the false paths pursued by those who contemplated God but who did so without using reason – this group included dualists and the deniers of the attributes of God as well as those that reject divine revelation in general. God had sent prophets using sound reason to Jews, Christians, and Muslims to clarify these matters and these groups then developed their own internal divisions which al-Yūsī names, but on which he does not comment.

Instead, he takes up the issue of the division of the sciences, which he categorizes in a number of ways, first focusing on the mathematical and natural sciences, and then giving two different breakdowns of the Islamic sciences.

Al-Yūsī's Division of the Sciences:[34]

- mathematical knowledge (*al-ʿilm al-riyāḍī*)
  - geometry (*ʿilm al-handasa*)
    - engineering (*ʿilm ʿuqūd al-abniya*)

[31] Ibid., 146–47. [32] Ibid., 147.

[33] This is a curious formulation, implying as it does that other religious communities have Sufis.

[34] Ibid., 151–64. I have given the Arabic for terms that may not be self-evident to a nonspecialist.

    - optics (*ʿilm al-manāẓir*)
    - science of burning mirrors (*ʿilm al-marāyā al-muḥriqa*)
    - science of weights (*ʿ ilm al-marākiz al-athqāl*)
    - surveying (*ʿilm al-misāḥa*)
    - science of discovering water (*ʿilm inbāṭ al-miyāh*) science of pulling weights (*ʿilm jarr al-athqāl*)
    - science of producing time-telling devices (*ʿilm al-bankāmāt*)
    - science of siege engines (*ʿilm al-ālāt al-ḥarbiyya*)
    - science of devices using air pressure (*ʿilm al-ālāt al-rūḥāniyya*)[35]
  - astronomy (*ʿilm al-hay'a*)
    - science of the astronomical almanac (*ʿilm al-zījāt*)
    - science of timekeeping (*ʿilm al-mawāqīt*)
    - science of stargazing (*ʿilm kayfiyyat al-arṣād*)
    - science of mapping the earth (*ʿilm tasṭīḥ al-kura*)
    - science of shadow measuring devices (*ʿilm al-ālāt al-ẓilliyya*)
  - arithmetic (*ʿilm al-ʿadad*)
    - science of open arithmetic (*ʿilm al-ḥisāb al-maftūḥ*)
    - science of calculating with a dust board and stylus (*ḥisāb al-takht wa l-mīl*)
    - science of algebra
    - calculation by double false position (*ḥisāb al-khaṭa'ayn*)[36]
    - the science of houses and bequests (*ʿilm al-dūr wa-l-waṣāyā*)
    - the arithmetic of dinars and dirhams
  - music
- natural science (*al-ʿilm al-ṭabīʿī*)
  - medicine
  - veterinary medicine
  - physiognomy
  - dream interpretation
  - astrology
  - magic
  - science of talismans
  - letter magic (*al-sīmiyā'*)
  - alchemy
  - agriculture
- First Division of Islamic sciences

[35] Al-Yūsī glosses this as a science that trains the soul through – one assumes mechanical – wonders (ibid., 153).

[36] On the history and nature of this science, I have benefited from a conference paper by Randy Schwartz, "Issues in the Origin and Development of *Hisab al-Khata'ayn* (Calculation by Double False Position)."

- Foundational Sciences Desired for Themselves
  - jurisprudence
    - the science of inheritance
  - Sufism
- Propaedeutic Sciences (*al-wasīla*) (compared to the role played by the sciences of arithmetic and timekeeping in the philosophical sciences)
  - Qur'anic exegesis
  - Prophetic Tradition
  - sciences that are subsidiary to these propaedeutic sciences (*wasīlat al-wasīla*)
    - science of recitations
    - science of writing
    - sciences of the Arabic language
    - the science of logic
- Second Division of the Islamic Sciences[37]
  - the core sciences
    - theology
    - jurisprudence
    - Sufism
    - Qur'anic exegesis
    - Prophetic tradition
    - legal theory
  - the sciences on which they rely (*al-musta'ān bi-hi*)
    - linguistics
    - science of vocalization
    - declension
    - semantics (*'ilm al-ma'ānī*)
    - exposition (*'ilm al-bayān*)
    - medicine
    - arithmetic
    - logic

Overall, and with a few minor differences, al-Yūsī's classification of the mathematical and natural sciences closely follows that of the eighth/fourteenth century Egyptian scholar Ibn al-Akfānī.[38] There are a few details in this classification that are interesting before we come to al-Yūsī's robust

[37] Beginning on ibid., 177.

[38] Ibn al-Akfānī, *Irshād al-qāṣid ilā asnā al-maqāṣid*, 168–70. Al-Yūsī lists ten natural sciences, omitting geomancy (*'ilm al-raml*), included by Ibn al-Akfānī (see ibid., 188). Al-Yūsī draws directly on some of Ibn al-Akfānī's definitions, but his own descriptions of the sciences are much briefer than those of his Egyptian predecessor.

defense of the study of logic and medicine. First, there is his succinct justification for his division of the types of natural science according to the presence of humors and souls: letter magic looks at what does not have humors (*mizāj*), what has humors but not a soul (*nafs*) is dealt with by alchemy, what has humors and a soul but is not sentient (*ghayr mudrika*) falls under agriculture, what is sentient but not speaking is dealt with by veterinary medicine, and that which speaks is dealt with by medicine. An external examination of one's morals is undertaken by phrenology and one's internal state is examined by dream interpretation.

The second point to make is al-Yūsī's repeated and extensive defense of astrology. Believing as he did that believers have been given the power of reason to study God's habitual actions in order to better understand His wisdom, al-Yūsī saw no reason to object to the study of the planet's movements to foretell the future as long as one remembered that the planets themselves have no influence on earthly events. To illustrate the effectiveness of astrology he gives the example of Muʿtamid b. ʿAbbād (d. 487/1095) and the founder of the Almoravid dynasty, Yūsuf b. Tāshfīn, and how the former put off the latter's order to attack the enemy until the proper time had come.[39]

A final, and vital, point is al-Yūsī's definition of Islamic science. Here, and as already discussed in the Introduction, al-Yūsī's reiterates his view that all knowledge that benefits the Muslim community should be understood as Islamic:

> All of these sciences [e.g., writing, grammar, logic] are generally understood to be Islamic, meaning that they are studied within the community of Islam or that they have been directly or indirectly beneficial to the religion of Islam. In this manner they are also revealed (*sharʿiyya*). It is commonly accepted to use the term revealed to both what is intended in its essence and that which is associated with it.[40]

There are fourteen different Islamic sciences, six of which are core, and which themselves rely on eight others, including medicine, calculation (*al-ḥisāb*), and logic.[41] Yet as already suggested, al-Yūsī's defense of the philosophical sciences goes far beyond medicine. He first makes a descriptive observation of the past practice of the Muslim community: when it came to such sciences as medicine, timekeeping, logic, and calculation, as well as

[39] Al-Yūsī, *al-Qānūn*, 160. Yet compare this anecdote with the one given by al-Yūsī in *The Discourses*, where he depicts the ʿAbbasid ruler al-Muʿtaṣim ignoring the advice of the astrologers and placing his faith in God (*The Discourses*, 287) and those who use astrology to deceive believers by pretending to be saints (ibid., 329).

[40] Ibid., 177. [41] Ibid., 178.

elements of astronomy and geometry, these had been incorporated into the community to their great benefit. Simultaneously, other sciences, perhaps due to worries regarding them, had remained in the hands of farmers, architects, sea captains, sorcerers, and others of the common people.[42] Al-Yūsī then turns to his own view of things:

> There is no harm in any of these sciences and I don't agree with those who forbid any part of them. In its essence, knowledge is sustenance for the mind, delight for the soul, and an attribute of perfection. Its fruits differ in terms of nobility according to subject and aim, and the judgment regarding them differs according to intention.[43]

Al-Yūsī emphasizes the key point that defines his understanding of the sciences: the desirability of their study is directly linked to the intention of the one studying them, not the content of the sciences themselves (thus even sorcery is a permissible subject of study, if for the purpose of defending oneself against it). This is a good moment to pivot to his long defense of the study of logic, a field in which he himself excelled.

Al-Yūsī's defense of logic and his harsh criticisms of those, such as al-Suyūṭī, who criticize its legitimacy as a discipline, are well documented.[44] In his division of sciences, he supported this assertion by including logic and medicine not only within philosophy, but, as seen above, also within the Islamic sciences, where they functioned as necessary propaedeutic ones. His justification of this inclusion was most explicit in his refutation of the views of the eighth/fourteenth-century Andalusi scholar Ibn Juzayy's condemnation of logic in his *al-Qawānīn al-Fiqhiyya*, where he also addresses how the Egyptian polymath al-Suyūṭī did not approve of the study of logic:

> Similar to this stance is that of al-Suyūṭī in his selection (*naqāyatihi*). He asserted the prohibition of the sciences of philosophy, and gave logic as an example of them. He had mentioned medicine as one of the important sciences, so it is to be said to him: If you hold that the sciences of philosophy are forbidden, then medicine is forbidden because it is one of them, so why did you mention it [favorably] when forbidding the others? If there was for you an indication that differentiated medicine from its like, then why didn't you draw attention to it and then explain it? It is for you to explain the prohibition of all of them. If it was due to their being innovations, that is the innovation of using them and their extraction to us, then the innovation (*al-mubtada'*) is more general than the forbidden, so a specific

[42] Ibid., 176. [43] Ibid., 177.

[44] See the discussion in the Introduction and El-Rouayheb, "Opening the Gate of Verification," 269–70.

indication is necessary. If it was due to their containing corruption, you had made known that many of them did not contain it, and details are necessary. Then the investigation comes to the content (*al-mushtamal*), as seen above. If this is the case, then medicine, with which you declared yourself content, is more worthy of prohibition than logic, for medicine contains the falsehoods of the natural philosophers and the philosophers concerning powers, spirits and the influence of the heavenly elements on the earthly ones, and so on. Concerning logic, there is nothing in it that is denied in the creed, as its investigation is into imaginations and confirmations of the mind, without objection to a specific image or a specific ruling. Yes, there occurs in logic the matter of classifying facts into imagined types and classes that do not resemble the bodies of the world, and this is different from the *mutakallimūn*, whose example entails inferring from one class to all of them. Addressing this is simple, for logic is based on abstracting facts, which does not prevent the representation of bodies in their essences. Arguing for abstractions (*al-mujarradāt*) has been taken up by many investigators (*al-muḥaqqiqīn*), while others maintained that it is sufficient to know about the creation of the world after the Decree: in this fashion, the existence of the chosen actor is proven. So, it truly strange that he would permit medicine and forbid logic.[45]

It is not only in his extended defense of the philosophical sciences that al-Yūsī is exceptional and distinguishes himself from his contemporaries (the organization and comprehensiveness of his thought alone is remarkable). The second and third sections of the *Qānūn* form together with the first a guide not only to how to understand and divide up knowledge, but how to live a life based on its pursuit.

In these latter sections, al-Yūsī argues that the teacher needs to be a model scholar and for the importance of remembering that first and foremost knowledge is an attribute of God and that the scholar should be pious in all his ways.[46] He should dress and behave properly, command the right and forbid the wrong, strive against his ego, and follow the advice set down in the books of the Sufis. The value of teaching is described as equal to that of the collective obligations of carrying out jihad against an enemy or preparing the bodies of the dead, and as such it is vital that knowledge be transmitted by the ethically upright. Such people are neither corrupt nor stupid and their character protects them from encounters with those who are in power. At the end of this section, al-Yūsī gives an overview of the books that during his day were considered the standard works in jurisprudence, Qur'anic exegesis, Prophetic Tradition, theology, legal theory, and grammar.[47] The final word is given to anecdotes on the value of books as physical objects, and the need to respect them, including instructions on

[45] Ibid., 299–300. [46] Ibid., 326. [47] Ibid., 363–70. See the list given in chapter 2.

how to stack them according to the nobility of their subject (the Qur'an is at the top, the rational sciences at the bottom).[48]

The third section, on the proper comportment of the student, is longer than the second. Al-Yūsī draws extensively here on the Sufi understanding of the relationship between master and disciple. The student's devotion to his studies and his teacher is a deep, internal, spiritual one and represents a commitment that the teacher should respect if the student's commitment is pure and not for material gain or due to fear of parents or rulers. Al-Yūsī offers standard Sufi depictions of the teacher-student relationship being one of a corpse in the hands of the washer, of a patient before a doctor, and of Moses and Khidr.[49]

When it comes to how one should pursue one's studies, after beginning with the Qur'an one should turn not to Prophetic Tradition – the social circumstances not being right for this, and the subject being so vast that many never advance beyond it – but to the necessary linguistic and rational principles (*qawā'id*) before proceeding to the religious sanctioned sciences in general.[50] Due to the limits of time it will not be possible to acquire all of these and the student will have to begin with his individual obligations, and to be cautious when approaching contentious issues, training his power of reason as one would an animal for the hunt.[51] The student must of course focus on the world to come and to seek knowledge constantly. Here too al-Yūsī compares the quest for knowledge with the carrying out of jihad, and includes a striking anecdote on the different abilities of students to persist in their studies that reflects his own belief in the effect a person's environment has on them in terms of conditioning their health:

> One person will need to relax and rest more than others. Let everyone proceed according to their nature and ability. In the days of my youth, I was in the study circle of our teacher Abū Bakr al-Taṭāfī, God have mercy on him. When the heat grew intense, they wanted to rest a little. I told them: "Let us work and pay the heat no mind." In this there was the energy of youth, and the strength of rural living. The aforementioned teacher said to me: "We will tell you a story about something like this so that you see the difference between you and us. There was one of the rulers of Egypt whom one of the doctors told about a herb that they had. If anyone ate it

48 Ibid., 372. 49 Ibid., 383.

50 These comments are intriguing and can be related to what was perhaps a broader lack of interest in Morocco in the *isnad* in general. In *The Discourses*, al-Yūsī notes: This attention to relations, events, and verifiable traditions is quite lacking among the Moroccans – in matters of knowledge they have given preference to intellectual appreciation over the transmission of knowledge, and for them nothing else is important" (al-Yūsī, *The Discourses*, vol. 1, 187).

51 Al-Yūsī, *al-Qānūn*, 394.

> they would die, if God permitted it. A Bedouin was present with them and he said: Bring this herb and I will eat it." And he ate it and did not feel pain, much less die. When the ruler saw this, he grew angry with the doctor and said: "You lied to me!" The doctor said: "O Ruler, we have not experienced this kind of person before," intending the Bedouin, "but if you want to know the truth of what I said, take this Bedouin, give him a house and a good bed so that a change takes place. Then give him the herb to eat and if he doesn't die, then kill me." The ruler did this with the Bedouin, and when the change in the comfort [of his lifestyle] had taken place, he ate it and died.[52]

The belief that a person's physical and cultural environment played a determining effect in how they responded to certain foods is characteristic of much of humoral medicine and al-Yūsī brought it up repeatedly in *The Discourses.* The most striking of these cases involved an account related to al-Yūsī by the prominent physician Abū ʿAbdallāh Muḥammad al-Darrāq al-Fāsī:

> "I went to Tangier," he said, "hoping to meet doctors – in particular, one who was teaching anatomy. One Christian doctor was surprised that we ate the aforementioned couscousūn, and laughed at us, saying, 'You're just filling your stomachs with dough!' Then one day, on this particular visit, I found him standing over a sick man with a severe fever. He was making him drink wine, so I said, 'What are you doing? What's the value of wine when treating fever? Both are "hot" in nature!' 'It won't hurt him,' he said, 'since he is used to it – he used to get it as an infant when he nursed at his mother's breast.' 'Good God!,' I said. 'We used to get couscousūn, the benefit of which you denied, in the same way, at our mothers' breasts when we were babes. How can it harm us now?' 'Too true,' he said, and was then at a loss for words. Peoples' natures differ according to habit and custom. Everyone enjoys eating food they are familiar with and everyone is averse to everything else. The Prophet, God bless and keep him, said about lizards: 'There are none in the land of my people, so you will find that I have an aversion to them.' He justified this opinion by their not being found where he lived."[53]

In the following pages of *The Decisive Rule*, al-Yūsī spends some time discussing the pros and cons of learning through books or from people, with some harsh comments regarding the abilities of book copyists of his own day.[54] He argues that one should be at least twenty years of age to live in and study at a college for it is dangerous to study in the proximity of

[52] Ibid., 427–28.

[53] Al-Yūsī, *The Discourses*, 209. Compare with a similar anecdote regarding al-Yūsī himself having been afflicted with diarrhea on ibid., 207, where al-Yūsī also explains his spelling of couscous.

[54] Ibid., 439. These comments should be read in the context of the argument of Khaled El-Rouayheb in "The Rise of 'Deep Reading' in Early Modern Ottoman Scholarly Culture."

beardless youths, and that study in a college is only legitimate if it is run properly. He then turns to the important question – especially considering the numerous times he has already referred to the example of al-Ghazālī – of whether one should devote oneself to the sciences or to mystical understanding, and includes a long anecdote about how he himself struggled with this choice:

During the days when I was in the companionship of our teacher the Imām Abū ʿAbdallāh b. Nāṣir, God the Exalted have mercy on him, my soul incited me to devoting myself solely to religious devotion (*tajarrud*), to travel for spiritual purposes (*siyāḥa*), and to giving up teaching. He did not agree to this and so I said to him one day: "Which is better, knowledge or mystical insight (*maʿrifa*)?" He said, "Mystical insight." I said to him: "Why then do we not busy ourselves with those things that bring it about?" He said, "Mystical insight is an allotment, and whoever has been allotted something of it will receive it. I have not seen anything better in our time than teaching knowledge." I was for a time in the city of Fes during the days of Rashīd b. al-Sharīf. I was teaching and receiving rewards. I rode to him, ate of his food, and dressed like the others. I became troubled by this, and planned to flee by myself and to travel the world, leaving my family to their Creator, Great and Mighty. I mentioned this to our teacher Abū Muḥammad ʿAbd al-Qādir b. ʿAlī al-Fāsī and he said to me: "If this had taken place in the proper fashion, it would be possible, but it is to be feared that you would do this out of vain desire, and so you will not find profit by doing so." For this reason, I gave up this plan. What he had been referring to is that if the worshipper moves toward God the Exalted, then he is a worshipper of God the Exalted, who will help him and care for him. If he moves due to his ego, then he is a worshipper of his ego and he will give himself over to it and will perish along with it.[55]

Al-Yūsī uses this anecdote as an occasion to stress the importance of having a teacher to guide one through complicated matters, and consoles himself by noting that while abandoning one's family to focus on God may have been appropriate for al-Ghazālī and al-Shādhilī, both the mystic and the scholar have roles to play and while that of the former is nobler, the latter is of greater use to God.

As the third section comes to a close, al-Yūsī offers a few insights into the nature and importance of memory and how to strengthen it. He stresses here the importance of avoiding sin, including avoiding the drinking of wine, and offers a number of strategies for strengthening one's memory, a faculty that he describes as a storehouse behind the brain from which anything one wishes can brought instantly.[56] He finishes the section

[55] Ibid., 452–53. [56] Ibid., 470.

stressing the importance of proper intention when it comes to seeking knowledge and takes up the dangers of employing jinn:

> I knew a young, reasonable, intelligent man who had grown up seeking knowledge, and he quickly acquired a good amount of it. If he had continued, he would have rapidly joined the great, established, accomplished scholars. Yet in this matter he became afflicted and we did not know what had happened to him, only that he was possessed. He was restless, deficient in both reason and faith. People asked about his state, and it was said that he had busied himself in seclusion with summoning them [i.e., jinn] to be employed. When he finished the task, they submitted to him and appeared before him, bringing him a great amount of money. They set the condition on him that he not miss the pre-dawn prayer (*lā yafūtuhu al-fajr*). He agreed to this and soon went to sleep until the sun rose and he found himself ritually unclean. He got up and went down to the river. He washed and returned. While he was approaching the house, he saw a tent (*khibā'*) above his head, pitched on top of him (*maḍrūb ʿalayhi*). This was the end of his sanity and he remained like this until he died. We seek well-being from God. There are many like him.[57]

This type of cautionary anecdote is a staple in al-Yūsī's writings, and numerous examples could be cited from *The Discourses* as well. It and several others like it bring the book to a fitting close in that they emphasize the importance of piety and of correct orientation toward to the Divine in the quest for knowledge. In many ways al-Yūsī was exceptional for his generation of Moroccan scholars: his taxonomy of knowledge, his often-detailed autobiographical asides in his writings, as well as, in the case of Moulay Ismāʿīl, his explicit criticisms of political authority. Yet his focus on the importance of proper ethical and spiritual comportment of teacher and student in the transmission of knowledge he was representative, if not somewhat conservative.

There are not many works from the Maghrib in the seventeenth–eighteenth century that address the same or similar subjects as *al-Qānūn*. The one that comes closest is ʿAbd al-Raḥmān al-Fāsī's *Kitāb al-uqnūm fī mabādi' al-ʿulūm*, an encyclopedia written in *rajaz* verse. The author, a prodigy, was the son of the prominent Fasi scholar ʿAbd al-Qādir al-Fāsī (d. 1091/1680). He was famed for his broad knowledge of innumerable sciences and described by his own father as the al-Suyūṭī of his age – high praise considering the respected status of the Egyptian scholar in the seventeenth century (pace al-Yūsī's criticisms). He is said to have excelled in astronomy (*al-hay'a*) and medicine and to have been acknowledged as an authority in every gathering he attended. He was known for

[57] Ibid., 479.

writing quickly, without drafts and without rereading. Having memorized the Qur'an by the time he was 7, he went on to write nearly 170 works during his career. The biographer al-Kattānī, whose entry in *Salwat al-anfās* I am relying on, notes that there was a good deal of difference of opinion regarding how many sciences he described in *al-Uqnūm*: some said 112 sciences, some 300, some 150. Al-Kattānī argues that the reason for this difference may have been the different copies of the manuscript, but it is also possible that the differences in number may result from variations in counting the entries' subheadings. He died in 1096/1685, six years before al-Yūsī, after having been bedridden for nearly six years, one source – *'Ināyat ulī al-majd* – claiming that his illness was due to his having employed jinn.[58]

In comparison to the *al-Qānūn*, *al-Uqnūm* is chaotic and rambling. A glance over the, by my estimation, 116 different sciences shows little in the way of an organizing principle.[59] There is no clear overall division between rational and transmitted, ancient and Islamic or permitted and forbidden sciences, and some sciences appear twice or more. Of interest is that the philosophical sciences are well represented, sometimes surprisingly so, as when six different kinds of astrolabes and quadrants are given their own sciences, or when astronomy in its broadest sense appears under eight different headings. The esoteric and spiritual sciences are given equal treatment and as for the question of the proper comportment of teachers and students, it falls under *'ilm ādāb al-qirā'a* ("the science of the proper study etiquette") where the paternalistic relationship of teachers to their students is stressed. While the poetic nature of the text may have made it most suited for being memorized as a reference for study – though its sheer length casts some doubt on this, the one manuscript I consulted runs to some 250 pages – the lack of internal structure and the overlapping chapters remain puzzling. We do learn that 'Abd al-Raḥmān was reading many of the same authors as al-Yūsī, seeing that he cites al-Sanūsī, al-Būnī, Ibn Saba'īn, Ibn Sīnā', al-Ghazālī, as well as Plato.[60]

By comparison, al-Ṭurunbāṭī's *Bulūgh aqṣā al-murād fī sharaf al-'ilm*, written roughly a century after *al-Qānūn*, is admirably well organized. Muḥammad b. Mas'ūd al-Ṭurunbāṭī was best known for his writings on grammar, although he also studied logic. Al-Kattānī's entry in *Salwat al-anfās* informs us that among his students was the 'Alawite Sultan Moulay Sulaymān b. Muḥammad (d. 1238/1822), and that he himself served as

[58] See al-Kattānī, *Salwat al-anfās*, vol. 1, 357–58.

[59] See Appendix 1 for a full list of these sciences.

[60] See 'Abd al-Raḥmān al-Fāsī, *Kitāb al-uqnūm fī mabādi' al-'ulūm*, at 112 (al-Sanūsī), al-Būnī (182), Ibn Saba'īn (184), Ibn Sīnā' (185), al-Ghazālī, as well as Plato (183).

judge for various periods of time in Sijilmāsa, and for six months in Essoueira. In his *Bulūgh* "according to what is said," he drew on several other works, among them al-Yūsī's *al-Qānūn*.[61] Al-Ṭurunbāṭī died in 1214/1799 of the plague in Fez.

The *Bulūgh* is divided into five chapters moving from the virtues of knowledge to the proper comportment of students, but the most relevant and interesting section is the introduction in which the author addresses the nature of knowledge more broadly. Drawing on al-Ghazālī's *Mustaṣfā*, al-Sanūsī's *Ṣughrā* and ʿAbd al-Raḥmān al-Fāsī's (the author of the *al-Uqnūm*) commentary on the *Sughrā*, he differentiates between eternal (*qadīm*) and created knowledge, the first belonging to God and the second being divided in turn into self-evident (as with the existence of cold and heat) (*ḍurūrī*) and speculative (*naẓarī*).[62] Al-Ṭurunbāṭī continues by explaining that God created the man's soul from the purest of light and clothed it with purity. He then took it and placed it in the vessel of the body (*qārūrat al-jism*), in His wisdom fashioning a strange and wondrous connection between the two. This connection ensured the soul would not abide in the body, due to the flesh of the body and the soul's own delicate nature. The purity of the soul is directly linked to knowledge, for without a pure soul there cannot be pure action and pure action is the heart of knowledge. It is for this reason that God sent prophets to mankind to educate him in how to purify his soul. The prophets came with two things to man, ritual practices (*ʿibādāt*) and knowledge of God's Habit (*ʿādāt*). The first of these was directed to sound beliefs and undertaking required ritual practices, the second of these was directed toward nourishing the body, providing medical treatment and means of life. Man has been given the responsibility of carrying out the first, but he can only do it by means of the second. The greatest teacher of all, the Prophet Muḥammad, carried out both together in order to guide the lives of his followers in this world and the next.[63]

Al-Ṭurunbāṭī then provides a brief history of knowledge: man's knowledge was sound at the time of the Prophet and his immediate successors, yet with time this knowledge was corrupted by man's vain desires, necessitating the recuperation of the original knowledge through speculative theology (*kalām*). This theology required knowledge of Arabic and if one wished to carry out the ritual practices of Islam, it necessitated

61 See al-Kattānī, *Salwat al-anfās*, vol. 2, 303–04.

62 Al-Ṭurunbāṭī, *Bulūgh aqṣā al-murād fī sharaf al-ʿilm*, 228–29.

63 Ibid., 230–31. For the first part of this paragraph al-Ṭurunbāṭī is relying on Abū ʿAbdallāh al-Sāḥilī's *Bughyat al-sālik fī ashraf al-masālik*, vol. 1, 121–22.

a knowledge of various rational sciences. Thus, he explains the origins of astronomy (*'ilm al-hay'a*) in the need to determine the qibla and the times of prayer, the origins of mathematics in creating portions of inheritance, the origin of logic in the need-to-know sound from false reasoning, and so on. In general, he argues, harkening back to al-Yūsī's own reasoning in his critique of al-Suyūṭī, that the wisdom of the Greeks was accepted by all the caliphs, as well as the Companions themselves, and that there is no reason to forbid any of these sciences, that they were all means to an end. Anyone who forbids one of them should forbid all of them so that if you say that theology and logic were reprehensible innovations (*bida'*), then you might as well go and call mathematics and medicine innovations as well. Perhaps taking a page from al-Sanūsī, al-Ṭurunbāṭī then goes on to emphasize that without proper knowledge a Muslim could not fulfill his ritual obligations. He gives an anecdote of someone who came to his own study circle only to have his own teacher explain to him that since he had not been performing his ablutions correctly, or the ritual observances during the pilgrimage, he would need to repeat them all again to make up what he had failed to do from the time he had reached puberty.

The subsequent five chapters show that al-Ṭurunbāṭī has also been reading al-Ghazālī, al-Rāzī, al-Suyūṭī, Sanūsī, and Ibn 'Aṭā' Allāh, but contain no explicit division of the sciences and aside from demonstrating a similar interpretation of Ash'arism as his seventeenth-century colleagues when it comes to the importance of understanding God's Habit, and of Sufism when it comes to submitting to one's teacher and focusing on knowledge for the sake of the world to come, there is little that is new. Al-Ṭurunbāṭī did include, however, a striking anecdote on the need for humility among teachers that he may have taken from 'Abd al-Wahhāb al-Sha'rānī's (d. 873/1565*) Lawāqiḥ al-anwār* and which deals with the great Sufi authority Ibn al-'Arabī:

> The supreme teacher Muhyī al-Dīn Ibn al-'Arabī related regarding himself that he had once been on a ship in the ocean and the winds had grown fierce. He said: "Become still O sea, for there is a sea of knowledge sailing upon you." A beast rose out of the sea and said to him: "We heard what you said. What do you say about the case of a wife whose husband is transformed (*mansūkh*)? Does she carry out the waiting period for the living or the dead?" The Shaykh did not know how to answer. The beast said to him: "Make me your teacher in this matter and I will teach you the answer." Ibn al-'Arabī assented. It said: "If he is transformed into an animal, she carries out the waiting period for the living, if into an inanimate object, the waiting period of the dead." This story

is recorded in the biographical accounts (*tarjama*) of his teachers of the jinn, humans, angels, and animals.

Considering that the point of law on which the great Sufi Ibn al-ʿArabī admits ignorance is a rather fanciful one with no place in manuals of jurisprudence, the reader would appreciate this anecdote for its sentiment that every scholar should practice humility rather than a criticism of Ibn al-ʿArabī himself.

The three works considered here organized knowledge, its transmission and study in light of the particular synthesis between Ashʿarī theology and (mostly) Shādhilī Sufism that predominated in early modern Morocco. Among the three, al-Yūsī's stands out not just for its comprehensiveness, but also for his extended defense of the rational sciences as part of God's revelation and as instrumental to the fulfillment of Muslims' obligations. Yet in al-Yūsī's emphasis on man's ability to investigate the workings of God's habitual actions he goes beyond al-Ṭurunbāṭī's more instrumentalist approach of seeing the rational sciences fulfilling only a supportive role for the Islamic sciences.

The occult sciences feature in these taxonomies not as an exception or a contrast category, but as an accepted part of the natural sciences. Even as this point bears emphasizing due to current narratives that undermine the rationality of the occult, it is also true that it can be difficult to precisely delineate the line between licit and illicit invocations of occult powers. The difference between a prayer and a spell is the addressee – God or jinn – not the form or mechanism with which it is carried out (e.g., talisman, letters, magic square). As al-Yūsī commented regarding sorcery, it is a scholar's intention that determines the nature of his study, and if the intention is directed toward the good of the community, the study is licit.

2.2 FRAMING THE NATURAL SCIENCES WITHIN INTELLECTUAL AUTOBIOGRAPHIES

The biobibliographical *ṭabaqāt* literature offers a general impression of how many scholars studied the natural sciences in a certain area or specific time period, and the literature describing and enumerating the sciences offers insight into how scholars framed and situated the natural and rational sciences more broadly. A third genre – one specific to the Maghrib – offers yet further insight into the overall place of the natural sciences in Moroccan scholarship of the period. The *fahrasa* or intellectual autobiography bears a resemblance to and at times overlaps with a scholar's list of his teachers, works studied, and collections of attestations of permission to teach certain works of subjects (*ijāza*, pl. *ijāzāt*), but by the sixteenth–eighteenth centuries at the latest had grown into a genre of its own.

As with many other Maghribi intellectual developments, the *fahrasa* has its origins in al-Andalus, where the first example appeared with Ibn al-Qālī (d. 356/967), a scholar of Eastern origin who had settled in al-Andalus in 330/942. It seems to have found immediate appeal, for a number of Ibn al-Qālī's contemporaries wrote their own, and the first example we know of in Morocco is that of Abū ʿImrān al-Fāsī (d. 430/1039), who wrote his half a century after Ibn al-Qālī. Abū ʿImrān's *fahrasa* not being extant, for the modern historian, the *fahrasa*'s Moroccan fate begins a century later with Qāḍī ʿIyāḍ (d. 544/1149).[64]

ʿAbdallāh al-Targhī, who has written the definitive work on the *fahrasa* in the Moroccan context, differentiates between those examples written between the sixth/eleventh and ninth/fifteenth centuries and those from the tenth/sixteenth to twelfth/eighteenth centuries. The break between the two, he argues, is due to the collapse of the Moroccan education system with the fall of the Merinids in the middle of the ninth/fifteenth century: the corollary of there being a lack of teaching and studying was a gap in the writing of intellectual biographies that recorded this transmission of knowledge. With the rise of the Saʿdī dynasty and their avid support of scholarship – here al-Targhī both draws on and is largely in agreement with Ḥajjī – the *fahrasa* experienced a revival, albeit with new characteristics.[65] With what was perhaps an awareness of the newly supportive environment for scholarship, authors included greater detail on both their studies and their teaching, and included greater discussion of those sciences that required rational acquisition (*ʿulūm al-dirāya*) as opposed to purely memorization (*ʿulūm al-riwāya*).[66] Finally, those examples of the genre written in

[64] Al-Targhī, *Fahāris ʿulamāʾ al-maghrib*, 100–13.

[65] See ibid., 144–46, with al-Targhī listing all examples of the *fahrasa* that he has found from this second period on ibid., 149–64. Al-Targhī emphasizes the importance of Fez and Ibn Ghāzī's circle specifically for the large number of *fahāris* written there. Similarly, the arrival of scholars from the Mashriq, especially students tracing themselves back to the famed ninth/fifteenth century Shāfiʿī scholar Ibn Ḥajar al-ʿAsqalānī (d. 852/1449), played an important role in the revival of education and the renewed importance of documenting the transmission of knowledge, particularly with regard to Prophetic Tradition (ibid., 145–46).

[66] For al-Yūsī's own emphasis on *dirāya* over *riwāya*, see El-Rouayheb, *Islamic Intellectual History*, 220, and compare with Brentjes' explanation of the term *dirāya* as referring to studying independently through reading instead of personal transmission (*Teaching and Learning the Sciences*, 151–52).

the eleventh/seventeenth–twelfth/eighteenth centuries differed markedly from earlier examples by including anecdotes their authors found striking or instructive, as well as poetry and legal opinions (*fatāwā*) – often by their authors.[67] Thus, while many intellectual autobiographies continued to confine themselves principally to a listing of a scholar's teachers, studies, and students, others were far more expansive and it is here where a number of this genre offer us insights onto the place of the natural sciences in their authors' intellectual lives.

In his survey of sources pertaining to the tenth/sixteenth–twelfth/eighteenth centuries, al-Targhī lists 112 Moroccan *fahāris* that he has found mentioned, of which 95 are extant in manuscripts in a variety of libraries, although the overwhelming majority remain unpublished. The roughly dozen examples I have been able to consult differ widely in length – from tens to hundreds of published pages – as well as in content, with some containing laconic accounts of teachers and students, or lists of scholars from whom the author transmitted *ḥadīth* – the famed astronomer Muḥammad al-Rūdānī's (d. 1095/1685) *Ṣilat al-khalaf bi mawsūl al-salaf* is a prime example – to extended collections of anecdotes and legal opinions. It is unsurprisingly in the latter that we find the most information concerning the natural sciences, and while these examples are not representative, they offer valuable examples of how individual scholars discussed the natural sciences within their own intellectual formation.

A number of scholars record in their *fahāris* that they studied one or more of the rational sciences, in accordance with the evidence offered by the *tabaqāt* literature. Ibn Ghāzī (d. 919/1513), whom we have already encountered as an important figure in the intellectual revival Morocco experienced under the Saʿdīs, was a prominent figure of his day, serving as judge in the town of his birth, Meknes, before becoming Imām of the Qarawiyyīn in Fez, where he also held teaching chairs in mathematics and inheritance law.[68] Aside from writing a short poem on mathematics with an extended prose commentary – *Bughyat al-ṭullāb fi sharḥ munyat al-ḥisāb* – he studied the influential plague treatise of the Egyptian Shāfiʿī jurist, Ibn Ḥajar al-ʿAsqalānī.[69] Half a century later, in the *fahrasa* of Aḥmad al-Manjūr (d. 995/1587), a renowned Mālikī jurist, we find that the author studied mathematics, timekeeping and logic not only with numerous teachers, but also with two scholars – Abū al-ʿAbbās Aḥmad b. Muḥammad al-Mawāsī

[67] Ibid., 176–81. [68] Ibn Ghāzī, *Fihris Ibn Ghāzī*, 8.

[69] Ibid., 12–15, 135. On the nature and significance of Ibn Ḥajar's plague treatise, see Stearns, *Infectious Ideas*, 85–89.

and Abū ʿAbdallāh Muḥammad al-Ṣaghīr b. Aḥmad b. al-Ḥājj al-Rajnī – whom he identifies as philosophers.[70] A student of Ibn Manjūr's, ʿAbd al-Wāḥid al-Sijilmāsī (d. 1003/1595), studied mathematics with him – al-Qazwīnī's *Talkhīṣ al-miftāḥ* and Ibn Bannā's *Talkhīṣ* – as well as introductory works on inheritance law and logic.[71] Muḥammad b. Aḥmad Mayyāra al-Fāsī (d. 1072/1661–62), a prominent Fassi scholar, records in his autobiography having studied among other subjects, mathematics, inheritance law, logic, and medicine, with Abū al-Qāsim b. Abī al-Nuʿaym al-Ghassānī (d. 1035/1632).[72] The head of the Nāṣirī lodge of his day, Ḥusayn b. Muḥammad Ibn Nāṣir al-Aghlānī al-Darʿī (d. 1091/1681) studied al-Jādirī's work on timekeeping and Ibn Ghāzī's *Munyat al-ḥisāb* with al-Mirghitī, whom we will discuss below, and with ʿAlī al-Zaʿtarī al-Maṣrī (d. 1120/1708) he studied two works on timekeeping, *al-Rubʿ al-mujayyab* and ʿAlī al-Zaʿtarī's own treatise *Niṣf al-dā'ira*.[73] And over a century later, Sulaymān al-Ḥawwāt al-Shafshāwnī (d. 1231/1816), studied *al-Muqniʿ* as well as Ibn al-Ṣaffār's (d. 426/1035) treatise on the astrolabe with the then head of the Nāṣirī order, Muḥammad bin Nāṣir al-Darʿī.[74] In addition, al-Ḥawwāt studied mathematics, timekeeping using instruments and without, and medicine with a certain Abū Rabīʿ Sīdī Sulaymān b. Aḥmad al-Fishtālī, who was known for his studies of ancient sciences, and "for being one of those who study philosophy (*al-ḥikma*) in a manner according with revealed law."[75] If this impressionistic list confirms in slightly greater detail what we had learned from the biobibliographical literature, the scholarly autobiographies of Abū Zayd ʿAbd al-Raḥmān al-Tamanārtī (d. 1060/1650), Abū ʿAbdallāh Muḥammad b. Saʿīd al-Mirghitī (d. 1089/1678), Abū Qāsim b. Saʿīd al-ʿUmayrī al-Jābirī al-Tādilī (d. 1178/1764), and Aḥmad ibn ʿAjība (d. 1224/1809) give much more detail regarding the natural sciences, including in the form of the types of anecdotes that al-Targhī described as typical of the *fahāris* of these centuries.

Abū Zayd ʿAbd al-Raḥmān al-Tamanārtī was one of the most important jurists of his time. He was born in 974/1545 in Tamanārt, a small village on the southern edge of the anti-Atlas on a tributary of the Draʿ River into

70 See Aḥmad al-Manjūr, *Fihris Aḥmad al-Manjūr*, 17. Ibn Manjūr shared a teacher with these scholars in timekeeping: the mathematician, timekeeper and specialist in inheritance law, ᶜAlī b. Mūsā ibn Hārūn al-Matgharī (ibid., 11–12, 40–41).

71 ʿAbd al-Wāḥid al-Sijilmāsī, *al-Ilmām bi-baʿḍ man laqaytuhu min ʿulamā' al-islām*, 100–06.

72 Muḥammad b. Aḥmad Mayyāra al-Fāsī, *Fahrasa*, 28–29.

73 Ibn Nāṣir al-Aghlānī al-Darʿī, *Fihris Ibn Nāsir*, 97–99.

74 I have not been able to locate this scholar's dates.

75 Sulaymān al-Ḥawwāt, *Thamrat anīs fī taʿrīf bi-nafsī li-abī al-rabīʿ Sulaymān al-Ḥawwāt al-Shafshāwnī (1160/1747–1231/1816)*, 62, 91.

a family with an illustrious Sufi heritage.[76] His father had studied with al-Tabbaʿ (d. 914/1508), the successor of the Sufi al-Jazūlī, whose movement had contributed to the rise of the Saʿdī dynasty at the beginning of the sixteenth century.[77] After studying in Tamanārt, al-Tamanārtī moved to Taroudant, the mosque of which had restored by the Saʿdī ruler Muḥammad al-Mahdī, and which the Saʿdīs exerted themselves to make an intellectual center on par with the northern cities of Fez and Marrakesh.[78] It was there that he continued his studies and became a respected judge for over thirty years, appointed in this position by three successive rulers during the tumultuous decades of the seventeenth century. The high quality of the intellectual environment of the Sūs valley and its regional capital Taroudant as well as the lodges in the anti-Atlas mountains to its south during this period is attested to by al-Tamanārtī attaining his reputation without ever leaving the south to study.

Al-Tamanārtī's *fahrasa*, *Collected Benefits Concerning the Muslim Community's Transmission of the Sciences* (*Fawā'id al-jamma fī isnad ʿulūm al-umma)* was a substantial work, containing extensive information regarding his studies, teachers and students, as well as several of the author's legal opinions, one of which will play a role in Chapter 3. We learn that he studied mathematics in both Taroudant and Tamanārt, but of greatest interest here are his studies with ʿAbd al-Raḥmān al-Buʿaqīlī (d. 1006/1598), who taught him both mathematics and astronomy.[79] Al-Tamanārtī introduces al-Buʿaqīlī as follows:

> Due to his excellence in astronomy (*ʿilm al-nujūm*), al-Manṣūr brought him to Marrakesh to carry out timekeeping and to teach this science. He has a useful commentary on [al-Jādirī's] *Rawḍat al-Azhār* on timekeeping and astrology (*al-tanjīm*) that he named *Qaṭf al-anwār min rawḍat al-azhār* and another on the *al-Yasāra* [of Ibn al-Bannā'] as well as a poem in *rajaz* on logic.

Al-Buʿaqīlī was originally from the Sūs valley, and it was in Taroudant that he taught al-Tamanārtī, who also offers an intriguing story on how his teacher was called to court in Marrakesh. The caliph al-Manṣūr had learned through astrological methods that armies were approaching him. Uncertain of what this meant, he turned to his confidant (*ṣāḥib sirrihi*) Abū

[76] For biographical information on al-Tamanārtī, I have drawn on the introduction to al-Tamanārtī, *Fawā'id al-jamma*, 11–41.

[77] See ibid., 80. [78] Ḥajjī, *al-Ḥaraka al-fikriyya*, vol. 1, 125–26.

[79] For his mathematical studies, see al-Tamanārtī, *Fawā'id al-jamma*, 120 and 133, and for his studies with al-Buʿaqīlī, see ibid., 115–16. There is some disagreement regarding the date of al-Buʿaqīlī's death, with al-Tamanārtī giving 1006/1598, whereas al-Qādirī's entry in *Nashr al-mathānī* has 1020/1611.

al-Ḥasan ʿAlī b. Sulaymān b. ʿAbdallāh al-Tamlī (d. 999/1590), who in turn wrote to his brother Sulaymān, who was studying with al-Buʿaqīlī in Taroudant. Al-Buʿaqīlī responded to his student's question regarding the significance of the foretold armies by explaining that they were armies of locusts. He wrote to al-Manṣūr to inform him of this and it did not take long before the locusts arrived and afflicted all parts of Morocco. Al-Manṣūr called him ʿAbd al-Raḥmān al-Jarrād (i.e., ʿAbd al-Raḥmān of the locusts), and brought him to Marrakesh because of his ability to correctly interpret what had been foretold for the caliph.

The anecdote leaves it unclear both precisely what astrological means al-Manṣūr had employed to foretell the future, just as it is vague how al-Buʿaqīlī was able to correctly interpret the prognostication. It may be telling, though, that this anecdote is related in close proximity to al-Tamanārtī's description of how al-Buʿaqīlī installed geographically tailored sundials in Taroudant's mosques:

> It was he [al-Buʿaqīlī] who placed in each one of Taroudant's two minarets a piece of marble on which was engraved the hour, the opened fingers (*al-aṣābiʿ al-mabsūṭa*), the azimuth, the line for sunset (*khaṭṭ al-zawāl*), the line for noon (*al-ẓuhr*), the line for the afternoon prayer (*khaṭṭ al-ʿaṣr*), and the line for the end of the afternoon prayer for the city of Taroudant and every town on its latitude. He fixed in the middle of it a nail, the shadow of which was considered in relation to each of these lines. If the shadow rested on the line of sunset, he knows it, and the same for the remaining ones, so that the one who calls to prayer (*muʾadhdhin*) does not need to take any other steps.[80]

Using the natural sciences to further Islamic ritual practice, and especially the establishment of prayer times and the beginning and end of months, has been studied extensively over the past decades by David King and others, and this work formed the basis for A. I. Sabra's influential thesis that the natural sciences had not declined in the Islamic world, so much as they had been naturalized and interest in them limited to what supported religious imperatives. Taken alone, al-Buʿaqīlī's sundials offer strong evidence for Sabra's view. I will touch on the naturalization thesis in Chapter 3 – here, I am interested in emphasizing al-Tamanārtī's framing of al-Buʿaqīlī as having excelled in both astrology and astronomy. In accordance with the later categorizations of the sciences discussed above, al-Tamanārtī

80 Al-Tamanārtī, *Fawāʾid al-jamma*, 116. The marking of the end of the afternoon prayer can be explained by Mālikī jurisprudence differentiating between optional and mandatory times to carry out prayer, with the second marking referring to the mandatory time. I am grateful to Mohammed Fadel for clarifying this point to me.

does not differentiate between the validity of the technical abilities needed to make a latitudinally accurate sundial and the occult knowledge needed to properly ascertain the relationship between the supra and sublunar worlds. Al-Buʿaqīlī is remarkable in this account not for having studied these sciences as much as he is for having excelled in them to the extent that after his death, he was lamented for not having left students of similar abilities to succeed him.[81]

Like al-Tamanārtī, but two generations later, Abū ʿAbdallāh Muḥammad b. Saʿīd al-Mirghitī (d. 1089/1678), with whom this chapter began, grew up in the Sūs valley, in this case in the small village of Mirghit, some twenty kilometers from Tiznīt.[82] Also like al-Tamanārtī, his family had a reputation for piety and Sufism, and he too left for Taroudant after his initial studies with his father. Unlike al-Tamanārtī, however, al-Mirghitī left the south and subsequently traveled to Marrakesh and then to the Dilā' lodge in the Middle Atlas. He studied astrology and letter magic (*al-sīmīyā'*) with Abū Qāsim al-Ghawl al-Fishtālī, learned astronomy from Abū ʿAbbās Aḥmad al-Walātī, and medicine from Abu al-Ḥasan ʿAlī b. Ibrāhīm al-Andalusī.[83] Among the medical authors whose works he knew were Ibn Sīnā, Ibn Zuhr, al-Wazīr al-Ghassānī, and along with astronomy and astrology, he studied numerology (*al-tarqīm*), letter magic, as well as al-Būnī's *Shams al-maʿārif.*[84] This last work was perhaps the single most famous work of the occult sciences in the Muslim world in the Early Modern period.[85] He wrote three works on astronomy, *al-Muqniʿ fi ikhtisar Abī Muqri'*, and then *al-Mumtiʿ fi sharh al-muqniʿ*, a commentary on the former, and finally *al-Muṭliʿ ʿalā masā'il al-muqniʿ*, a summary of the former commentary. In addition, he authored a poem on timekeeping, *Maʿūnat al-ḥaysūb fī ʿamal al-tawqīt bi-l-jayyūb*, a poem in 110 lines, and a treatise on how to defend

[81] Al-Tamanārtī quotes an Abū Muḥammad ʿAbdallāh b. al-Mubārak to the effect that the latter was never as saddened by the death of a scholar as he was by al-Buʿaqīlī and the knowledge that was lost with him, as he left no one in Morocco as skilled in this as him since few had studied with him – the majority having withdrawn from him (*li-ghulbat al-inqibāḍ ʿalayhi*) (ibid., 116).

[82] For background on al-Mirghitī – whose *nisba* is given in a variety of ways, including in al-Yūsī's *fahrasa* as al-Mīrghithī – I have relied on the editor's introduction to al-Mirghitī, *al-ʿAwā'id al-muzriyya,* vol. 1, 17–55.

[83] Al-Mirghitī, *al-ʿAwā'id al-muzriyya*, vol. 1, 49. I have not been able to identify these latter two scholars.

[84] Ibid., 61–62.

[85] On the identity and context of al-Būnī and his work, as well as his having been credited with multiple works that began with the title *Shams al-maʿārif*, see the impressive article by Noah Gardiner, "Forbidden Knowledge? Notes on the Production, Transmission, and Reception of the Major Works of Aḥmad al-Būnī."

oneself against sorcery, *Risāla fī ibṭāl al-siḥr.*[86] Aside from teaching, which he did, among other places, at the Dilā' lodge, where he had al-Yūsī as a student – who, in turn, spoke highly of his teacher – he also worked at times as a physician.[87] He died of the plague in 1089/1678 at the age of seventy-two and was buried in Aghmāt, near Marrakesh.

Al-Mirghitī's autobiography is not only quite long – its published edition comprises three volumes and the text itself covers over seven hundred pages – but contains numerous anecdotes concerning medicine, astronomy, and magic that range from a recipe for wheat soap to calculating eclipses to quotes from Ibn Ghāzī on how to boil an egg properly and to Ibn al-Nafīs on forgetting.[88] The vast majority of these anecdotes – there is no clear organization of what al-Mirghitī calls "beneficial accounts" (*fā'ida*, pl. *fawā'id*) – admittedly do not address the natural sciences, concerning themselves instead with issues of spirituality, and ritual. But in a work of such a size, more than enough of it takes up the natural and rational sciences to show that for al-Mirghitī there was no qualitative difference in validity between astronomical observation, talismanic magic, Galenic medicine, and invocations and spells. All were legitimate and effective ways of addressing the challenges posed by the natural world and were ways to gain insight into its workings.

The majority of the material related to the natural sciences included by al-Mirghitī relates to medicine, although this term itself requires some discussion. Students of medicine in the Islamic world have in the past differentiated between Graeco-Islamic or Galenic and Prophetic medicine, with the former being framed as largely free of religious influences and the latter seen as primarily a mixture of religious and folk practices.[89] Additionally, in studies of the sixteenth–eighteenth centuries in the Ottoman Empire, historians have more recently focused on the reception of Paracelsus' medical theories by Muslim physicians, giving us a third distinct medical tradition practiced in Mediterranean

[86] Al-Mirghitī, *al-'Awā'id al-muzriyya*, 73–74. The poem *Ma'ūnat al-ḥaysūb* was printed in Fez in 1305 along with al-Māridīnī's treatise on *al-rub' al-mujayyab*. I hope to address his treatise against sorcery in a future publication.

[87] Ibid., 51. For al-Yūsī's impressions of him, see al-Yūsī, *Fahrasat al-Yūsī*, 68–70.

[88] See, respectively, al-Mirghitī, *al-'Awā'id al-muzriyya*, vol. 1, 144, 338, 200, and vol. 2, 789.

[89] For two critical estimations of Prophetic medicine, see Ullmann, *Islamic Medicine*, 4–5; Savage-Smith, "Medicine," 927–30. A more nuanced and productive reading of the tradition is offered by Imerli Perho, *The Prophet's Medicine – A Creation of the Muslim Traditionalist Scholars.*

Muslim societies.[90] As we will see in Chapter 4, this last medical tradition attracted little attention in Morocco during the period in question. Most relevant here, instead, is the distinction between Galenic and Prophetic medicine, which has in part been maintained by the famous modern trinity of science, religion, and magic. More specifically, when we look at the medical material provided by al-Mirghitī and many of his contemporaries, we are easily tempted to differentiate between legitimate material – that which reflects Galenic humoral theory – and illegitimate accretions in the form of prayers, invocations, or spells. To give into this temptation would, however, not only distort how al-Mirghitī and his colleagues understood the science of medicine by subjecting it to a modern definition in which science is science precisely because it is not magic or religion and does not invoke occult forces. It would also submit to the desire to locate parallels and precedents to modern medicine in premodern Muslim societies in order to ground broader generalizations between the relationship of Islam and Science, however problematically defined. Al-Mirghitī's understanding of medicine, as reflected by the material included in his *fahrasa*, is not legible when framed in this fashion, as for him magic squares, attention to diet, and an awareness of the imminence of God lay on a continuum.

Given the nature of the source as a series of disjointed "benefits," any attempt to reconstruct al-Mirghitī's understanding of any science is tendentious insofar as it requires cobbling together disparate passages into a composite whole. Nonetheless, his general understanding of medicine and the care of man's body and soul is clear:

> In matters of medicine, know that curing the sick person is not a condition or to be made conditional for the expert doctor, much less to prolong his life. He is required to look into illness and the state of the sick person. If he finds a way to treat it, let him do so. Health is dependent on the decision of God, Most High. If the cause is determined to lead to the sick person's death, then stop the treatment. There are three reasons for death: One: Blows, collapse, falling, drowning, burning, and so on. When this happens, the spirit (*al-rūḥ*) contracts entirely to the heart and then exits all at once. Two: An increase of one of the four humors over its opposite, overpowering it, putting an end to the radical moisture (*al-ruṭūba al-aṣliyya*) little by little until the pain becomes intense and the spirit exits at last. Three: Death in accordance with the end of the natural lifespan.[91]

[90] On this, see especially the work of Natalia Bachour, *Oswaldus Crollius und Daniel Sennert im frühneuzeitlichen Istanbul: Studien zur Rezeption des Paracelsismus im Werk des osmanischen Arztes Ṣāliḥ b. Nasrullah Ibn Sallūm al-Ḥalabī.*

[91] Al-Mirghitī, *al-ʿAwā'id al-muzriyya*, vol. 1, 253. Compare with the later inclusion of a poem on the four natures (*ṭabā'i*ʿ), which al-Mirghitī attributes to the Granadan

While adhering to a belief in humoral medicine, al-Mirghitī takes pains to stress the limits God has placed on man's ability to cure and heal the afflicted body. A great deal of what one can do is related to diet. The author recommends risen bread, male sheep, almonds, sugar, milk, turnips, apples, pomegranates, fresh cheese, and honey and warns against consuming beef, goat (*maʿz*), female sheep, all fish (*ḥītān*), fruits other than those mentioned, cabbage, eggplant, and aged cheese.[92] Al-Mirghitī mentions bread again later when in connection with observing that scholars have warned against befriending people with physical deformities, he notes that philosophers (*al-falāsifa*) agreed that whoever eats mostly bread, especially of the sweet kind, is smarter and quicker of understanding.[93]

Beyond general advice, al-Mirghitī also offers a number of specific remedies. To cure an affliction of the chest or throat, for example, one should take one measure (*ūqiya*) of saffron, one of cinnamon bark, one of olibanum, and four of spikenard (*sunbul*). Sift these together, make tablets out of them and eat with thickened licorice juice.[94] Other recipes provide cures for hemorrhoids, pimples, scabies, and toothache, excessive choler as well as differing ones for sciatica and forgetfulness.[95] While some of these recipes entail mixtures of herbs, others, such as the first for sciatica (*ʿirq al-nasā*) involve Qur'anic invocations:

Go before dawn on Thursday or Saturday to the bush called henbane, which intoxicates and which can be used to intoxicate fish by putting it in water. Then cut off a twig, cut off the end of the root from which the twig grew and dig around it until a hand width and three fingers are visible. Order the sick person to place the hurting foot on the root. Take a cutting blade and put it in the earth and recite the 15th verse of Sūrat al-Ḥajj seven times, then cut off the root and say, "I have cut

polymath Ibn al-Khaṭīb (d. 776/1374), but which the editor notes is actually that of the famed jurist Aḥmad al-Wansharīsī. The poem links the natures to their attributes and is followed by a graph (ibid., 297). On the development of the notion of radical moisture in the Medieval Mediterranean Galenic tradition, see Michael McVaugh, "The 'Humidum Radicale' in Thirteenth-Century Medicine."

[92] Ibid., vol. 1, 169. The author later repeats his warning against eating cabbage (ibid., vol. 2, 676).

[93] Ibid., vol. 1, 245. Other authors in this period also wrote of an aversion to people with deformities: see the anecdotes regarding al-Shāfiʿī related by al-Yūsī in *The Discourses*, 251–55.

[94] Al-Mirghitī, *al-ʿAwāʾid al-muzriyya*, vol. 1, 233.

[95] Ibid., vol. 2, 574 (hemorrhoids), 577 (pimples and scabies), 577 (toothache), vol. 2, 538 (purification of a choleric temperament [*ṣufarāʾ*]), vol. 1, 169–70, 321–22 and vol. 2, 571–73 (sciatica), vol. 2, 576 and 789 (remedies for forgetfulness, ascribed to the famed doctors Ibn Zuhr [d. 557/1162] and Ibn al-Nafīs [d. 687/1288], respectively).

your root, o son of so and so [presumably the name of the patient]." Place it on the rock and take it like a corpse on a bier to the grave.[96]

This is not the only place where al-Mirghitī describes driving a knife into the earth to alleviate pain. In a later passage, he states that for all pains in the body, place a knife in the middle of a natural triangular drawing (*jadwal*) on the edges of which nine Arabic letters are written, place your left hand on the knife and your right hand on the pain and recite verse thirty-five of *Sūrat al-nūr.* If the pain moves, move the knife to the side that the pain moves in the body until it stops. The patient should then lick the letter where the pain stopped, and if it had been written on the earth, he should take it and spread it over the place of the pain.[97] Talismans and Qur'anic invocations are not uncommon in other cases as well. The two most striking are perhaps al-Mirghitī's specifications on how to prevent pregnancy and his advice on how to avoid being afflicted by epidemic disease. Regarding the first he writes:

To prevent pregnancy (*li-qatʿ al-ḥaml*), write this talisman with this seal (*al-khātim*) as you see here: ḥ-ḥ, ḥ-ḥ-ḥ, ṭ, ṭ, ā, ḥ-k-ā-waw 66 sins l-h, and place it in an amulet. The woman places it on herself, and if she wishes to become pregnant, she takes it off.[98]

The varied approaches he offers to counter the dangers of epidemics are of particular interest considering the repeated outbreaks of plague in Morocco in the seventeenth century, including that of 1603 in which the ruler Aḥmad al-Manṣūr died and that of 1678 in which al-Mirghitī himself perished. These including constantly reciting *ḥayy* (alive) *ḥalīm* (clement) *ḥannān* (tender) *ḥakīm* (wise) – these being four of the names of God; reciting the Throne verse eighteen times a day; using a prayer against the plague transmitted from the Algerian theological authority al-Sanūsī (d. 895/1490); and, on the authority of al-Shāfiʿī (d. 204/820), the founder of the Shāfiʿī law school, covering oneself with purple paint.[99] Two further

[96] Ibid., vol. 1, 169–70. The description of how the root is treated once in the grave is subsequently described in detail.

[97] Ibid., vol. 1, 317–18. The relevant Arabic letters follow the description in three lines: bā'-ṭa'-dāl/zā'-hā'-jīm/ waw-alif-ḥā'. Compare with another remedy for pain given a few pages later for: "These letters contain a wondrous secret: ṣad 'ayn alif-lam kaf alif-ḥa alif-ḥa ra-ḥa to end pain in a tried-and-true manner: write them on three chips (*barawāt*) with saffron and rosewater, mash one in a bowl and give it to the sick person and he will recover, God willing; if he does not, mash the second, and then the third" (ibid., vol. 1, 335).

[98] Ibid., vol. 1, 317. For an overview on talismans in the Islamic tradition see Constant Hamès' edited volume *Coran et talismans: Textes et pratiques magiques en milieu musulman.*

[99] Al-Mirghitī, *al-ʿAwā'id al-muzriyya,* vol. 1, 322. Compare with Dols, *The Black Death in the Middle East*, 121–42.

related, if distinct, examples of such remedies include treating dog bites with the blood of a descendant of the Prophet, and drinking water in which a paper with the names of the Companions of the Prophet had been submerged in order to avoid both sinning and using it to wash the face to avoid being afflicted with blindness.[100]

As we will discuss further in Chapter 4, al-Mirghitī's use of humoral medicine combined with spells and prayers was common in the Islamic West during the seventeenth century. His interest in astronomical topics in his *fahrasa*, while not as extensive, combined technical issues with more general cosmological ones. Early on, drawing on – though not citing – al-Jādarī's (d. 839/1416) *Iqtiṭāf al-anwār*, al-Mirghitī describes how to divide the heavenly spheres (*al-falak*) according to the days of the year, and later he gives a brief overview of how to calculate the trajectories of the planets (*tarḥīl al-dārārī*).[101] Yet, the most striking case is that mentioned at the beginning of this chapter, when he introduces the issue of the trepidation of one of the outer spheres (*ḥarakat iqbāl al-falak wa idbārihi*) – though which of the spheres he is referring to is not clear.[102] Al-Mirghitī opens his discussion of the issue by stating that Ptolemy and his followers said that the sphere's (*al-falak*) trepidation is every hundred years. Others said it is eighty, and al-Jādirī, in his *Rawḍat al-azhār*, a standard astronomical reference in Morocco in this period, said it was sixty years while most of the scholars in the East say seventy-two Persian years.[103] Regardless of the disagreements mentioned, the issue already appears confused, as trepidation, in Ptolemaic astronomy generally the accession and regression of the fixed stars resting on the eighth sphere, was defined by astronomers in terms of how many years it took the stars to move a single degree.[104]

Al-Mirghitī continues by explaining that Abū Zayd al-Jazūlī al-Buʿaqīlī – whom we met above as having installed sundials in the minarets of the mosques of Taroudant – said in his commentary on the *Rawḍa* that what is accordance with the attested practice of timekeeping is that the amount of approach is 14 degrees. After some further explanation, al-Mirghitī notes, referring to his collaboration with his own teacher in astronomy:

[100] Al-Mirghitī, *al-ʿAwāʾid al-muzriyya*, vol. 1, 353 and 362, respectively.

[101] Ibid., vol. 1, 163 and 338, respectively. [102] Ibid., vol. 2, 533.

[103] The Persian calendar was based on a standard year of 365 days. See the entry on "Ta'rīkh" in EI3.

[104] On the early history of trepidation in Islamic astronomical thought, see Jamil Ragep, "al-Battānī, Cosmology, and the Early History of Trepidation in Islam." I am grateful to Robert Morrison for this reference.

The calculations of our companions in Marrakesh today is that it is 15 degrees. Know this. But in my opinion and that of the teacher Abū ʿAbbās b. Muḥammad al-Walātī, based on observations with the astrolabe carried out in 1049/1639–40, we found that it was 16 and two thirds degrees, approximately (*bi-taqrīb yasīr*).

What observations this entailed are unclear, as is exactly what was being measured. Nonetheless, we learn from this at the very least that al-Mirghitī had the technical training to use an astrolabe and the curiosity to test earlier observations of which he had read.

There are a handful of passages in which al-Mirghitī takes up the matters of the heavens. The first is an extended discussion of a group of nine or twelve stars or comets (*dhawāt al-dhawāʾib*), that he describes Ptolemy having referred to as prisoners of the sun due to their usually being hidden by its rays. If seen at dawn in the east, these portend war and strife. He discusses each of these stars in detail, referring in part to the writings of the Brethren of Purity.[105] In another, more cosmological passage, he reflects on the immensity of the universe:

Know that the world and all that is on it in the way of seas and mountains in the nine spheres is like a dot on edge of a circle (*kurr al-dāʾira*). In the spheres (*al-falak*) there are 1029 stars, the smallest of which is eighteen times smaller than the earth, the largest of which is a hundred and seven times larger, appearing to you like a cast pearl on a green carpet. If you contemplate this immensity, the wisdom of the Creator and his Grandeur will become clear to you, rousing you from the sleep of unawareness and ignorance.[106]

The connection between the heavens and a proper appreciation of God's creation continues in a short comment al-Mirghitī makes on the mention of a "piercing star" in the third verse of the eighty-sixth Qur'anic sura al-Ṭāriq, in which he identifies the star with Saturn, the light of which pierces through the seven spheres between it and Earth.[107] The astronomical, astrological, and cosmological are part of a continuum in al-Mirghitī's text, which, organized as it is in generally unrelated anecdotal form, does not lend itself to presenting clear taxonomies of the sciences as found in the work of al-Yūsī. Rounding out the material he included on

[105] Ibid., vol. 2, 647–50. The term *dhū dhuʾaba,* "possessing locks of hair," usually refers to comets when designating planets or stars, and some of the names that al-Mirghitī provides are the same mentioned in a sixteenth century Persian astrological manuscript referring to comets (E. S. Kennedy, "Astronomical Events from a Persian Astrological Manuscript," 164).

[106] Ibid., vol. 1, 183. Compare with al-Mirghitī, *al-Mumtiʿ fī sharḥ al-muqniʿ*, 12.

[107] Ibid., vol. 1, 313–14. Al-Mirghitī adds a quote from al-Ṣafadī that proposes the Pleiades as an alternative identification.

medicine, astronomy, and astrology, are several passages that deal with other spells and invocations. They address, among other things, how to diminish the heat of the sun; how, on the authority of Ibn Waḥshiyya, one can draw a circle around oneself and with certain Qur'anic passages protect oneself from thieves; how, with a combination of fasting and the recitation of Sūrat al-Anʿām 96 one hundred times after every prayer for seven days one can open locks.[108] The underlying structure of causality of the universe, both sub- and supralunar, is, as noted by al-Mirghitī in his discussion of medicine, entirely dependent on the will of God. Yet, it also clearly has an ascertainable structure that can be observed and measured with the senses and the help of instruments, and manipulated through proper knowledge of the attributes of plants, the names of God, the use of the Qur'an, and insight into their esoteric sympathetic connections with physical reality as well as such sympathetic relations more generally. Put differently, al-Mirghitī takes pains to adhere to the standard Ashʿarī understanding of occasionalism that denies causative power to anything but God and has God recreating the world at each moment while also stressing the potential of man to understand this world through God's habit of recreating it in a regular fashion. Thus, close to the beginning of his work he presents a long quote from the Moroccan legal and spiritual authority Qāḍī ʿIyāḍ (d. 544/1149) that decries the claims of soothsayers and astrologers to be able to independently ascertain the causes of future occurrences.[109] Considering al-Mirghitī's inclusion of several astrological benefits in his *fahrasa*, it is probable that he did not share Qāḍī ʿIyāḍ's more categorical rejection of astrology, and that like al-Yūsī, he was more concerned with a correct understanding of how God's Habit played out in the supralunar world.[110]

A proper understanding of causality and repeated discussion of astrology continued within the *fahrasa* literature of the twelfth/eighteenth century. Abū Qāsim b. Saʿīd b. Abī al-Qāsim al-ʿUmayrī al-Jābarī al-Tādilī (d. 1178/1764) was born in Fez in 1103/1691–92 (only a year after al-Yūsī's death), although his family soon moved to the royal capital in Meknes, where his father worked at the court of Moulay Ismāʿīl.[111] His father had studied with both al-Yūsī and al-Mirghitī and al-ʿUmayrī quoted repeatedly and extensively with appreciation from al-Yūsī's *The Discourses* in his own

[108] See ibid., vol. 1. 317, 329–31, and 333, respectively. On the enigmatic figure of Ibn Waḥshiyya, see the EI3 article by Jaakko Hämeen-Anttila.
[109] Ibid., vol. 1, 122–23.
[110] See the discussion of al-Yūsī's views above and compare with al-Yūsī, *al-Qānūn*, 147.
[111] Abū Qāsim al-ʿUmayrī, *Fahrasat al-ʿUmayrī*, 36.

fahrasa. Professionally, he led a career as a judge and suffered persecution during the reign of Moulay ʿAbdallāh (d. 1170/1757), one of Moulay Ismāʿīl's sons and principal successor. In his *fahrasa,* al-ʿUmayrī is nostalgic for the stability and piety of Moulay Ismāʿīl's reign, lamenting the political turmoil into which Morocco had fallen after his death.[112] Although al-ʿUmayrī did not study the natural sciences himself – the only rational science he describes himself studying was logic – his deep interest in omens and divination runs through his writing.

Al-ʿUmayrī began his *fahrasa* with a chapter on the nature of reason. According to the philosophers (*al-ḥukamāʾ*), there are four levels of reason: "material reason" (*al-ʿaql al-hayūlānī*), which is a possessed by children; "reason of aptitude" (*al-ʿaql bi-l-malaka*), through which one acquires knowledge of the necessities; "reason through action" (*al-ʿaql bi-l-fiʿl*), by which one attains speculative insights from necessary knowledge; and, finally, "profitable reason" (*al-ʿaql al-mustafād*), by which all the speculative sciences are preserved.[113] As for the physical vessel of reason, al-ʿUmayrī presents two slightly different possibilities: the first is that reason consists of subtle bodies diffused through the body disappear when one dies; the second is that it is a subtle body infused in the body like water in green wood.[114]

In the second chapter, ostensibly on the author's birth and his background, al-ʿUmayrī takes up the issue of omens and associated sciences. Here he takes pains to distinguish between legitimate and illegitimate divinatory sciences. Al-ʿUmayrī begins with a favorable discussion of the stars that that foretell the coming of rain (*naw'*, pl. *anwā'*): this occurs, when, in accordance with Habit of God, one such star rises on the eastern horizon while another sets in the west.[115] Another science like it is geomancy (*ʿilm al-raml*), which entails coming to an insight through lines in the sand and through other esoteric matters (*wa ghayru-hu ʿalā umūr ghaybiyya*). Al-ʿUmayrī notes people have gone as far as claiming that it is a divine science (*qīla anna-hu min al-ʿulūm al-ilāhiyya*), and he cites a Prophetic tradition referring to it as a prophetic science. An entirely different case is that of seeing omens in the flight of birds (*ʿilm al-zajr*) – al-ʿUmayrī describes this as a false science based on tricks, as is the science of guessing unknown matters correctly (*ʿilm al-zakn*), which the

[112] See ibid., 121 (edited text). The complaint was made with good reason. For an overview of the political turmoil of the period, see Stephen Cory, "Sharīfian rule in Morocco (tenth–twelfth/sixteenth–eighteenth centuries)," 469–71.

[113] Ibid., 3 (edited text).

[114] Ibid., 4 (edited text).

[115] Ibid., 18 (edited text). For incorrect or heretical views of causality, see ibid., 54.

Judge Iyās b. Muʿāwiya (d. 121/739) was famous for, but which was really only close observation.[116]

More striking than refuting the power of evil omens is al-ʿUmayrī's observation a few pages later that some scholars consult tables that have information about future events and then use things like the barking of dogs and so on to invoke their knowledge of these events. That this comment is not meant as a reference to trickery as much as accurate insight is borne out by al-ʿUmayrī providing the reader with a series of tables in the following pages. He explains in detail how to determine under which sign of the zodiac one falls on any given day:

> You look at the line of the days, how many days of the month you are in have passed in the Arab calendar, then place the finger of your right hand on it, and you will know which of the foreign (*al-ʿajam*) months the Arab month begins in (*istahalla*). Then place the finger of your left hand on of the months, and move the finger of your right hand horizontally (*ʿarḍ^{an}*) and the finger of your left hand vertically (*ṭūl^{an}*), and where they meet, the moon is in that house, God willing. Then look at what is below the mansion that the moon is in to find the omen in the table and what will happen according to the table of the moon.[117]

The last sentence refers to the reader turning to a subsequent series of tables while remembering the sign of zodiac obtained from the first table in order to cross-reference this sign with a list of omens (creaking of wood, croaking of a crow, dogs barking, etc.) in order to discover what a particular omen on that specific day could portend (sickness, buy a robe, a man will die in the night, etc.).[118] Al-ʿUmayrī does not cite the origins of these tables or comment on their underlying causality. Instead, he immediately launches into a discussion of gematria (*ḥisāb al-nīm*), which he attributes to Aristotle's *Politics*, and which he notes was a science discussed by Ibn Khaldūn and versified by Ibn Ghāzī.[119] Al-ʿUmayrī takes the legitimacy of this science for granted, and ends the chapter with a brief discussion of how to calculate the night of power (*laylat al-qadr*) during Ramadan, his last substantive engagement with the natural sciences.

The final *fahrasa* discussed here, by the celebrated Sufi and Qur'an commentator Aḥmad Ibn ʿAjība (d. 1222/1809), shows an ongoing engagement with the question of causality and the esoteric natural

116 Ibid., 19 (edited text). Iyās b. Muʿāwiya was a judge in Basra during the Umayyad period, who grew famous for his powers of close observation and who occurs in this fashion in numerous anecdotes in Arabic literature (see Charles Pellat's entry in EI2).

117 Ibid., 28. 118 Ibid., 30–34. 119 Ibid., 35–36.

sciences.[120] Born near Tangier in 1160/1747–48, Ibn ʿAjība grew up in a rural environment where he memorized the Qur'an and according to his *fahrasa*, exhibited his own pious bent early on, exhorting his mother to pray when he was still a child. When he was eighteen or nineteen, a certain Sīdī Muḥammad al-Sūsī al-Samlālī, passing through his village, was impressed by the questions posed to him by this young man, and took him with him to Qaṣr al-Kabīr, where Ibn ʿAjība began his studies in Islamic jurisprudence. After two years, he moved to Tetuan, which would be his home for most of the rest of his life, and applied himself rigorously to study and to religious devotions. His studies included *tafsīr* (Qur'anic exegesis), *ḥadīth* (Prophetic Tradition) and its commentaries, logic, rhetoric, and theoretical writings on Sufism.

After an encounter with the founder of the Darqāwī order, Shaykh al-ʿArabī al-Darqāwī (d. 1239/1825), and with al-Darqāwī's student Muḥammad al-Būzīdī (d. 1230–31/1814), who became Ibn ʿAjība's spiritual guide, he undertook writing influential commentaries on foundational works for the Shādhilī order, including most famously one on Ibn ʿAṭā' Allāh's *Ḥikam*. He also took up ascetic exercises, including public begging and the carrying of garbage, that caused a local furor and resulted in embarrassment for his family. His activities and those of his followers attracted the attention of the authorities to the point where he and some of his companions were imprisoned for a short time and ordered to renounce Sufism. As an extension of these exercises, between 1209/1796 and 1213/1798 he left his family and traveled with a group of followers through the Rif in Northern Morocco and as far south as Rabat, calling Muslims to God and attracting new adherents to the Darqāwī order. By 1214/1799 he had returned to Tetuan and witnessed the plague's devastating effects on the city's inhabitants, which included the death of all his children. In the plague treatise he wrote the following year, Ibn ʿAjība took pains to refute the contagiousness of plague by giving the example of fellow Sufis whom he had witnessed tending to the sick and wearing the clothes of those who had died of the plague.[121] As with earlier scholars such as the Granadan jurist Ibn Lubb (d. 782/1381), and the famed Egyptian scholar Ibn Ḥajar al-ʿAsqalānī (d. 852/1448), who himself witnessed the death of two of his children to plague, Ibn ʿAjība was invested in the responsibility of the Muslim community to each other in times of

[120] For the following biographical sketch of Ibn 'Ajiba, see Jean-Louis Michon, *The Autobiography of the Moroccan Sufi Ibn 'Ajība.*

[121] On this plague treatise, see Stearns, *Infectious Ideas*, 149–50.

epidemic disease and believed that jinn were the vehicle through which God caused plague to afflict mankind. Ibn ʿAjība subsequently left the city of Tetuan for the surrounding countryside, where he settled in Djimmīj, where he was buried, following his own death to the plague in 1224/1809.

Ibn ʿAjība discusses the sciences in his *fahrasa*, and while he himself was not a student of the natural sciences, he was particularly concerned with a proper understanding of sorcery, which he differentiated from medicine by its affecting the psyche and not the body.[122] Ibn ʿAjība, who repeatedly praised al-Yūsī and referred to al-Yūsī's own *fahrasa* in this work, largely follows the taxonomy of the sciences laid out by the earlier scholar in his *al-Qānūn*.[123] When it came to sorcery (*siḥr*), which he listed as one of the natural or physical sciences along with the science of talismans (*ṭilismāt*), and hypnotism (*sīmiyā*), Ibn ʿAjība listed its division by the Mālikī jurist al-Qarāfī (d. 684/1285) into sensory illusions (*sīmiyā*), astrology (*hīmyā*), and that which is related to the properties of things.[124] It is Ibn ʿAjība's discussion of talismans that is of the greatest interest, for it shows his general acceptance of sympathetic magic (as opposed to all forms of divination) and elucidates the examples of magical squares given by al-Mirghitī:[125]

> To make talismans, specific names are written down that are in relation to the heavenly bodies and planets, and also, according to what adepts of this science claim, to physical bodies, like metals and so forth. Certain virtues correspond to these talismans, which custom attributes to them. All necessarily carry the three specific names [mentioned below]; moreover, they should evoke certain astral conjunctions and should be placed in a material object. It is of the utmost importance, also, that they be made by a man who has intact psychic powers, since not everyone has a talent for these operations. Magic squares (*awfāq*; sing. *wafq*) belong to this category. These are based on numerical correspondences written in a square figure: the square is divided into boxes – sixteen, for example – in each of which a number is written, such that the same of all the numbers laid out, whether they be horizontal, vertical, or diagonal, is constant and identical – twenty, for example. If the result equals one hundred, the square has a special virtue [or protection], which is why it is especially used in amulets (*ḥurūz*; sing. *ḥirz*) and for [getting] help from those who are in a state of holiness. When the sum obtained in all directions is equal to fifteen, the talisman is especially effective for removing

[122] See Jean-Louis Michon, *The Autobiography of the Moroccan Sufi*, 146.
[123] See ibid., 141–44, 161. [124] Ibid., 146.
[125] For Ibn ʿAjība's dismissal of divination, including (*zajr*), as well as *ʿilm al-anwāʾ* and *ʿilm al-raml*, see ibid., 161.

difficulties, freeing a prisoner, easing childbirth, and all analogous situations. Al-Ghazzālī was so well versed in this science that he is considered its father. The three words that enter into the compositions of talismans are, according to what the specialized works say: *BṬD, ZHJ, WĀḤ*.[126]

To understand the final sentence in which Ibn ʿAjība lists words, the letters of which are used in the creation of talismans, one must know that Arabic letters have numerical values, which allows them to be used as a shorthand for the recording of astronomical or geographical coordinates as well as more esoteric notations.[127] While Ibn ʿAjība was opposed to divination and skeptical of astrology, he provided the science of talismans with an illustrious genealogy, linking it with arguably the most influential Sufi author in his day, Muḥammad al-Ghazālī (d. 505/1111), author of *The Revival of the Religious Sciences*.[128] The identification of al-Ghazālī with talismans is not unique – until today he is famed as an authority of this science in Mauritania – although talismans only appear once in his own writings and then in his spiritual autobiography as a rhetorical foil for the power of prayer.[129] Ibn ʿAjība, similar to his seventeenth-century predecessors, believed in the sympathetic powers of causality resting in the correct arrangement of letters that linked language with the physical world that could be employed by those with esoteric knowledge who were respectful of God's causative role.

2.3 CONCLUSION: NATURAL SCIENCES, RELIGIOUS SCIENCES, MAGIC

As detailed in the Introduction, during the nineteenth century and for much of the twentieth century European and Arab scholars embraced a narrative of the intellectual decline of the Muslim world during the Early Modern period. One staple in this narrative was that scholars who were famed for

126 Ibid., 161.

127 For an introduction to the science of letters in Islamic scholarship, see Pierre Lory, *La science des lettres en Islam* but for the broader context see now Sebastian Günther and Dorothee Pielow (eds.), *Die Geheimnisse der oberen und der unteren Welt: Magie im Islam zwischen Glaube und Wissenschaft*.

128 For Ibn ʿAjība's regret on having studied astrology, see Michon, *The Autobiography of the Moroccan Sufi*, 163.

129 For the sole mention of a talisman in the writings of al-Ghazālī, see the reference in *The Deliverance from Error* in al-Ghazālī, *Majmūʿa rasāʾil al-imām al-Ghazālī*, 79; on al-Ghazālī being associated with talismans in West Africa, see Constant Hamès, "Entre recette magique d'al-Bûnî et prière islamique d'al-Ghazālī: Textes talismaniques d'Afrique occidentale."

their knowledge of rational and philosophical sciences were persecuted by a religious establishment that viewed engagement with natural science as illicit if not heretical. In an influential survey of the fate of the philosophical sciences in the premodern period, Ignaz Goldziher (d. 1921), the most important European scholar of Islam of his time, used the persecution of the Cairene Hanbalite scholar Sayf al-Dīn al-Āmidī (d. 631/1233) to support his argument that following al-Ghazālī (d. 505/1111) a "dark orthodoxy" opposed to philosophy dominated the Muslim world.[130] Recent scholarship, and in particular the careful historiographical work of Sonja Brentjes, has shown that Goldziher, Dominique Sourdel (d. 2014) and many of the scholars who followed in their wake were mistaken in this interpretation of al-Āmidī's career, whose fall from grace was due to political reasons.[131] More directly relevant for us here is that, as shown in the epigraph at the beginning of this chapter, al-Yūsī had already recognized in the seventeenth century that it was scholarly envy and ignorance, not principled objection against philosophy that had led to al-Āmidī's fall. This is, of course, not to say that there was not opposition to the philosophical and natural sciences in some quarters, but rather to stress that such opposition cannot be assumed to be general or pervasive. As this chapter has argued, a careful reading of a range of biographical, classificatory, and autobiographical materials – as opposed to generalizations based on a single scholar's career – offer us a contextualized understanding of the place of the natural sciences in a specific Muslim society at a particular point in time. The quote from al-Yūsī's *The Discourses* can be read productively alongside his other writings discussed above: not only should all sciences that benefit the Muslim community be understood as revealed and licit according to Islamic jurisprudence, but opposition to these sciences is rooted in the smallness of the human ego or in a failure of reasoning. Thus, the opposition faced by the founder of the Almohads, Ibn Tūmart, champion of the thought of al-Ghazālī and initiator of a dynasty that would support both the philosophers Ibn Ṭufayl and Ibn Rushd (Averroes), can be grouped together in al-Yūsī's writings with al-Suyūṭī's misguided opposition to logic, and for that matter,

[130] See Goldziher, *Stellung der alten islamischen Orthodoxies zu den antiken Wissenschaften*, 39.

[131] See Sonja Brentjes, "On Four Sciences and Their Audiences in Ayyubid and Mamluk Societies," 157–58, 164. In his discussion of Muslim scholars who opposed the "sciences of the ancients," Goldziher relied on many more examples than just al-Āmidī – one other notable case being al-Suyūṭī's opposition to logic (see Goldziher, *Stellung der alten islamischen Orthodoxies*, 41). Sourdel wrote the entry on al-Āmidī for the second edition of the *Encyclopedia of Islam*.

the jealousy al-Yūsī himself faced following his arrival in Fez in the late 1660s.[132]

Yet, this chapter has moved past attempts to frame the story of the natural sciences in the Muslim world within a binary of flourishing and oppression to investigate the nature of these sciences within sixteenth–eighteenth century Moroccan scholarship. It has shown that the biobiographical literature that functioned to defend and represent genealogies of Moroccan scholarship included, acknowledged, and occasionally celebrated scholars who specialized in the rational and natural sciences. To be sure, we should not understand the *ṭabaqāt* literature as a comprehensive vision of all the scholarship being written in Morocco during these centuries – the example of the physician ʿAbd al-Ghanī al-Zammūrī, author during the reign of Aḥmad al-Manṣūr of an important treatise on gallstones, is but one example of a scholar missing from this literature.[133] Nonetheless, the evidence gathered here suggests compellingly that studying and writing on the natural sciences and the philosophical sciences more broadly was an established and accepted, if not encouraged practice in the scholarly circles examined here. The classificatory writings of al-Yūsī, al-Fāsī, and al-Ṭurunbāṭī bear out this suggestion and also draw our attention to the presence of astrology and letter magic and other occult sciences within the list of studied and practiced natural sciences along with their authors preoccupation with the nature of causality in the natural world and the imminence of the Divine. Finally, the discussion of the *fahrasa* literature fleshes out the first two sections by giving examples of how these sciences were practiced and incorporated with scholars' intellectual biographies. Taken together, these sections provide the necessary context for the texts that will be discussed in Chapter 4.

Historiographically, the texts presented here speak to current efforts by scholars to reinsert the occult into Islamic intellectual history. The student of Early Modern European history can easily see parallels in the ways in which magic was rehabilitated within European intellectual history a half a century ago.[134] The lesson to take from this comparison is not to measure developments in the Muslim world by their European counterparts, but as

[132] On the last case, see al-Yūsī, *The Discourses*, vol. 1, 221–23.

[133] Physicians are not the only ones missing from the *ṭabaqāt* literature. Caitlyn Olson had drawn my attention to the case of Muḥammad b. ʿUmar Ibn Abī Maḥallī, a theologian at the heart of a major theological debate during the lifetime of al-Yūsī, who is also not mentioned in this literature.

[134] See Matthew Melvin-Koushki, "Conjuncting Astrology and Lettrism, Islam and Judaism." Wouter Hanegraaff, "Beyond the Yates Paradigm: The Study of Western Esotericism Between Counterculture and New Complexity," and Liana Saif, "What is Islamic Esotericism?"

Wouter Hanegraaff as noted regarding the European context, to understand the esoteric sciences not as an essentialized countertradition to a variously defined orthodoxy, but rather to recover a previously neglected aspect of a complex intellectual milieu that had been written out of a particular teleological narrative of the rise of modern science.[135] Doing so in the Middle Eastern Islamic context involves a particular challenge, the genealogy of which I have addressed in the Introduction, but which bears reiterating here: seeing that the history of Islamic science is frequently given the burden of articulating either a contribution to the rise of modern science as it was birthed in Europe, or of representing the apex of Muslim rationality in contrast to a variously defined later rigid orthodoxy, accentuating the role of the esoteric in this history could be seen as tendentious if not provocative. And yet denying the importance of the esoteric sciences within the broader study and transmission of the natural sciences severely distorts our understanding of the nature, role, and pervasiveness of these very bodies of knowledge.

[135] Hanegraaff, "Beyond the Yates Paradigm," 30.

EXCURSUS

The Horizons of Causality or How to Think about Causes, Nature, and Ghosts of Scientific Methods

Does any rational person deny the wisdom of this, claiming for instance, that "The fact that foods exist has no influence, its presence or absence makes no difference," and considering the person who eats in order to become full ignorant? This is also true of the benefits God created in medicines, remedies, and their properties, knowledge of which He granted to doctors and those with experience. Can anyone deny this? This is true of every matter that habitually occurs in association with something else. This habitual occurrence is not to be denied; rather, it should be considered divine wisdom, in accordance with true faith in God's unity. One should not attribute influence to it beyond its existence as an occasion for the manifestation of the eternal decree – not by its very means but attendant with it. If someone attributes to something other than God's influence for the presence or absence of something else, he is a polytheist. And whoever denies the wisdom God placed in the forms of all that exists is ignorant and blind. If such ignorance consisted merely in his refusal to perceive this, and habitual events and experimental knowledge would inform him correctly, then it would be an easy affair to correct.

But this is a denial of the wisdom of the Lord, praise Him, and of the splendor of His Dispensation, which guides one to knowledge, and provides cognizance of His Will for what is good and harmful. It is looking out of a single eye without using the other. When experience shows, for example, that a person is unhappy on a given day because of someone's departure, or someone's marriage, or because of any other particular circumstance, or conversely that he is happy because of these things, then when we hear this, we do not rush to deny it, saying that that person has associated something with Exalted God.

On the contrary, there is nothing wrong with acknowledging this with proper faith, and when taking this divine Habit into account, to attribute influence to a day or to anything else. Concerning this, people are of three kinds: (1) A person who considers something a reason to do or not to do something, while ignoring Exalted God. He does so either while attributing influence to the occasion – these are the polytheists – or without doing so, but overly emphasizing the occasions and relying on what he set aside for himself. Such people are heedless. (2) A person who does not consider it at all and immerses himself in believing in the oneness of God, relying upon Exalted God, ignoring the occasions while not denying God's wisdom. There is no harm in this. If his reliance on God and his ridding himself of a belief in causes are sound, then this itself is a cause for his salvation by the grace of Exalted God through the necessary contingency of His Habit. Even if a snake bites his foot, it will not hurt him, for when he breaks the Habit for himself by separating it from customary things and frivolous things, Exalted God breaks the Habit for him by curing him from what habitually takes place and from His Habit's normal effects. (3) The best position to take is that of the person who considers this an education in how to preserve correct belief in light of God's ceaseless wisdom, and how to exercise proper reliance upon God in attendance with the occasions, not relying on them.

Al-Yūsī, *The Discourses*[1]

Causality frustrates. This is as true for the historian attempting to discover how populations in the past understood the workings of the natural world around them, as it is for the scholar parsing the influence of one thing upon another. And while it is a commonplace to contrast correlation and causation, it is far more difficult to distinguish clearly between the two – both in history and in the world around us. In the case of the narratives related to the history of science in Muslim societies, causality has played a role in two substantial ways. The first relates to what the premodern Islamicate world shares with other parts of the premodern world, however constituted: the absence of a modern science that first articulated its distinctive nature around a putative "scientific method" rooted in a rigorous understanding of causality in stark opposition to the nominally weak grasp of causal connections that characterized any other system of explanation. Here, particularly stark counter examples were the occult and magical sciences (the stability of magic itself relying to a large extent on that of the contrast category of science). And while we, the well-educated lay audience, may be willing to accommodate some difference in terms of what this scientific method actually is – it does not usually trouble us that scientists of different stripes may tell us that they have different ways of doing the things they do at different times – we rarely doubt that modern science has given us a better and deeper understanding of the causal mechanisms of our world. Even without a scientific training, we can recognize the iconic examples of science's modern success such as airplanes, vaccinations, computers, and genetically modified crops, as well as the degree that we

[1] Al-Yūsī, *The Discourses*, vol. 1, 247–49.

profit from them. Belief in other causal matrices or underlying structures of the seen and unseen worlds, it would follow, have less validity, if any. Almost everything about the narrative sketched here is problematic, of course, from the "we" that is unevenly applied to the populations of the world today, with a decided bias toward Europe and its former colonies, to the persistence of multiple understandings of causality within those societies where the scientific method enjoys prominence. Yet our inability to cleanly differentiate between the presence or prevalence of scientific and other forms of causality, or to rid ourselves of various incarnations of modernization theory does not prevent us from equating the premodern with the prescientific or structuring our historical narratives of scientific thinking around this opposition.

The second way in which talk of causality has impacted the narrative of the history of science in Muslim societies is particular to them and relates to an important tenet of the school of Islamic theology known as Ashʿarism: occasionalism, which it shared with other schools of *kalām* to various degrees. In brief, the Ashʿaris argued that there was no necessary link between any two phenomena and that the appearance of a web of underlying secondary causes with their own ability to exert influence was actually the result of God's recreating the world in every instant in a predictable fashion. Occasionalism permitted the Ashʿaris to safeguard the central role of the Divine in directly causing every aspect of Creation while explaining the appearance of regularity and God's ability to change His habitual creation at any point, which in turn clarified the nature of miracles. Ashʿarism was not the only school of Islamic theology, nor did all Muslims believe that the discipline of speculative theology itself was a valid way of talking about God and they chose to do so in distinct ways that supported secondary causality. But it was a prominent enough one – the dominant one in West and North Africa in the Early Modern period – that, as noted in the Introduction, it became a target of both Muslim reformers and Orientalist observers in the late nineteenth century who were looking for causes of intellectual decline in the Islamic world. For supporters of what was then becoming the increasingly culturally prominent identification of Western civilization and modernity with scientific progress, the problem with occasionalism was that it undermined scientific understandings of causality by denying the existence of secondary causes. Historically speaking, for these critics, it impeded the emergence of the proper understanding of causality that was necessary for the emergence of modern scientific thought.

There is – and has been – much to say about how these two different ways of measuring past understandings of causality against a putative correct modern scientific one have shaped our evaluations of the rationality of past societies, and have led to anachronistic if not teleological readings of past texts and worldviews. Let us consider instead, how an Ashʿārī understanding of causality provided a causal model that promoted a deeper understanding of natural phenomena. The reflections on causality quoted above and taken from al-Yūsī's *Discourses* reveal in an unusual level of detail how a scholar could argue passionately from within Ashʿarism for the importance and validity of contemplating God's habitual arrangement of nature (al-Yūsī's very passion for the subject speaks to the subject's contested nature during his lifetime). The arrangement was, critically, directly dependent upon God, who could alter it at any moment (including in order to reward the piety of an exceptionally devoted worshiper). But the arrangement was also an intelligible pattern that the believer would be negligent to ignore, and from which he could learn much about the structure of God's creation. For al-Yūsī as for other prominent seventeenth-century scholars, such as his prominent Damascene near contemporary ʿAbd al-Ghānī al-Nābulusī, what mattered was less how one talked about causation – it was clear that there was a structure of causality in natural phenomena that could be observed and predicted – so much as preserving the believer's proper attitude towards God's imminent role in Creation. This structure included the predictability of astrological and astronomical phenomena and the responses of patients to medicines as well as the power of alchemical procedures in effecting the changing of the elements. It is important to stress again that despite God's ability to change the nature of his intervention at any moment, the human observer of nature should assume that God will reliably follow the course of His Habit and that He only departs from this Habit under extraordinary circumstances.

It is of course curious that this point has to be made, considering the subject of this book and its demonstration of the prevalence and prominence of the study and transmission of the natural sciences in seventeenth-century Morocco. To engage in any natural science the practitioner has to have not just an implicit but to some extent also an explicitly formulated understanding of the predictable processes of nature. Occasionalism can only be understood as a potential impediment to understanding the workings of nature if, haunted by a putative singular modern scientific method, one limits valid understandings of causality to one that functions according to natural laws that are independent from any supernatural (though not necessarily occult) power. While such an observation may seem self-evident, it also speaks to the centrality of causality to modern science's self-understanding and its differentiation of itself from prior traditions of natural science. Accepting alternate conceptions of causality opens a door to considering both the validity of divergent understandings of the world and the contingent nature of our own received assumptions. Causality is thus a thread that when pulled (or followed through a maze depending on one's metaphor) can destabilize "science," "modernity," and other associated concepts that undergird our ability to qualitatively differentiate between ourselves and those relegated to premodernity. One arguably productive way to frame occasionalism alongside contemporary scientific understandings of secondary causality in nature would be to posit their defining distinct and nonsequential paradigms – a nod to the writings of Thomas Kuhn that I will take up in the excursus on Kuhn. This may well be a necessary move, to preserve nonmodern understandings of causality from being preemptively thought invalid. But it is also a limiting one, as it pulls the discussion back into an oppositional framing instead of allowing its own logic to play out: as a coherent description of nature in light of God's immanence.

3

Legalizing Science: The Authority of the Natural Sciences in Islamic Law

The answer to this is that there are philosophical sciences that are practiced in Islam, and it is correct that they be counted among the religious sciences due to the Law's benefit from them (*li-l-intifā' bihi fī-l-sharī'a*). The Shaykh Sīdī al-Ḥasan al-Yūsī, may God have mercy on him, has mentioned in his book *al-Qānūn* the science of timekeeping (*'ilm al-tawqīt*) as one of the Islamic sciences. Concerning it he said: It is one of the sciences of the ancients (*min 'ulūm al-awā'il*) such as the science of logic and the like. Regarding it being an Islamic science (*fī ma'nā kawnihā islāmiyya*): it is practiced in the Islamic community, which benefits from it in its religious practice (*fī dīn al-islām*), either directly or in a mediated fashion. It is also legally valid (*shar'iyya*), and what is established (*al-mashhūr*) is to grant the title of legal validity to a matter in both its essence and associated subjects.

He says in another place . . . I count those [sciences] as belonging to Islam whose benefit has spread and the utility of which has grown great: along with the aforementioned [sciences] such as logic and accounting, [I count] and what is needed of astronomy (*'ilm al-hay'a*) and geometry . . ..

The speech of this shaykh, God be content with him, has indicated that what benefits Islam is not lessened by having a source other than Islam (*la yaqdaḥ fī-hi kawn aṣl waḍi'hi bi-ghayr al-islām*). It has been said: the inventor of the astrolabe was the prophet of God, our Lord Idrīs, may the prayer and peace [of God] be upon him and our Prophet. So, let he who has no knowledge of this beware that he not place his tongue in a place it should not be, as we have related regarding one of the ignorant deniers, that he said concerning the rejection (*tanfīr*) of these time-keeping instruments that they are of the science of the Christians, may God destroy them, and that mechanical clocks (*al-majānāt*) and their like are similar.

> Nothing of this is to be trusted. Such a one as he is ignorant of the fact that reliance upon something is [to be evaluated] according to its benefit, not according to its location or its creator.
>
> Al-Wazzānī, *al-Mi'yār al-jadīd al-jāmi' al-mu'rib 'an fatāwā al-muta'akhkhirīn min 'ulamā' al-maghrib*

> Concerning opium, it is permitted to ingest an amount that does not intoxicate or produce languor as this does not affect the temperament (*al-ṭab'*), although the judicious do not approach it. And concerning tobacco, it is a recent arrival and there is no explicit textual ruling regarding it from the early legal scholars, only from recent ones. The best approach to take is that it is not forbidden. I hold that it harms one's honor (*muruwwa*) and that one should not accept the testimony of one who smokes it . . .. All the Sufis (*min ahl al-ṭarīq*) that I know hate it . . ..
>
> Al-Yūsī, *As'ila wa ajwiba*

3.1 INTRODUCTION

Law has long occupied a central place in Islamic Studies and the various strands of Euro-American Orientalism, where it has often been a metonym for Islam itself. This process of reducing Islam to a body of legal texts had a number of genealogies. First, European colonial officials in Muslim countries in the nineteenth and twentieth centuries took great pains to take a diverse and intricate Islamic legal tradition and to codify it into a straightforward code that they could apply in their administrative efforts (thereby removing interpretive authority from Muslim jurists themselves).[1] Then, during the nineteenth century in the Ottoman Empire, South Asia, and North Africa, Muslim reformers paid special attention to Islamic law and the importance of reforming it in order to strengthen their societies in the face of European colonialism. A corollary of this attention was the debate among nineteenth- and twentieth-century Muslim and European scholars about the "closing of the gates of interpretation," a reference to much misunderstood debates in the post-formative period between Muslim jurists about juridical authority that many modern scholars interpreted as the end of creative legal thinking in the Muslim world.[2]

[1] See David Powers, "Orientalism, Colonialism, and Legal History," and Iza Hussin, *The Politics of Islamic Law: Local Elites, Colonial Authority, and the Making of the Muslim State.*

[2] There is a considerable body of literature on this subject. For some good entryways into this literature, see Wael Hallaq, "Was the Gate of Ijtihad Closed?"; Sherman Jackson, "*Taqlīd*, Legal Scaffolding and the Scope of Legal Injunctions in Post-Formative Theory *Muṭlaq* and

Recent critiques of reified discussions of Islam in modern scholarship – encompassing both European and Salafi reformist scholarship – have extensively parsed the consequences of equating Islam with its legal traditions to the detriment of the spiritual, theological, artistic, and literary aspects of Muslim societies.[3] These critiques have rightly emphasized the ways in which this limiting of Islamic intellectual history to jurisprudence has impoverished our understanding of the complex richness of Muslim scholarship and have functioned specifically to marginalize the claim that Sufism has been an integral and authentic part of the Islamic religious experience.

Without then overstating the role and importance of Islamic jurisprudence, this chapter turns to the place of the natural sciences in Moroccan legal writings in the sixteenth–eighteenth centuries. The Islamic legal tradition, and with it particularly the legal opinions *(fatwā,* pl. *fatāwā)* that scholars have focused on during the past decades as a fertile source for the social context of Islamic legal history, is especially valuable as a window onto the broader relevance of the natural sciences for both scholars who had likely themselves not specialized in them, and for the broader Muslim public whose lives could be affected by these legal debates. It is important to read this chapter in the light of the previous two to remember that the texts discussed here are but one of many genres of scholarly writing during this period. One attraction of jurisprudence in general is that it offers clarity by framing topics in terms of what types of actions or activities are permitted or not. This observation is true for Islamic law as well, despite a long tradition of scholarship that has explored how Islamic jurisprudence was an ethical discourse that went far beyond differentiating between mandatory and forbidden acts to determine which were encouraged, discouraged, or permitted.[4] The legal texts discussed here are of interest not because they speak to whether jurists declared the natural sciences were permitted or not according to God's law – the question was rarely posed in this fashion – but because they illustrate the nature of the authority of the natural and rational sciences

ʿĀmm in the Jurisprudence of Shihāb al-Dīn al-Qarāfī"; Mohammad Fadel, "The Social Logic of *Taqlīd* and the Rise of the *Mukhtaṣar*"; Ahmed Fekry Ibrahim, "Rethinking the Taqlīd Hegemony: An Institutional, Longue-Durée Approach."

[3] See the above references to Ahmed, *What is Islam?: The Importance of Being Islamic* and Bauer, *Die Kultur der Ambiguität* on Islam being equated with law.

[4] A good place to begin with here is the classic article of A. Kevin Reinhart, "Islamic Law as Islamic Ethics," and the edited volumes of Jonathan Brockopp (ed.), *Islamic Ethics of Life: Abortion, War, and Euthanasia*, and together with Thomas Eich (eds.), *Muslim Medical Ethics*.

within Islamic jurisprudence. They show, in other words, how the natural sciences permeated Islamic discourses and shaped the intellectual worlds of many Muslim scholars who had never studied them. Building on this point, I will argue that certain areas of Islamic law were predicated on knowledge of aspects of natural science and that the significance of these sciences for Moroccan scholarship during these centuries was far greater that the comparatively small number of scholars who specialized in them.

The chapter falls into two main parts. In the first, I survey the place of the natural sciences in a number of collections of legal opinions from Morocco from the fifteenth to nineteenth centuries, showing how these sciences were defined and how Muslim jurists both circumscribed and deployed their authority. This survey also offers an opportunity to review some of the most widespread paradigms for the statues of the natural sciences in the Muslim world. In the second, I take one of the most vehemently argued legal debates of the Muslim Mediterranean in the seventeenth century – that of smoking tobacco – as a case study for the varied ways in which the natural science of medicine was of importance to Islamic law. In conclusion, I return to the limitations and possibilities of using Islamic jurisprudence as a window onto the study and practice of the natural sciences.

3.2 LEGAL OPINIONS AND THE NATURAL SCIENCES[5]

The three most important collections of legal opinions in the Muslim West during the late medieval and early modern periods were those of al-Burzulī (d. 841/1438), al-Wansharīsī (d. 914/1508), and al-Wazzānī (d. 1342/1923). Instead of representing editions of the authors' own legal opinions, these collections contain selections from the legal decisions of hundreds of jurists over a substantial period of time. As such they offer valuable windows into the nature and variety of legal thought in the Muslim West in the post-formative period.[6] Legal opinions were, it should be stressed, just that: they were not rulings by a judge that could be enforced by the power of a ruler or state, although judges did at times request legal opinions from prominent jurists. Because Islamic law covered a wide

[5] Much of this section of the chapter is drawn from my "The Legal Status of Science in the Muslim World in the Early Modern Period: An Initial Consideration of fatwās from Three Maghribi Sources."

[6] The formative period of Sunni Islamic law is generally held to have ended with the formation of the principal law schools by the fifth/eleventh century (Melchert, *The Formation of the Sunni Schools of Law, 9th–10th Centuries C.E.*).

range of subjects that ranged well beyond questions of criminal and ritual law, as a discourse it contained what might today seem a surprising range of topics, and it was within legal opinions that jurists had the space to consider these topics – including the natural sciences – at length.

In Chapter 2, I reviewed classifications of the sciences by al-Yūsī, al-Fāsī, and al-Ṭurunbāṭī that elucidated how Moroccan scholars in the period in question situated and defined the natural and rational sciences. Here, I additionally describe these sciences as a varied body of specialist knowledge the nature of which was not primarily defined by the same interpretive practices that were the purview of jurists, which for its part drew on both scriptural sources – Qur'an, Prophetic Tradition, legal precedent – and interpretive practices. I recognize that this is a broad interpretation, but have chosen to cast as wide a net as possible as I am primarily interested in how legal scholars evaluated, framed, and controlled an authority defined by criteria qualitatively different from their own – in this case specifically the authority of those who cultivated the natural sciences and their scientific practices. As reflected in the questions posed to *muftī*s (those jurists who hand out *fatāwā*), the main bodies of scientific knowledge considered in the collections of legal opinions examined here were astronomy, astrology, and medicine. In the past decades a substantial amount of scholarship has been published on the practice of astronomy and medicine in the Muslim world, as well as on the potential of writing social history through an investigation of legal opinions.[7] Little to date has been written on science as an alternative discourse of authority within Islamic law, and in large part, this is understandable, for Muslim jurists comparatively seldom addressed the natural sciences.[8] The following discussion, due to the scope of the material it

[7] Of the many recent publications, I have found the following to be especially useful (this is hardly a representative, much less a comprehensive, list of recent works of importance): for astronomy, the work of David King in general, but especially King, *In Synchrony with the Heavens: Studies in Astronomical Timekeeping and Instrumentation in Medieval Islamic Civilization* and three articles of A. I. Sabra's ("The Appropriation and Subsequent Naturalization of Greek Science in Medieval Islam: A Preliminary Statement"; "Science and Philosophy in Medieval Islamic Theology: The Evidence of the Fourteenth Century"; "Situating Arabic Science: Locality Versus Essence"). For medicine, I have been especially influenced by Imerli Perho's work on Prophetic Medicine, and have found Peter Pormann and Emilie Savage-Smith's synthesis quite useful (Pormann and Savage-Smith, *Medieval Islamic Medicine*). For notable efforts to write social history through an examination of legal opinions, special attention should be drawn to Masud et al., *Islamic Legal Interpretation: Muftis and their Fatwas* and Powers *Law, Society, and Culture in the Maghrib, 1300–1500*.

[8] Taken together, the works of al-Burzulī, al-Wansharīsī, and al-Wazzānī contain more than 8,000 printed pages. In a survey of the indices to these volumes, I found no more than a few dozen references to astronomy, alchemy, and medicine. For passing references in the

surveys, while arguably representative of the Mālikī legal tradition that dominated Muslim North and West Africa during this period, is certainly not exhaustive.[9]

A final issue should be briefly addressed before turning to the legal decisions themselves. As discussed in the Preface and Introduction, the tension between science and religion has often been characterized as either one between reason and revelation, or between empirical evidence and scriptural authority. Such characterizations are insufficient and misleading. Instead, we find that, when challenging the authority of the natural sciences, Muslim legal scholars, with some notable exceptions, take issue with how empirical evidence should be interpreted, and with which empirical evidence is relevant to the question at hand, and not with the value of empirical evidence itself. In doing so, they are motivated by what can be called an ethical concern for both the spiritual and the physical well-being of the Muslim community.

3.3 ASTROLABES AND REVEALED LAW

While some Muslim jurists in the early modern period may have disapproved of interaction with astrologers and soothsayers – as seen in Chapter 2, this was certainly not true for all – they valued the technical abilities of astronomers, whose calculations facilitated both establishing prayer times and the beginning of the lunar months, especially Ramadan.[10] The work of David King in particular has shown how, with the creation of the office of timekeeper (*muwaqqit*) from – at the latest – the seventh/thirteenth century onward, individual mosques in Egypt, al-Andalus and North Africa

writings of al-Qarāfī to the "rational sciences" remaining outside the purview of law properly understood, and "scientific observation" playing an important role in establishing proper *taqlīd*, see Jackson, *Islamic Law and the State: The Constitutional Jurisprudence of Shihāb al-Dīn al-Qarāfī*, 115, 128.

[9] This is especially true when it is considered that not only are solely Mālikī sources being discussed here, but that large numbers of Mālikī *fatwā* collections remain to be published (see al-Hīlah, "Classification of Andalusian and Maghribī Books of Nawāzil from the Middle of the Fifth to the End of the Ninth Century AH" and al-Ḥarbī, "Namādhij min juhūd fuqahā' al-mālikiyya al-maghāriba fī tadwīn al-nawāzil al-fiqhiyya." I am indebted to Jocelyn Hendrickson for both of these references).

[10] For an overview on the place of astrology in premodern Islamdom, see Saliba, "The Role of the Astrologer in Medieval Islamic Society." Al-Wansharīsī includes an opinion of the respected Syrian scholar al-Nawawī (d. 676/1278) on visiting astrologers (*ityān al-munajjimīn*) being forbidden (*al-Mi'yār*, vol. 12, 366–67). See also Ibn Khaldūn's remarks on astrology as discussed in King, "On the History of Astronomy in the Medieval Maghrib." I am grateful to Professor King for having sent me a copy of this article.

generally had religious officials who possessed sufficient astronomical knowledge to calculate the daily times of prayer, which varied according to time and place, and which could be notoriously difficult to establish in the case of the noon and afternoon prayer, as well as on cloudy days.[11] Based on King's work, A. I. Sabra suggested that the study of astronomy in general may have declined in the early modern period even as it was "naturalized" and its study became restricted to an "instrumentalist" view to aid with the correct establishment of ritual activities such as prayer and fasting (Sabra, 1987: 240).[12] I will return to some of the implications of Sabra's – admittedly tentative – suggestion below, but the first case discussed here would initially seem to bear out his argument. Neither King nor Sabra, however, offer an extended consideration of the legal status of scientific inquiry in post-formative Islamic societies.

At some point in the late nineteenth/early twentieth century, al-Wazzānī, the Moroccan author and compiler of the twentieth-century *fatwā* collection *al-Miʿyār al-Jadīd*, was party to a serious disagreement regarding the role and authority of astronomical instruments in Islamic ritual.[13] Covering the entire twenty-four page chapter on the call to prayer (*nawāzil al-adhān*), al-Wazzānī's *fatwā* with its numerous citations of Mālikī authorities offers the reader a long and somewhat convoluted account of the difficulties of ascertaining the proper time for the call to prayer.[14] As such, it can be profitably read against Ebrahim Moosa's discussion and translation of Aḥmad Muḥammad Shākir's (1309–77/1892–1958) 1939 treatise on the permissibility of beginning the month on the basis of scientific calculation and not actual visual sighting of the moon.[15] Both al-Wazzānī and Shākir conclude that astronomical calculations are necessary for the establishment of Islamic ritual, and cite a long list of, respectively, Mālikī and Shāfiʿī authorities to substantiate their position. Both authors also confront and refute contrary opinions in previous scholarship. Where the two differ is

[11] See King, "On the Role of the Muezzin and Muwaqqit in Medieval Islamic Societies."

[12] Sabra's article was cited by Huff in support of his argument, and in his critique of Huff, Saliba argued that neither Sabra nor Huff had advanced enough evidence to prove that the "naturalization" of Greek science led to its decline. See Saliba, "Seeking the Origins of Modern Science?" and compare with Saliba, *Islamic Science and the Making of the European Renaissance*, 125ff.

[13] I have not been able to find any references to this episode in the historical sources of the period. On al-Wazzānī and his juridical project, see Etty Terem, *Old Texts, New Practices: Islamic Reform in Modern Morocco.*

[14] Al-Wazzānī, *al-Miʿyār al-jadīd*, vol. 1, 215–39.

[15] See Moosa, "Shaykh Aḥmad Shākir and the Adoption of a Scientifically-Based Lunar Calendar." I am grateful to David Powers for this reference.

that, while Shākir was writing for a decidedly Salafi audience, al-Wazzānī, who was writing at the other end of the Arab world some decades before Shākir, situates himself within a purely Mālikī environment.[16] While Moosa argued that Shākir was to some degree trying to come to terms with the technological challenge of modernity, there is no explicit sign that al-Wazzānī was primarily reacting to anything more than a local disturbance in Morocco for which he wished to find the appropriate legal response.[17]

Al-Wazzānī's *fatwā* is long and, as is common enough for the genre, interspersed with numerous quotations from the works of jurists of previous generations. The first sections related to the legal status of astronomy begins as follows:

> There was a debate (*mudhākara*) between a group of legal scholars regarding Friday Prayer. One of them said: It takes place very late in Fes: it is desirable for its prayer to be at the beginning of the time, even if only in one mosque, so that the preacher begin the sermon following the *adhān* of only one *mu'adhdhin* [presumably as opposed to waiting for all calls to end]. He stated that this was the way it had been in the Prophet's time. The others of the group did not agree with him.[18]

After defending the custom of the people of Fez, and citing several authorities on the proper manner of sounding the call the prayer, al-Wazzānī relates that this argument became widespread among both scholars (*al-khāṣṣ*) and commoners *(al-ʿāmm*), and took on broader significance. He

[16] While the term Salafi is quite complex in that it carries several distinct meanings, in this case if refers generally to those Muslims who in the nineteenth and early twentieth centuries believed that Islam needed to be reformed by rejecting the consensus of the four Sunnī law schools (Ḥanafī, Mālikī, Shāfiʿī, and Ḥanbalī) and in engaging in a critical reexamination of Islam's scriptural sources. See the discussion in the Introduction.

[17] "In this instance, the existing social context not only impinges on the self-understanding of Muslims, but also forces jurists to re-read the religious texts in order to derive new meanings that are in harmony with the new context. Surreptitiously, a new juridical logic evolves and transplants itself onto existing practice, without hardly any acknowledgment of the occurrence of such changes – in itself a phenomenon insufficiently documented in the history of Islamic Law. Aḥmad Shākir, in the present instance, and other jurists who wrote on other issues, consciously or unconsciously interpolated Islamic law with the technology of modernity, creating thereby a desire for regularity and consistency in *fatāwā* or juridical responsa, a subject that needs to be addressed elsewhere." (Moosa, "Shaykh Aḥmad Shākir and the Adoption of a Scientifically-Based Lunar Calendar," 62.) While I agree with Moosa's general observation on the challenge modernity has posed to Muslim scholars, I am wary of seeing every reinterpretation of Muslim law in the modern period as only or even primarily the result of their interaction with modernity. Doing so could easily obscure the fact that premodern Islamic jurisprudence contained numerous opinions that while of use to Muslims today are not the product of an engagement with modern concerns.

[18] Al-Wazzānī, *al-Miʿyār al-jadīd*, vol. 1, 215.

describes the party opposed to his own views as having no knowledge of the science of timekeeping (*ʿilm al-tawqīt*), and the situation only growing worse when an unnamed scholar rose up, seeking the approbation of the masses and supporting their stance:

> He spoke to them with words from Khalīl's *Mukhtaṣar*, deluded that there was no knowledge, not even a little, greater than his own. He said to them: everyone ascertains the passing of dusk (*maghīb al-shafaq*); only those who are stubborn and in denial could ignore its obvious nature. The timekeeping devices such as the astrolabe and others cannot be depended upon (*lā yuʿawwal ʿalayhā*), nor should one turn to them for the knowledge of when the time [of prayer] begins.[19]

The *Mukhtaṣar* of the Egyptian Mālikī jurist Khalīl b. Isḥāq al-Jundī (d. 749/1348 or 767/1365) was the standard legal reference work of the Mālikī school after the eighth/fourteenth century, and while its importance is clearly reflected by the substantial commentaries devoted to it, it was primarily important due to its recording of accepted consensus.[20] As such, it may well have been the first reference a Mālikī jurist would turn to, but for an experienced jurist faced with a thorny problem, it would rarely be the last. Although Khalīl fails to mention astrolabes or astronomy, in his chapter on prayer his description of how to ascertain the prayer times refers only to a staff (*al-qāma*) and its shadow, while in his discussion of fasting, he warns against trusting astronomers (*al-munajjim*).[21] Al-Wazzānī quickly notes that he tried to find this individual to set him straight, but, being unable to locate him, he decided to express his thoughts in writing in the *fatwā* at hand.

Knowledge of the times of prayer, it is clear, is mandatory for all believers, but following the authority of al-Ḥaṭṭāb (d. 954/1547), author of a commentary on Khalīl's *Mukhtaṣar*, al-Wazzānī clarifies that, due to the difficulty of the matter, establishing these times is a collective and not an individual obligation.[22] Knowledge of how to calculate the time accurately, he argues, belongs to those who have a command of the astrolabe and

[19] Ibid., vol. 1, 217.

[20] On *mukhtaṣar Khalīl*, see Mohammad Fadel, "The Social Logic of *Taqlīd*," and ibid., "Adjudication in the Mālikī Madhhab: A Study of Legal Process in Medieval Islamic Law," 262–65. Rüdiger Lohlker has argued that Khalīl's *Mukhtaṣar* didn't achieve its widely accepted status until the sixteenth century (Lohlker, *Islamisches Völkerrecht: Studien am Beispiel Granda*, 124).

[21] See al-Azharī, *Jawāhir al-iklīl ʿalā mukhtaṣar al-imām Khalīl*, vol. 1, 32 (prayer), 145 (fasting). Since, as David King has shown, the institution of *muwaqqit* was widespread in Mamlūk Egypt, it is curious that Khalīl did not, at least in passing, refer to a more exact form of timekeeping.

[22] Al-Wazzānī, *al-Miʿyār al-jadīd*, vol. 1, 219.

who are familiar with the science of timekeeping.[23] Al-Wazzānī backs this claim up, appropriately enough, by citing the empirical observations of the renowned Fāsī *muwaqqit* al-Jādarī (d. 818/1415 or 839/1435), whom we met previously in Chapter 2.[24] Al-Wazzānī does not simply rely upon the authority of specialists such as al-Jādarī, but enters into the discussion of specifics, explaining to his reader why the views of Abū ʿAbdallāh Ibn al-Ḥabbāk (d. 867/1463) and Muḥammad b. Yūsuf al-Sanūsī on the subject of the changing length of dusk are to be preferred to those of al-Burzulī.[25] Al-Wazzānī's sources reflect the complexity of the relationship between law, science, and theology (*kalām*) in the early modern Muslim world. Al-Sanūsī's creeds and his summation of Ashʿārī theology, *al-Muqaddimāt*, were widely read in the Maghrib until the modern period, and he also wrote on the use of the astrolabe, as well as on medicine. While al-Sanūsī adamantly denied the existence of secondary causality in his theological writings – even as he affirmed God's habit (*ʿāda*) of acting in a the fashion of Ashʿarī theologians we are now familiar with – in his writings on medicine and astronomy, he supported their legitimacy.[26] The compatibility of these various scientific pursuits, or perhaps more accurately the ability of early modern Muslim scholars to engage simultaneously in multiple intellectual discourses, should be emphasized, for it is precisely their incompatibility that some scholars have recently alleged.[27]

Al-Wazzānī interrupts the list of quoted authorities to note that one of those opposed to him wrote to a scholar living in Rabat, Aḥmad b. ʿAbdallāh al-Ghurfī, distorting the issue (*lam yufṣiḥ ʿan sharḥ ḥaqīqati-hā wa-kunhi-hā bal awhama fī su'āli-hi*) in order to receive an answer that would condemn the use of the astrolabe.[28] Al-Wazzānī wrote to this scholar, and clarified the nature of the argument to him, after which he received an answer which contained the following passage:

[23] Ibid., vol. 1, 221.

[24] Compare Kaḥḥāla, *Muʿjam al-mu'allifīn*, vol. 2, 106 with vol. 2, 113–14. al-Wazzānī also cites a treatise on the astrolabe by a certain Abū Faḍl Dāniyāl al-Shāfiʿī, whom I have not been able to identify.

[25] For Ibn al-Ḥabbāk, see Kaḥḥāla, *Muʿjam al-mu'allifīn*, vol. 3, 114. The reference here is to his poem on the use of the astrolabe, *Bughyat al-ṭullāb fī ʿilm al-asturlāb*, on which al-Sanūsī wrote the commentary to which al-Wazzānī is referring.

[26] On al-Sanūsī's creeds see now Olson, "Beyond the Avicennian Turn: The Creeds of Muḥammad b. Yusūf al-Sanūsī (d. 895/1490)." Historians have struggled with evaluating the influence that the general Ashʿārī denial of secondary causation may have had on evaluations of empirical evidence.

[27] See Huff, *The Rise of Early Modern Science: Islam, China, and the West*, 69–72.

[28] I have not been able to identify this scholar.

From the time that these instruments, the sine quadrant, the astrolabe and others, appeared in Islam and among its people, they were investigated, and it was found that the one knowledgeable in them, when proficient in their use (*idhā utqinat fī nafsihā*) benefited from the certainty [they brought]. It is necessary for him who has no knowledge of them to follow him who does have such knowledge (*taqlīd al-ʿārif bihā*). One should act according to what he says, though it is necessary for him to exercise some caution when the sky and the horizon are cloudy. The people of both east and west act in this fashion, as al-Ḥaṭṭāb has related from al-Qarāfī (d. 684/1285) and others. No one condemns the reliance on instruments, save the ignoramus (*al-jāhil*) whose opinion has no value (*lā ʿibra bi-qawlihi*).[29]

Al-Wazzānī notes that this is similar to what the Imām Sīdī al-Tāwudī b. Sūda (d. 1208/1793) had said on the use of the astrolabe having a basis in the revealed law, and in the next few pages he presents an array of authorities who similarly support the use of the astrolabe.[30] It is clear that there is no scholar of note of the Mālikī or any other school who does not support the use of astronomy and the astrolabe for the purpose of bringing certainty to ritual practice.[31]

Before finishing with his opponents, al-Wazzānī has one final objection to deal with: the blanket assertion that the use of the astrolabe has no basis in law, and that it is suspect due to its association with philosophy. To deal conclusively with this accusation, he turns to the work of the prominent Moroccan scholar al-Ḥasan al-Yūsī, from whose *al-Qānūn* he quotes the passage that I have used as an epigraph for this chapter, and which is largely the same one I introduced in detail in Section 3.1 above. Al-Wazzānī follows al-Yūsī in adopting an instrumentalist logic to blur the boundaries between Islamic and non-Islamic sciences:[32]

[29] Al-Wazzānī, *al-Miʿyār al-jadīd*, vol. 1, 223.

[30] A scholar of Fez, Ibn Sūda wrote a marginal commentary on al-Zurqānī's (d. 1122/1710) commentary on Khalīl's *Mukhtaṣar*. See Kaḥḥāla *Muʿjam al-muʾallifīn*, vol. 3, 363. The authorities cited by al-Wazzānī, besides those already mentioned, include ʿIzz al-Dīn Ibn ʿAbd al-Salām (d. 660/1262), al-Ghazzālī (d. 505/1111) and al-Māzarī (d. 536/1141). Later in the *fatwā* the support of the following scholars is also cited: Ibn ʿArafa (d. 803/1401), Muḥammad al-Raṣṣāʿ (d. 886/1481), and al-Maqqarī (d. 1041/1631). All but Ibn ʿAbd al-Salām and al-Ghazzālī were of the Mālikī school.

[31] While I do not doubt that the majority of Mālikī authorities shared al-Wazzānī's view, he may well have omitted inconvenient exceptions. Compare with Moosa's discussion of Shākir's treatise, where he notes that, while the famed Shāfiʿī scholar Taqī al-Dīn al-Subkī (d. 756/1355) had accepted the use of calculations, the possibly even more famous Ibn Ḥajar al-ʿAsqalānī (d. 852/1448) and Ibn Taymiyya (d. 728/1328) had rejected them (Moosa, "Shaykh Aḥmad Shākir and the Adoption of a Scientifically-Based Lunar Calendar," 62, 76).

[32] I have not been able to locate in al-Yūsī's own writing all of the passages quoted here, but see the above discussion of al-Yūsī's *al-Qānūn* in Section 3.1.

The answer to this is that there are philosophical sciences that are practiced in Islam, and it is correct that they be counted among the religious sciences due to the Law's benefit from them (*li-l-intifāʿ bihi fī-l-sharīʿa*). The Shaykh Sīdī al-Ḥasan al-Yūsī, may God have mercy on him, has mentioned in his book *al-Qānūn* the science of timekeeping (*ʿilm al-tawqīt*) as one of the Islamic sciences. Concerning it he said: It is one of the sciences of the ancients (*min ʿulūm al-awāʾil*) such as the science of logic and the like. Regarding it being an Islamic science (*fī maʿnā kawnihā islāmiyya*): it is practiced in the Islamic community, which benefits from it in its religious practice (*fī dīn al-islām*), either directly or in a mediated fashion. It is also legally valid (*sharʿiyya*), and what is established (*al-mashhūr*) is to grant the title of legal validity to a matter in both its essence and associated subjects.

He says in another place . . . I count those [sciences] as belonging to Islam whose benefit has spread and the utility of which has grown great: along with the aforementioned [sciences] such as logic and accounting, [I count] what is needed of astronomy (*ʿilm al-hayʾa*) and engineering . . . .

The speech of this shaykh, God be content with him, has indicated that what benefits Islam is not lessened by having a source other than Islam (*la yaqdaḥ fī-hi kawn aṣl waḍʿihi bi-ghayr al-islām*). It has been said: the inventor of the astrolabe was the prophet of God, our Lord Idrīs, may the prayer and peace [of God] be upon him and our Prophet. So, let he who has no knowledge of this beware that he not place his tongue in a place it should not be, as we have related regarding one of the ignorant deniers, that he said concerning the rejection (*tanfīr*) of these timekeeping instruments that they are of the science of the Christians, may God destroy them, and that mechanical clocks (*al-majānāt*) and their like are similar. Nothing of this is to be trusted. Such a one as he is ignorant of the fact that reliance upon something is [to be evaluated] according to its benefit, not according to its location or its creator.[33]

Al-Wazzānī's final paragraph shows that he is well aware that, by adopting technology in late nineteenth/early twentieth century Morocco, one opens oneself to the accusation of mimicking the colonialist Europeans.[34] Nonetheless, he clearly rejects such a shallow, not to mention ironic,

[33] Al-Wazzānī, *al-Miʿyār al-jadīd*, vol.1, 227–28. It is striking that after marshaling the above arguments, al-Wazzānī still finds it necessary to posit a Muslim origin for the astrolabe. For a comprehensive discussion of other origins ascribed for the astrolabe in Medieval Islamic sources, see King, "The Origin of the Astrolabe According to the Medieval Islamic Sources" (for the popularity of attributing the invention of the astrolabe to Idris (Enoch), see ibid., 45). Finally, for a discussion of the term *al-majāna*, see García Gómez, *Foco de antigua luz sobre la Alhambra*, 82–85. I am grateful to Mercè Comes and David King for this reference.

[34] An interesting parallel to al-Wazzānī's dilemna can be found in the Algerian Ḥamdān b. ʿUthmān Khoja's (d. ca. 1258/1842) treatise on the plague, *Itḥāf al-munṣifīn wa-l-udabāʾ fī al-iḥtirāz ʿan al-wabāʾ*, in which he argues for the necessity of the quarantine, despite its alleged European provenance (Stearns, *Infectious Ideas*, 152–57).

accusation – considering that astrolabes had been used in the Muslim world for over a thousand years before the twentieth century – and in the last pages of his opinion emphasizes that for testimony concerning the correct prayer time to be accepted, the observer must be experienced.[35]

3.4 THE LIMITS OF BENEFICIARY SCIENCE

In the fifteenth-century *Miʿyār* of al-Wansharīsī, we find a *fatwā* of the renowned scholar Qāḍī ʿIyāḍ (d. 544/1149) that can productively be read against that of al-Wazzānī and which is distinctly reminiscent of a passage al-Ghazālī placed in his introduction to *The Incoherence of the Philosophers (Tahāfūt al-falāsifa)* in which he discussed eclipses.[36] It differs from the above discussion of astronomy in that instead of addressing a situation in which science is being used directly in the service of Islamic Law, it instead examines a case where the authority of astronomers problematizes ritual practice.

In Khalīl's *Mukhtaṣar*, compiled some two centuries after Qāḍī ʿIyāḍ's death, the author described the necessity of praying during an eclipse, a practice that had a firm basis in the Mālikī school as well as in Prophetic Tradition.[37] With this subject in mind, ʿIyāḍ's questioner expressed a certain amount of anxiety regarding the claims made by astronomers that they could foretell both eclipses and their length:

> He [Qāḍī ʿIyāḍ] was asked about the eclipses of the sun, regarding the fact that the Prophet, God bless him and grant him peace, had ordered at their occurrence prayer, invocations, the freeing of slaves, giving of alms, fear, and supplication. He [had] said: "[Pray] until what is in you [of fear] is removed (*ḥattā yukshaf mā bikum*)". There is no doubt that what he ordered is the truth, and that it [the eclipse] is one of the signs of God. This is what the people of the Sunna follow. Yet we see astronomers (*ahl al-ḥisāb wa l-nujūm*) stating that they perceive the eclipse before it occurs, saying: "It occurs itself at such a time." It is well known that this is knowledge acquired through calculation, and that it happens due to an association (*iqtirān*), which they claim is between the stars in their respective spheres (*fī aflākihā baʿḍuhā bi-baʿḍ*). How can this be reconciled for the astronomers with the ordained fear and supplication [at the time of the eclipse], with it being an affliction that has descended upon people, and with their being ordered to engage

[35] Al-Wazzānī, *al-Miʿyār al-jadīd*, vol. 1, 231–32.

[36] See al-Ghazālī, *The Incoherence of the Philosophers*, 6–7.

[37] Al-Azharī, *Jawāhir al-iklīl*, 104–5. There are numerous traditions in both al-Bukhārī and Muslim which link eclipses to visions of the day of judgment, while denying that they are related to the death of individuals.

in supplication until they are free of reprehensible types of things? There is no fright in them at such a time, no fear, for they say: We know when it will occur. Is there a way to reconcile the two matters? If we declared their statement void in its entirety, something would remain in the soul due to their getting it right (*iṣābati-him*) ... and if what they say is not declared entirely false, and we support them in this matter being reachable through calculation, then where is our fear for our safety? Where is the fear that was ordered for us? .... Indeed, what is sought after from this is how can we pray for the occurrence of the end of the eclipse (*fī injilā' al-kashf*), with them telling us the time when it will occur and end, and with them possibly being correct in this. How can this be reconciled? Clarify for us what knowledge you have regarding this, may you be rewarded, God willing.[38]

The query puts Qāḍī ʿIyāḍ in a difficult position, for not only does there seem to be tension between Prophetic Tradition and astronomers, with the former arguing that eclipses evoke fear, and the latter claiming to be able to predict their occurrence, but there is also the potential that the astronomers' claims could adversely affect the piety of the Muslim community.[39] At the beginning of his response, he seems to dispute the validity of astronomical predictions of eclipses in an absolute fashion, denying that the astronomers' proofs could reflect what God alone knows (*ʿilm al-ghayb*). Yet this is a discursive move, one that functions to set his questioner at ease as to the authority of astronomers: it is at best conditional, and not absolute.[40] ʿIyāḍ goes on to cautiously admit that astronomers can rely upon the habitual experience of a competent observer (*ʿāda jarrabahā mutamakkin*) in their observations, though their proofs do constrict the common good of the Muslim community (*al-maṣāliḥ*). In this context, the latter presumably refers to the effects of prayer and the act of relying upon God. While ʿIyāḍ notes that leading scholars generally dispute the knowledge of astronomers regarding eclipses, he also cites the exception of the prominent jurist Abū al-Walīd Ibn Rushd (d. 520/1126), whose *al-Bayān wa l-taḥṣīl* was possibly the single most comprehensive reformulation of Mālikī jurisprudence in the post-formative period of

[38] Al-Wansharīsī, *al-Miʿyār*, vol. 11, 259.

[39] It is worth noting that the case here is distinctly different from that in the eighth/fourteenth century when Ibn al-Khaṭīb (d. 776/1374) famously challenged the validity of any Prophetic Tradition that would endanger the Muslim community. See Stearns, "Contagion in Theology and Law: Ethical Considerations in the Writings of 14th-Century Scholars of Nasrid Granada."

[40] The parallels with the case of contagion are again striking, for while few if any Muslim jurists admitted the existence of "contagion," many affirmed the transmission of disease. See Stearns, *Infectious Ideas*, chapter 5, and Conrad, "A Ninth-Century Muslim Scholar's Discussion of Contagion."

Islamic jurisprudence.[41] Ibn Rushd had no objection to the views of the astronomers' argues 'Iyāḍ, because he saw no contradiction between them and Prophetic Tradition. After all, notes ʿIyāḍ, we certainly do not believe that the appointed prayers cause the eclipse to end, though the ending of the eclipse is the goal of the prayer. No, the true benefit of the prayer is the acknowledgment of the eclipse as one of the signs of God, the intensity of fear felt during the prayer, and the recognition of God's omnipotence.[42] In the end, ʿIyāḍ finds a balance between assuring his questioner that astronomical knowledge does not threaten religious practice, and acknowledging the ability of astronomers to accurately perceive God's habit, the existence of such habits being a central concept in Ashʿārī theology. To be sure, the *fatwā* is not a ringing endorsement of the need for astronomical research; yet, strikingly considering the way in which the question was framed, it supports the compatibility of religious law and astronomical knowledge.

3.5 LAW AND MEDICINE: OF ALCHEMISTS, LEPERS, AND TOBACCO

The practice of medicine appears in the *fatwā* collections in a decidedly different fashion than astronomy, relating in part to a concern for the legitimacy of doctors as qualified witnesses. Not only is the authority of medical knowledge addressed, but also the question of whether the pursuit of medicine affects the status of the practitioner as a witness, i.e., does an interest in a specific art in and of itself constitute a moral flaw which invalidates anything the practitioner might say? Let us remember that jurists saw themselves as the guardians of Muslim society, and that they were well aware of the activities of charlatans, astrologers, and tricksters, who would falsely present themselves to society as legitimate authorities.[43] Still, we should not imagine jurists as invariant killjoys, as on occasion they tolerated illusionists and while they were careful when discussing sorcery, they allowed the writing of amulets.[44]

[41] On the status of *al-Bayān wa l-taḥṣīl* in the Mālikī school, see Fernández Félix, *Cuestiones legales del Islam temprano: La ʿutbiyya y el proceso de formación de la sociedad islámica andalusí.*

[42] Al-Wansharīsī, *al-Miʿyār*, vol. 11, 260–61. Included in this fear is the awareness that the eclipse may be a sign of the end of time. In structuring his argument, ʿIyāḍ cites both Abū Bakr Ibn al-ʿArabī (d. 543/1148) and Abū al-Qāsim al-Muhallab (d. 436/1044). On al-Muhallab see 'Iyāḍ, *Tartīb al-Madārik*, vol. 2, 751–52.

[43] See Saliba, "The Role of the Astrologer in Medieval Islamic Society" and Pormann, "The Physician and the Other: Images of the Charlatan in Medieval Islam."

[44] See al-Burzulī, *Fatāwā*, vol. 1, 380–82, for an example of juridical tolerance of entertainment: in this case, a group of players who would pretend cutting off the head of an actor, and then having the head speak to them.

As the boundaries between astrology and astronomy, between sorcery and medicine, were not always clear, jurists had to be careful not only to protect believers' health and wealth, but also to shield them from all forms of polytheism.[45]

In this context, the character of the practitioner and the way in which he used his knowledge were understood to be as important as the status of the art itself. In al-Wansharīsī's chapter on giving testimony (*al-shahādāt*) the following question addressed to ʿIyāḍ illustrates this:

> He [ʿIyāḍ] was asked about the practice of alchemy (*ṣināʿat al-kīmiyāʾ*), whether it is permitted (*hal hiya min bāb al-jāʾiz aw min bāb al-mustaḥīl*), and whether or not the one practicing it (*ṭālibu-hā*) was to be forbidden [from doing so], and whether or not the practice of it invalidated (*yaqdaḥ*) the testimony of its practitioner?[46]

ʿIyāḍ immediately affirms that alchemy is indeed a legitimate science, dealing with the making of glass and the analysis of pearls. His authorities here are unnamed doctors, and he notes that he has dealt with this subject at greater length in a separate work.[47] Turning to what he perceives to be the root of the question, ʿIyāḍ denies that alchemy can be linked in its essence to the practice of forging money using alchemical knowledge. Such examples of fraud do not invalidate the efforts of expert practitioners of the science, who in general test the abilities of possible charlatans or hucksters in their profession. Al-Wansharīsī, the editor, juxtaposes this clear endorsement of alchemy's legitimacy with the opinions of two prominent scholars, Ibn ʿArafa (d. 803/1401) and Abū al-Ḥasan ʿAlī b. al-Muntaṣir (d. 742/1341), the first of whom had stated that the testimony of practitioners of alchemy was similar to that of those who sell sets of backgammon, drums, and pipes, and that it was therefore not valid; al-Muntaṣir, for his part, had stated that such a person was not permitted to lead the prayer.[48] The reader of the opinion is left in the not unusual position of having to choose between conflicting opinions of prominent

45 The central issue here is causality, and the importance of not-believing that anything besides God could cause anything to occur. How difficult this could be is seen in the example of prayers or spells: if they were written in Arabic, they were considered safe, and reminiscent of the discussion of al-Mirghitī's work in Chapter 2, al-Burzulī mentions that his own teacher (possibly Ibn ʿArafa) would write certain names on paper, mix the paper with chicken eggs, cook the mixture and eat it. If, however, the words were written in a foreign language, then the object of invocation was unclear, and the act was disapproved, if not forbidden (al-Burzulī, *Fatāwā*, ibid.).

46 Al-Wansharīsī, *al-Miʿyār*, vol. 10, 155.

47 I have not been able to identify this source.

48 Al-Wansharīsī, *al-Miʿyār*, vol. 10, 155.

jurists, two of whom had cast doubt on the character of the practitioner of alchemy, and thus implicitly on the art itself.

In the case of medicine, the situation was clearer. In general, jurists acknowledged, though often only in passing, the expert testimony of a doctor regarding the nature of a disease or the state of a person's health. I have shown elsewhere how jurists generally acknowledged the transmission of disease in the case of leprosy, while often denying it in the case of the plague.[49] In the latter case, instead of rejecting the value of empirical evidence or the authority of doctors, jurists argued for alternative explanations of empirical observations that reflected the complexity of contemporary medical plague etiologies.[50] This was the case with the two legal opinions, also collected by al-Wansharīsī, of the eighth/fourteenth century Granadan jurist Ibn Lubb (d. 782/1381) who argued that Muslims should not abandon the sick in times of plague, as there was no conclusive evidence that it was transmitted by contact.[51] To be sure, there were jurists such as Tāj al-Dīn al-Subkī (d. 771/1369), who argued that if two doctors testified that a specific plague victim was a potential cause of harm to others, he should be avoided. In giving doctors such authority in the case of plague, al-Subkī was, however, in the minority.[52] With leprosy, on the other hand, a disease that was widely held to be a legitimate cause for divorce, the opinion of doctors was seldom disputed.

A particularly striking example of the acknowledgment of medical authority is found in al-Burzulī's collection of legal opinions, where he includes a *fatwā* from Ibn al-Ḥājj (d. 737/1336) on the value of a doctor's opinion on the presence or absence of leprosy:

> Of the questions [treated by] Ibn al-Ḥājj is also the following case: Doctors testify regarding leprosy that is present before the date of the marriage contract, as [other witnesses] testify concerning a sale. There is no oath needed from the spouse to confirm this, nor is the matter open to dispute, as [is the case] when dispute occurs with his oath at the giving of testimony. (This is) because the judgment provided by

[49] See Stearns, *Infectious Ideas*, *passim*. As always, there were notable exceptions, with Ibn Rushd (d. 520/1126) denying the contagious nature of leprosy (ibid., 114–16).

[50] There were notable exceptions, including the case of the renowned and influential Egyptian jurist Ibn Ḥajar al-ʿAsqalānī, who wrote a long treatise on the virtues of the plague – dying of it could lead to martyrdom – and on how it was caused by *jinn* (Stearns, *Infectious Ideas*, chapter 5).

[51] I have discussed these legal opinions at length in Stearns, "Contagion in Theology and Law."

[52] See Stearns, *Infectious Ideas*, 88–89. Al-Subkī was also decisively refuted by Ibn Ḥajar in the latter's treatise less than a century later. This refutation may – I speculate – relate to al-Subkī's treatise on the plague remaining extant only in citations in other works.

the testifying of doctors has long been a definitive judgment (*ḥukm bi-l-qaṭʿ bihi ʿalā-l-qidam*).[53]

Yet, even while acknowledging the legal authority and value of medical testimony, jurists were concerned with controlling this testimony, either by investigating the qualifications of individual practitioners or by presenting themselves as medical authorities in their own right. Many jurists had also studied medicine – though few of them may have had any actual clinical experience – but here I am interested in the ability of jurists to present medical knowledge as transmissible within legal circles without the presence of an actual doctor.

In al-Burzulī's *Fatāwā* we find an interesting example of how, while acknowledging the authority of medicine, a jurist could challenge that of the doctor. The scholar involved was al-Māzarī (d. 536/1141), a respected Mālikī scholar.[54] The query reads as follows:

Al-Māzarī was asked about someone who had married his virgin daughter [to a man]. The husband asked to have sex with her (*al-dukhūl bi-hā*) and the father claimed that on his [the husband's] body there was leprosy (*baraṣ*). They took the case to a judge, and he sent two doctors, one of them a *dhimmī* [a Christian or a Jew living under Muslim rule] who testified that there was on his body leprosy regarding which there could be no doubt (*lā yashukkūna fī-hi*). Does the wife have a choice [of sleeping with her husband] or not? And is the word of a non-Muslim to be accepted?[55]

In his answer, al-Māzarī stresses the necessity of the examining doctor being competent, and carrying out his investigation of the alleged leper's body in a thorough fashion. If he is incompetent (*qaṣīr al-bāʿ*), then this will obviously not do. Al-Māzarī is clear about how one can diagnose leprosy:

Is it or is it not a case in which there is a smell, which in connection with sitting or lying together has a clearly harmful effect? If they say: there is no smell, test his

[53] See al-Burzulī, *Fatāwā*, vol. 2, 287. Compare with the view of Ibn Rushd: "The word of the doctor is valid in what is asked of him by the judge, concerning what is specific to the knowledge of the doctors even if he is not of moral probity (*ghayr ʿadl*), or is Christian if there is no other to be found. The preference is for two of moral probity" (al-Wansharīsī, *al-Miʿyār*, vol. 10, 17). A similar (anonymous) opinion acknowledging the authority of doctors to ascertain leprosy can be found ibid., vol. 7, 341–42. Al-Wansharīsī also records the opinion of Ibn Lubāba (d. 330/942), who cites the ability of doctors to ascertain whether a sickness is life-threatening or not (ibid., vol. 10, 294). On the position of lepers in Medieval Islamic society in general, see Dols, "The Leper in Medieval Islamic Society."

[54] Al-Māzarī had also written an influential work on prayer, in which he had defended the use of the astrolabe. See al-Māzarī, *Sharḥ al-Talqīn*, vol. 1, 386–88.

[55] Al-Burzulī, *Fatāwā*, vol. 2, 338.

place (of illness) with the head of a needle. If it changes and its color becomes red, blood appearing in the area, then it is not leprosy, and the woman has no say in it. This is the word of the authorities among the doctors (*qudamā' al-aṭibbā'*), and I don't know of a stronger (*awthaq*) position than this.[56]

Immediately after affirming the authority of doctors of previous generations, al-Māzarī attacks the practitioners of his own day, claiming that there are no longer any doctors of any note. In part, this is certainly representative of a broader anxiety regarding the possibility of charlatans exploiting believers by posing as doctors. Yet, al-Māzarī's concerns are more specific. For him, it is not so much a question of charlatans, or a doctor's religion, as whether a given doctor acknowledges the medical position presented by al-Māzarī:[57] if a doctor follows the criteria laid out by al-Māzarī, then it is simply a matter of sensory perception and deduction (*amr ḥissī ḍarūrī*). If a doctor gives different criteria, backing it up with citations, then he is to be permitted to do so, but his trustworthiness must be investigated by a judge. Al-Māzarī then proceeds to deny the phenomenon of contagion, while affirming that proximity to lepers was harmful, a differentiation made at least as early as Ibn Qutayba (d. 276/889).[58] While affirming the veracity and authority of medical opinion, al-Māzarī has managed to appropriate medicine's authority as a discipline to an extent not seen in al-Wazzānī's discussion of astronomy above.[59]

Though the legal opinions of the scholars here are far from univocal, repeatedly their discussions of astronomy, alchemy, and medicine show an ability and desire to support science's legitimate authority, while ensuring that it not be understood in a fashion that would challenge proper piety.

56 Ibid.

57 Not all jurists were as generous with regard to religious identity. Ibn al-Ḥājj, whose support of the authority of doctors was cited above, was particularly vicious on the subject of Jewish doctors (Ibn al-Ḥājj, *al-Madkhal al-shar'ī al-sharīf 'alā l-madhāhib*, vol. 4, 107–15, 140–50).

58 Ibid. On Ibn Qutayba, see Conrad, "A Ninth-Century Muslim Scholar's Discussion of Contagion." For a discussion of why the concept of contagion (as opposed to disease transmission) was rejected by many jurists, see Stearns, *Infectious Ideas*, chapter 1.

59 Another example of a legal scholar appropriating the authority of medical knowledge can be found in al-Wansharīsī's denial of the permissibility of abortion (*isqāṭ al-janīn*), in which, while giving a detailed discussion of the process of gestation, he cities only legal authorities (al-Wansharīsī, *al-Mi'yār*, vol. 3, 370). This example should be compared with the opinion of al-Mawwāq (d. 897/1492) on both *coitus interruptus* and abortion (ibid., vol. 4, 235–36). In both cases, al-Wansharīsī appends the views of Ibn al-'Arabī (d. 543/1148) to the initial opinion, and refrains from citing the views of doctors. On abortion in Islamic thought, see Katz "The Problem of Abortion in Classical Sunni fiqh" (in which al-Mawwāq's *fatwā* is briefly discussed) and Musallam, *Sex and Society in Islam: Birth Control before the Nineteenth Century.*

No conclusive generalizations can be drawn from these few opinions, however intriguing they may be. Above all, the substantial chronological range of material drawn upon here, comparing *fatāwā* from over almost a thousand years, suggests that this broad survey needs to be supplemented with collections and debates from the sixteenth to eighteenth centuries. This section concludes with an examination of one such collection before turning to a legal debate that shook the Muslim juridical community in the eleventh/seventeenth century.

Maḥammad[60] b. Muḥammad b. Aḥmad b. Muḥammad b. Ḥusayn b. Nāṣir b. ʿAmr b. ʿUthmān al-Darʿī (d. 1085/1674) was the founder of the Nāṣirī branch of the Shādhīlī Sūfī order in Tamgrout in the Dar'ah valley, which I have discussed in Chapter 1. As such he was an influential scholar and had many students, including al-Yūsī, who received from him his formal induction into the Shādhīlī order. His legal opinions were gathered into a book, *Kitāb al-ajwiba al-Nāṣiriyya* by his student, Muḥammad b. al-Qāsim al-Ṣanhājī (d. 1103/1692).[61] Ibn Nāṣir, as he was commonly referred to, was known for his profound piety and not for having studied the natural sciences. Yet in his opinions, he did address a small number of matters related to timekeeping, epidemics, and the origin of rain.

The question on timekeeping takes up the calculations set down in a poem by the seventh/thirteenth century Moroccan scholar Abū Muqriʿ, specifically the series of letters "ṭ-z-h-j-b-a-a-b-d-h-ḥ-ī."[62] This poem was commented upon by al-Mirghitī, who explains that these letters allow one to know the time to combine the noon and afternoon prayer, but does not specify how exactly.[63] Ibn Nāṣir responds to the query that one should follow the calculations of Abu Muqri' to calculate the time according to minutes, and not to follow the common method of praying the two prayers together after a fourth of the stick, presumably a reference to a period of time measured by shadow.[64] In the same

60 The variant pronunciation (and writing) of Muḥamamad as Maḥamamad is common in the south of Morocco.

61 I consulted a manuscript of this text held by the King 'Abd al-Aziz Foundation library in Casablanca, number. MS203_M3. I follow here the page numbering that is written in pencil on the manuscript itself.

62 On the difficulty of identifying Abū Muqriʿ see G. S. Colin and H. P. J. Renaud, "Note sur le 'muwaqqit' marocain Abû Muqri' – ou mieux Abû Miqra' – al-Baṭṭîwî (XIIIe siècle J.-C.)."

63 See al-Mirghitī, *al-Mumtiʿ*, 72.

64 Ibn Nāṣir, *Kitāb al-ajwiba al-Nāṣiriyya*, 21. For a detailed consideration of the use of popular astronomical methods at the other end of the Muslim world, see Petra Schmidl, *Volkstümliche Astronomie im islamischen Mittelalter. Zur Bestimmung der Gebetszeiten und der Qibla bei al-Aṣbaḥī, Ibn Raḥīq und al-Fārisī*, especially at 96–99.

answer, he briefly gives advice on how to tell time by positioning oneself toward to the sun in various ways, and alludes to the long standing debate about the correct orientation of mosques in Morocco, noting that to find the location of prayer in the south of Morocco one should place the north pole over one's left shoulder.[65] Ibn Nāṣir's answers on the subject are brief and do not even reach the limited detail included in al-Mirghitī's *fahrasa*, but they speak to his considering the proper interpretation of astronomy vital to ritual matters.

It is in the context of the difficult issue of visiting the sick and washing the dead that Ibn Nāṣir takes up the question of the transmission of illness, one Muslim scholars had been debating for centuries in medical and legal writings as well as in commentaries on Prophetic Tradition.[66] As alluded to in Chapter 2, epidemic disease, and particularly the plague, occurred repeatedly in North Africa in the seventeenth century and with devastating effect.[67] How to balance one's fear of the sick with the moral responsibility to care for them would have been a challenge for many Muslims. Ibn Nāṣir, like the eighth/fourteenth-century Granadan jurist Ibn Lubb (d. 782/1381), whose legal opinion was preserved by al-Wansharīsī in *al-Mi'yār*, and Ibn 'Ajība, whose *fahrasa* was discussed in Chapter 2, focuses on the duty of each Muslim to take care of his fellow believer. Specifically, Ibn Nāṣir was asked about the belief that some sicknesses are acquired by the healthy when a sick person looks at them, and that for this reason the healthy avoid the sick. How can such a belief be reconciled, the questioner continued, when the Prophet had related that there was no evil omen and no contagion, and had also said not to approach a place where an affliction (*al-balā'*) had broken out?[68]

In his answer, Ibn Nāṣir addressed the apparent contradiction between various sayings of the Prophet by stating that it would be bad (*qabīḥ*) to neglect visiting the sick because of these traditions. The proper understanding of this apparent contradiction, and the reason for the Prophet advising Muslims not to approach a disease-struck area was so that they would not be afflicted by the whispering of devils, i.e., doubt and fear of

65 Ibn Nāṣir, *Kitāb al-ajwiba*, 22. For background on this debate, see Monica Rius, *La Alquibla en al-Andalus y al-Magrib al-Aqṣà*.

66 Ibn Nāṣir, *Kitāb al-ajwiba*, 26.

67 For an overview of the plague treatises written in Morocco during this period, see Stearns, *Infectious Ideas*, chapter 5.

68 On the body of traditions attributed to the Prophet that dealt with contagion, plague, and leprosy and the how commentators on hadith collections reconciled them, see ibid., chapter 1.

death. To bolster his argument, he cites the *Iḥyā'* of al-Ghazālī and *al-ʿĀqiba* of ʿAbd al-Ḥaqq (d. 582/1186), dwelling at length on the afterlife.[69]

Confirming that legal opinions can move well beyond ritual and social matters, Ibn Nāṣir's interest in the phenomena of the natural world continues in some paragraphs reflecting on the nature of clouds: after being asked about the cause of gusts of wind, he answered that the air is a subtle body. When it moves like the waves in the sea, the air becomes stirred up, save that one does not see its body in the way one sees the body of the sea. When asked to explain the reason for the differences in the air, he said if the air is stirred up on one side it moves in other directions, pushing the air in those directions, creating wind. Turning to the origins of rain, he explained that God created it from clouds and caused it to fall in a known amount and quantity (*mikyāl maʿlūm wa mīzān maʿlūm*) at the hands of angels entrusted with this task. If God caused it to all come down at once, it would cause harm, hence the importance of the work of the angels. Ibn Nāṣir rejected the view of those who implied that clouds acquired their water from pausing above the sea and drawing water up from it. There is nothing to this (*fa-laysa bi-shay*).[70] While Ibn Nāṣir does not mention the sources that he is drawing on for his understanding of meteorology, they are clearly drawn from Prophetic Tradition and not the natural sciences.

The final issue that Ibn Nāṣir is asked about that relates to the natural sciences is that which will occupy the second half of this chapter: the status of smoking tobacco. Ibn Nāṣir's discussion of the subject raises some of the major issues that characterized this debate, even as he includes certain issues – the connection of smoking and dowsing for water among them – I have seen nowhere else. Asked about the herb tobacco, whether it is impure or not and what the ruling is on smoking it, Ibn Nāṣir answers that its purity is like other herbs that are not intoxicating and those who consider it forbidden do not think you should smoke it, while those who see it as permissible, do.[71] Clearly unsatisfied by this general description of the debate on the subject, his student and the compiler of the book

[69] Abū Muḥammad ʿAbd al-Ḥaqq b. ʿAbd al-Raḥmān al-Ishbīlī al-Azdī's *al-ʿĀqiba* is a famous collection of Prophetic Traditions on the afterlife. Al-Ghazālī discussed the disease and the Muslim's duty of tending to the sick in the book of the *Revival* on reliance (*tawakkul*) on God (al-Ghazālī, *Faith in Divine Unity and Trust in Divine Providence*, 143). When Ibn Nāṣir quotes a long tradition attributed to Saʿīd b. al-Musayyib about people coming to Abū Bakr when he was about to die and asking him for a sermon, and then gives the sermon, he may be drawing on book 40 of the *Revival* that deals with the afterlife (al-Ghazālī, *The Remembrance of Death and the Afterlife*, 75–76).

[70] Ibn Nāṣir, *Kitāb al-ajwiba*, 89. [71] Ibid., 56.

Muḥammad b. al-Qāsim al-Ṣanhājī interjects "O Teacher (*Yā Sīdī*), but what do you say?" Ibn Nāṣir responds that renowned scholars with knowledge of both external and internal matters have agreed that it is forbidden, and the only people who have said that it is permitted are following their own bodily desires. Already in 1083/1672 he had ordered his students to burn all that they had of this herb. Al-Ṣanhājī then brings up the fact that some figures renowned for their piety are known to smoke: "Yā Sīdī, the saints smoke it and they draw out (*yan ʿathuna)* water from under the earth to the people. This has been proven true by experience." In line with his student al-Yūsī's comments on saints standing outside of the legal and moral orders that define proper practice for other Muslims, Ibn Nāṣir explains that there is no problem with such respected figures smoking it since their hearts were filled with light.[72] Their ability to discern water under the earth was an art like other arts, and for that matter unbelievers were able to do so as well. One must judge people by their behavior and their adherence to the Prophet's example. Al-Ṣanhājī brings the subject to a close, writing: "I was pleased by what he said . . . and disturbed by what I saw of their violating of the Revealed Law (*shari ʿa*) and their surrendering to their passions."

The natural sciences were clearly not central to Ibn Nāṣir's preoccupations, nor had he studied them, but matters that they evoked reoccurred through his collection of legal opinions. Similarly, the legal opinions from the three major collections discussed above demonstrate not a prevalence of legal matters addressing the natural sciences so much as they do the importance of the authority of these sciences for dealing with a discrete set of legal issues. This comparatively narrow remit contained matters in which these sciences play an important role, as was shown by one of the most divisive legal issues of the seventeenth–eighteenth centuries: The Great Tobacco Debate.

3.6 STATES OF CONSCIOUSNESS IN THE GREAT TOBACCO DEBATE

As alluded to in Ibn Nāṣir's legal opinion on the subject, during the seventeenth and eighteenth centuries, a bitter and inconclusive debate on the

[72] See al-Yūsī, *The Discourses*, 313, where he writes: "Everything we have warned about and cautioned against applies to disciples and commoners on the path vulnerable to errors and lapses. About accomplished gnostics, even if they are not protected from error and not exempt from caution, we have nothing to say, as they know best their spiritual states, what they have been granted and can communicate, what they can express openly, and what they must conceal. It would be ludicrous as well as bad form for one such as me to counsel them."

permissibility of smoking raged throughout the Muslim Mediterranean. Tobacco, which had been introduced into Muslim societies in the late sixteenth century, proved to be a battleground for a struggle not only between the Ottoman and Saʿdī states and a public fascinated with new forms of consumption and sociality, but also between Muslim jurisprudents, Sufis, and doctors, who argued about whether tobacco was beneficial or harmful for the health, and whether its effects on the mind should count as intoxicating or not.[73] Social and intellectual historians such as Aziz Batran, James Grehan and Lutz Berger have shown how this debate reflected the ways in which Muslim societies responded and reacted to previously unknown drugs such as tobacco and coffee, a reception that built on earlier debates in Muslim scholarship that had addressed hashish and opium.[74] While by the mid seventeenth century the majority of Muslim scholars in both the Mashriq and the Maghrib had clearly stated in numerous treatises and legal opinions that the consumption of tobacco – by smoking or in the form of snuff – should be avoided if not shunned, and Ottoman and Saʿdī rulers Aḥmad I (r. 1603–17) and Aḥmad al-Manṣūr (d. 1603) had issued decrees forbidding its use, Muslims continued to smoke in public and private and to find prominent champions for their habit, such as the Damascene Sufi and theologian ʿAbd al-Ghānī al-Nābulusī (d. 1143/1731).[75] Precisely because of the continued popularity of smoking, it is not surprising that Muslim scholars down to the twentieth century would keep returning to the Great Tobacco Debate in order to argue against what most of them saw to be a distasteful if not outright forbidden activity. In the Maghrib, two examples of these twentieth-century condemnations of smoking are Muhammad b. Jaʿfar

[73] The nature and significance of this debate for both the Ottoman Empire and the Maghrib has been discussed in a substantial body of scholarship. See Muhammad Ḥajjī, *al-Haraka al-fikriyya*, vol. 1, 246–66; Aziz Batran, *Tobacco Smoking under Islamic Law: Controversy Over its Introduction*, which contains the translation of six *fatāwā* on smoking; an earlier version without the translations appeared as Baṭrān, "Ḥarb Fatāwī al-Tadkhīn Bayna al-ʿUlamāʾ wa-l-Muslimīn min Shamāl wa Gharb Ifrīqiyya fī-l-Aqdayn al-Awwal wa-l-Thānī min Ẓuhūr al-Tabagh"; Lutz Berger, "Ein Herz wie ein trockener Schwamm: Laqanis und Nabulusis Schriften über den Tabakrauch"; James Grehan, "Smoking and 'Early Modern' Sociability: The Great Tobacco Debate in the Ottoman Middle East (Seventeenth to Eighteenth Centuries)"; Dorrit van Dalen, "This Filthy Plant: The Inspiration of a Central Sudanic Scholar in the Debate on Tobacco," and compare with Rudi Matthee, *The Pursuit of Pleasure: Drugs and Stimulants in Iranian History 1500–1900*, chapter 5.

[74] See the cited sources and compare with Franz Rosenthal, *The Herb: Hashish versus Medieval Muslim Society*; Ibn Taymiyya, *Le haschich et l'extase*; Yahya Michot, *L'opium et le café*; Kafadar, "How Dark Is the History of the Night, How Black the Story of Coffee."

[75] This was but the first of several attempts of the Ottoman state to control smoking. In 1633 Murad IV (r. 1623–40) attempted to ban it outright on the pain of death with offenders being summarily executed (Grehan, "Smoking and 'Early Modern' Sociability," 1363–64).

al-Kattānī's (d.1345/1927)'s *The Ruling of the Four Jurisprudents and Others on Smoking* (*Ḥukm al-tadkhīn ʿinda al-aʾimma al-arbaʿ wa ghayri-him*), which brought together all of the sources the author could find from the proceeding centuries that provided arguments against smoking, and sections of *Medicine in the Light of Faith* (*al-Ṭibb fī daw' al-imān*), a series of articles and legal opinions by the prominent contemporary Tunisian mufti Muḥammad al-Mukhtār al-Salāmī.[76]

This section of the chapter, while it draws from and builds on these earlier studies, is not primarily interested in either the social history of consumption or in the legal arguments marshaled for or against tobacco. Instead, I will show that the Great Tobacco Debate is a productive site for us to consider how the body and human consciousness were constructed and deployed within the ethical and legal discourse that is Islamic jurisprudence. In the first part of the chapter, I argued that Muslim jurists in the post-formative period drew on the authority of astronomy and medicine, among other sciences, in order to enhance the authority of their own juridical opinions. Here, I am interested not only in how by bringing medical knowledge into legal debates, Muslim jurists not only acknowledged and at times appropriated the authority of the natural sciences, but additionally how in the process they argued about how and when human consciousness changed, and what the implications of these changes were for the believer.

3.7 THE ARRIVAL OF TOBACCO IN MOROCCO

In his comprehensive historical chronicle of Morocco's history, Aḥmad b. Khālid al-Nāṣirī (d. 1315/1897) quotes al-Ifrānī (d. ca. 1151/1738) to describe how tobacco arrived in Marrakesh, the capital of the Saʿdī sultanate, at the end of the sixteenth century along with a caravan of elephants that had been sent to the ruler Aḥmad al-Manṣūr (d. 1012/1603), and which caused great astonishment among the city's inhabitants:[77]

Because of these elephants entering the Maghrib, this vile herb called tobacco (*al-tābagh*) appeared. For the people of Sudan [sub-Saharan Africa], who were the

[76] See al-Kattānī, *Ḥukm al-tadkhīn ʿinda al-aʾimma al-arbaʿ wa ghayri-him* and Muḥammad al-Mukhtār al-Salāmī, *al-Ṭibb fī daw' al-īmān*, 259–79.

[77] Batran has shown convincingly, based on other contemporary sources, that the 1001/1592–93 date given by al-Ifrānī is unreliable and that the evidence points to the elephant's arrival in Marrakesh being in 1006/1598 and in Fez the following year, with tobacco having previously appeared in Timbuktu in 1005/1596–97 (Batran, *Tobacco Smoking*, 19–25). Compare with al-Laqānī, *Kitāb naṣīḥat al-ikhwān*, 59, where he states that tobacco first appeared in Timbuktu in 1005/1597.

elephant-handlers, brought it with them to smoke, claiming that it contained benefits. From them it spread out in the areas of Darʿa, Marrakesh, and elsewhere throughout the Maghrib. Concerning it the legal opinions of the scholars, God be pleased with them, conflicted. There were some who forbade it, others who permitted it, and others who suspended judgment (*mutawaqqif*). The knowledge of this matter is with God, praise be to Him.[78]

Al-Nāṣirī immediately followed this quote with his own opinion regarding smoking, which was that ingesting tobacco was forbidden according to Islamic law, as it was clearly one of the repulsive substances [*khabā'ith*] that Muḥammad had been ordered in the Qur'an to forbid to his purified community.[79] The issue was, however, not nearly so clear to the scholars who witnessed the initial spread of tobacco and smoking in the Saʿdī sultanate, which itself faced considerable challenges at the beginning of the eleventh/seventeenth century. In 1603 Aḥmad al-Manṣūr died, and control of the Far Maghrib was bitterly contested by his sons. The political turmoil that followed al-Manṣūr's death is of some interest to us here, for one of the most prominent rebels who challenged the right of the Saʿdī dynasty to rule during this period, Ibn Abī Maḥallī (d. 1022/1613), had previously also been one of the most prominent defenders of the smoking of tobacco.[80] Ibn Abī Maḥallī was a prominent legal scholar and Sufi who claimed descent from the Prophet's uncle ʿAbbās b. ʿAbd al-Muṭṭalib, and who justified his uprising, which began south of the Atlas in Sijilmāsa, by lamenting the devastation that the struggle between the sons of al-Manṣūr had brought to the Maghrib.[81] After an initial series of victories against the forces of Moulay Zaydān, Ibn Abī Maḥallī was defeated and killed outside of Marrakesh in 1613. His status as a rebel during a tumultuous time of Maghribi history served his views poorly in the debate on the permissibility of tobacco, as his failed attempt at rebellion may have rendered his judgment on this issue questionable.[82] The importance of Ibn Abī Maḥallī's contribution to the Great Tobacco debate lies in the earliness of his writing and the attention he paid to tobacco's effects on the

[78] Al-Nāṣirī, *Kitab al-istiqsa*, vol. 5, 131. The passage in al-Ifrānī can be found in al-Ifrānī, *Nuzhat al-ḥādī bi-akhbār mulūk al-qarn al-ḥādī*, 248.

[79] See ibid. The verse is Q7:157, which reads in part: "... the Messenger ... who makes good things lawful to them, and repulsive things unlawful." I follow here the translation of M. A. S. Abdel Haleem, changing only the translation of *khabā'ith*.

[80] The best overview of the life of Ibn Abī Maḥallī is that of al-Majid al-Qaddūrī, *Ibn Abī Maḥallī al-faqīh,* which also contains a partial edition of his *fahrasa* and, most relevant for our purposes, a long discussion of smoking.

[81] See al-Nāṣirī, *Kitab al-istiqsa'*, vol. 5, 244–47.

[82] For al-Laqānī's attack on Ibn Abī Maḥallī, see Batran, *Tobacco Smoking*, 26, and the section of his partial translation of al-Laqānī's treatise, on 192–93.

body. Ibn Abī Maḥallī wrote his opinion on smoking in 1014/1605 while in Fazzān, in modern south-western Libya, while on pilgrimage to Mecca, and in response to two earlier opinions that had forbidden smoking, those of the head of the Mālikīs in Egypt, Sālim al-Sanhūrī (d. 1015/1606), and the judge of Timbuktu, Muḥammad ibn Aḥmad ibn ʿAbd al-Raḥmān (d. 1016/1608).[83] Ibn Abī Maḥallī was not the only scholar in the Maghrib, however, to defend smoking in an extensive legal opinion – another extensive opinion in its defense was authored by the prominent West African jurist Aḥmad Bābā al-Tinbuktī (d. 1036/1627), who was asked about smoking in 1016/1607 by the Chief Judge (*qāḍī al-jamāʿa*) of Fez, ʿAli b. ʿImrān al-Salāsī (d. 1018/1609).[84] Aḥmad Bābā was a remarkable individual and scholar who has attracted a good deal of attention in previous scholarship both due to his biography and his writings against Moroccan enslavement of West Africans. A leading scholar of Timbuktu, Aḥmad Bābā was brought to Marrakesh after the Moroccan invasion of West Africa in 1591 due to fears that he could lead local resistance against the Moroccan occupation. Once he arrived in Marrakesh, he was put into what was essentially house arrest for the subsequent fourteen years, during which he became famous for his erudition and taught an entire generation of Moroccan scholars.[85] In 1607, he was granted permission by Moulay Zaydān to return to Timbuktu and it was in Tamgrout on his way home that al-Salāsī's request reached him and he wrote his *fatwā* defending the practice of smoking.[86]

3.8 THE TERMS OF THE DEBATE

What was it about tobacco that drew the attention of Muslim jurists? In the last years of the reign of al-Manṣūr we already find two scholars, Muḥammad

[83] Tobacco had been banned by the ruler of Fazzān around 1014/1605. For the broader context of Ibn Abī Maḥallī's *fatwā*, see Batran, *Tobacco Smoking*, 43–44. Batran translated the entire section of Ibn Abī Maḥallī's *al-Iṣlīt al-khirrīt* that dealt with smoking (ibid., 100–41).

[84] My discussion here of Aḥmad Bābā's opinion differs from that found in Batran, (ibid., 58–62, and translated on 169–90), as he referred to only to that section of Aḥmad Bābā's *fatwā* included by Ibn Abī Maḥallī and not to the entire legal decision as preserved in the manuscript I consulted. Batran situates Aḥmad Bābā's *fatwā* as a rebuttal of the view of the judge of Timbuktu, Muḥammad ibn Aḥmad ibn ʿAbd al-Raḥmān, and gives a useful overview of the tense relations between their two families (ibid., 58–62). In the process, however, he didn't take account of Aḥmad Bābā's opinion being primarily addressed to the Moroccan context he was leaving.

[85] See Mahmoud Zouber, *Aḥmad Bābā de Tombouctou (1556–1627)*, 26–31.

[86] See Zouber, *Ahmad Bābā*, 184–87 and John Hunwick, "A New Source for the Biography of Ahmad Baba al-Tinbukti."

b. Qāsim al-Qaṣṣār (d. 1012/1603) and ʿAbd Allāh b. Ḥassūn (d. 1013/1604) urging the Saʿdī ruler to outlaw smoking, and they were the first of many in the Maghrib.[87] Opponents of tobacco established, collectively, four major rationales for declaring both it and smoking to be forbidden: 1) in analogy with either wine or hashish, it was an intoxicant; 2) smoking was repulsive and tobacco a substance that violated ritual purity; 3) smoking was bad for the health; and 4) it was a European custom.[88] Authors of treatises against smoking would usually adduce several if not all of these themes simultaneously. Ibrahim al-Laqānī (d. 1041/1631), an Egyptian jurist and theologian and author of one of the most famous early anti-smoking treatises, effectively combined the last two reasons by presenting the spread of smoking in the Ottoman Empire as a European plot:

> I was informed by one who associates with Christian Englishmen that they didn't bring tobacco to the lands of the Muslims until their doctors had agreed on forbidding them from using it on a regular basis. The doctors agreed that the Englishmen should only use an amount of it that would cause no harm. They had taken a man after his death who had died of an infection of the liver (*iḥtirāq al-kibd*) and who had been a regular user of it, and had dissected him. They found it having spread into his veins and nerves (*sāriy*[an] *fī ʿurūqi-hi wa ʿaṣabi-hi*), to the point that his bone marrow had turned black. They discovered his heart to be like a dry sponge with many holes, some small and others large, and his liver was as if it had been grilled over a fire. From this time onward they forbade its habitual use and ordered for it to be sold to Muslims.[89]

Al-Laqānī's treatise was influential in both East and West, and when the Maghribi scholar ʿAbd al-Qādir al-Fāsī (d. 1091/1680) was asked about smoking, he quoted it at length before adding himself that he had been reliably informed that Christians in the city of Rabat mixed their urine or the urine of horses with tobacco before selling tobacco to Muslims.[90]

[87] See Baṭrān, "Ḥarb Fatāwī al-Tadkhīn," 189.

[88] Other reasons advanced against smoking included wasting money and consorting with men of low morals, criticisms that were similar to those advanced against coffee.

[89] Ibrāhīm al-Laqānī, *Kitāb naṣīḥat al-ikhwān*, 93. Compare with the translations given in Lutz Berger, "Ein Herz wie ein trockener Schwamm," 261, and Aḥmad al-Aqḥiṣārī, *An Ottoman Manifesto Against Smoking*, 59.

[90] See al-Shaykh ʿAbd al-Qādir al-Fāsī, *Jawāb su'āl ḥawl al-tabagh "ṭāba,"* 6. On the importance of the al-Fāsī family in sixteenth–seventeenth Morocco and their history see Fernando Mediano, *Familias de Fex (ss. XV–XVII)*, 143–60, where ʿAbd al-Qādir al-Fāsī is mentioned as *shaykh al-jamāʿa* of Fez (ibid., 159). Al-Laqānī had similarly been anxious regarding tobacco being mixed with human urine, horse dung, vinegar, wine, hashish, or simply being replaced with another vile, foul-smelling plant (al-Laqānī, *Kitāb naṣīḥat al-ikhwān*, 86–89).

While such fears of foreign contamination are comparatively rare, they point not only to the anxiety of some Muslim jurists regarding the introduction of a new and foreign substance into Muslim societies, but also to the heightened fears throughout the Muslim Mediterranean in the early seventeenth century of expanding European influence. These fears were hardly abstract for scholars in the Maghrib, where Castile and then Spain had established colonies on their northern Mediterranean shore and where the Portuguese had spent much of the sixteenth century establishing a series of fortified ports on the Maghribi Atlantic coast.[91] The main thrust of ʿAbd al-Qādir's treatise was, however, not tobacco's European origin, but that as a substance either as leaf or as snuff it was impure and thus could not be used without violating Islamic laws of ritual purity.[92] It was, admittedly, in his opinion permissible (though not desirable) for Muslims to consume tobacco that was imported from the Western Sudan or which originated in the Maghreb itself.[93]

Maghribi legal opinions from earlier in the seventeenth century debated the issue largely without references to European contamination, although ritual purity played an important role. The debate spoke to the close intellectual and commercial connections between the Western Sudan, Maghrib, Egypt, and the Hijaz, with the greatest opponents of smoking counting al-Laqānī's Egyptian teacher Sālim al-Sanhūrī (d. 1015/1606) and al-Laqānī himself, and prominent defenders including Aḥmad Bābā, Ibn Abī Maḥallī, the Egyptian scholar ʿAli al-Ujhūrī (d. 1066/1657), and later the prominent Damascene Sufi and theologian ʿ Abd al-Ghānī al-Nābulusī (d. 1143/1731). A third group of scholars, including the Palestinian Marʿī al-Karmī (d. 1032/1623–24) and the Maghribi Abū Bakr al-Suktānī (d. 1063/1652) held that while they were personally disposed against smoking, they could not find enough evidence to justify banning it outright.[94] The problem that this last group faced, and which had been addressed in varying ways by the first two, was that the disagreeability of the smell of tobacco was largely a matter of taste, that the medical profession seemed to be uncertain as to whether it was beneficial or harmful, and that it was quite unclear whether tobacco was an intoxicant or not.[95] The debate on this issue remained

[91] For Portuguese aspirations in the Maghrib, see Matthew T. Racine, *A Most Opulent Iliad: Expansion, Confrontation & Cooperation on the Southern Moroccan Frontier 1505–1542.*

[92] See ʿAbd al-Qādir al-Fāsī, *Jawāb su'āl ḥawl al-tabagh "ṭāba,"* 7. [93] See ibid., 6.

[94] See Grehan, "Smoking and 'Early Modern' Sociability," 1369, and on al-Suktānī see al-Wazzānī, *al-Miʿyār al-jadīd*, vol. 1, 106.

[95] The debate on tobacco has a striking similarity with a different drug that was introduced into the Muslim Mediterranean roughly a century earlier: coffee. For the discussion on whether coffee was an intoxicant see the discussions in Ralph S. Hottox, *Coffee and*

vigorous throughout the seventeenth century and involved the broad diffusion of medical terminology and concepts within Muslim legal circles.

3.9 SITUATING SMOKING BETWEEN INTOXICATION, LANGUOR, AND ANESTHESIA

Aḥmad Bābā's opinion on the question of smoking tobacco is striking for its systematic approach to the question, the extensive attention it devotes to the formal legal aspects of the debate, and his consideration of tobacco's effects on the consciousness. For our purposes, it is a productive doorway into the debate. He begins by describing the appearance of tobacco, which appeared "at the beginning of the eleventh century" and which had an unnamed scholar of the age in Cairo and elsewhere asking another for clarification for another on whether smoking was licit or not. The scholar questioned did not answer, but another did with the conclusion that using it was licit as it was not an anesthetic (*murqid*), nor a narcotic (*mufsid*), nor an intoxicant (*muskir*), three categories that we will come to below. As an aside, Aḥmad Bābā notes that he had previously written an opinion on this issue when he was still in Marrakesh in which he answered the judge and mufti of Fez, Abū al-Ḥasan ʿAlī b. ʿImrān (d. 1018/1609), and the Judge of Darʿa, the jurist Aḥmad Abū Saʿīd, when they asked him about the matter.[96]

Aḥmad Bābā's subsequent discussion of the effects of tobacco on the body and the smoker's consciousness drew repeatedly on the influential legal writings of the seventh/thirteenth century Egyptian Mālikī scholar Shihāb al-Dīn al-Qarāfī (d. 684/1285). Al-Qarāfī's writings in the realm of legal theory, especially among Mālikīs, have continued to be authoritative down until today.[97] A Berber of the Sanhaja, he was born near Cairo and achieved fame as a scholar in Egypt, with Mālikī students coming to study with him from as far as the Maghrib.[98] Before continuing with Aḥmad Bābā's own argument, it is worth reviewing what al-Qarāfī had written regarding a different herb – hashish – on which he took a provocative position.

Coffeehouses: The Origins of a Social Beverage in the Medieval Near East, and Michot, *L'opium et le café*, 106. Compare with ibid., 79–80, where Michot quotes ʿAbd al-Qādir al-Jazīrī (d. ca, 977/1570) as to the "toxicity" (*mafsada*) of coffee.

96 I know of no extant copies of this previous opinion of Aḥmad Bābā's.

97 Batran has noted the centrality of al-Qarāfī to the tobacco debate (Batran, *Tobacco Smoking*, 49–50).

98 On al-Qarāfī, see Sherman Jackson, *Islamic Law and the State*.

In his *Book of Legal Distinctions* (*Kitāb al-furūq*), in the fortieth distinction (*farq*) he addresses the differences between substances that intoxicate (*al-muskirāt*), those that numb (*al-murqidāt*), and those that corrupt (*al-mufsidāt*).[99] Noting that differentiating between these three is confusing for many jurists, al-Qarāfī explains that substances that cloud (*tughīb*) the senses are anesthetics (*al-murqidāt*), and that those that do not do so either bring about elation (*nashwa*) and pleasure, or do not. The former are intoxicants (*al-muskirāt*), the latter narcotics (*al-mufsidāt*). The difference between anesthetics and the other two categories lies in the first affecting the senses, while the second two affect the mind. Examples that help clarify the difference between the second two categories include wine, beer (*al-mizr*), honey wine (*al-bat'*), and wheat wine (*al-sukurka*), which all cloud the mind and induce pleasure and elation, and henbane (*saykurān*), and hashish, which cloud the mind without bringing about pleasure. A further differentiation between intoxicants and narcotics is that the first lead the one who consumes them to feeling courageous, to becoming violent, and to undertaking acts of extraordinary generosity, whereas the second do not.

Al-Qarāfī turns to explain in detail why hashish is not an intoxicant, and draws on humoral medicine to do so: hashish "excites the hidden humor (*al-khilṭ al-kāmin*)" in each person's body in the following fashion according to the well-known four humors. The person in whom yellow bile is prominent feels heat after using it, it causes lethargy in the person in whom phlegm is strong, and brings about crying and sorrow in a person in whom black bile is prominent it, while, finally, in the person in whom blood is prominent, it causes a corresponding amount of pleasure. In sum, there are those who cry a great deal, while others are quiet. By contrast, al-Qarāfī notes, with wine and other intoxicants, you rarely find anyone who drinks it who is not delirious and happy. Those who drink wine often become quarrelsome and attack each other with weapons and carry out attacks of the kind that they would never do while sober. Those who consume hashish do not do any of this, and even if provoked they do not exhibit as much violence as wine drinkers. Instead they are quiet like beasts. In light of these arguments, al-Qarāfī considers hashish a narcotic and not an intoxicant. This is important primarily because it means that it does not invalidate ritual purity, specifically in the case of prayer. It also means that he does not impose the Qur'anic *ḥadd* punishment for it, but proscribes instead the *ta'zīr* (discretionary) punishment. Finally, it means that whereas it is forbidden to consume even a small

99 The following discussion is based on al-Qarāfī, *Kitāb al-furūq*, vol. 1, 363–65.

amount of an intoxicant, small amounts of hashish and opium, even just a grain or two, can be consumed without legal implications. Thus, al-Qarāfī concludes, it should now be clear how jurists should differentiate in their rulings regarding these substances.[100]

It is worth pausing briefly to consider the implications of al-Qarāfī's categorization. He differentiates between intoxicants on the one hand and narcotics and anesthetics on the other by invoking their effects on the body and the mind, criteria that bring the human body and its physical and mental states into legal theory along with the medical knowledge required to establish the value of the body's testimony. As Franz Rosenthal has shown in his exemplary study of hashish in premodern Islamdom, al-Qarāfī's view that hashish was not an intoxicant, while not unique, was not widely shared and was bitterly opposed by the Ḥanbalī Ibn Taymiyah (d. 728/1328) and the Shafiʿī al-Zarkāshī (d. 794/1392).[101] Yet, while the jurists may have argued over how the intoxicated or body under the influence of narcotics should be read and interpreted, the notion that it had a place in the legal and thus social discourse was widely accepted.[102] Medicine, either as a discourse or a profession, had an important role in converting the drugged body into legal testimony.[103] The terms set out by al-Qarāfī, which notably did not include languor-inducing substances (*al-mufattir*), provided the legal basis for how the body and human consciousness were defined in the debate on smoking between Maghribi scholars in the seventeenth century.

Returning to the legal opinion of Aḥmad Bābā, we find him beginning by stating the uncontroversial position that all intoxicants are forbidden. He proceeds by invoking al-Qarāfī's differentiation between intoxicants obscuring man's reason but not his senses and causing pleasure, while the situation is reversed with anesthetics and narcotics.[104] Why is this? It is because it is reason that separates mankind from animals and the

[100] See also the discussion of this passage in Rosenthal, *The Herb*, 108–10.

[101] See ibid., 111–12, and compare with ibid., 161–62, and Ibn Taymiyya, *Le haschich*, 79–80.

[102] For al-Zarkashī's use of medical sources as well as the categories of *muskir and mufsid*, see his treatise as edited in Rosenthal, *The Herb*, 180.

[103] For one contemporary discussion of how testimony, be it rational, empirical, or textual should be evaluated, see al-Manjūr, *Sharḥ al-manhaj al-muntakhab ilā qawāʾid al-madhhab*, vol. 1, 167–77. The importance of medical testimony in a wide variety of legal matters in the premodern period, while not his main interest, has been surveyed by Ron Shaham in *The Expert Witness in Islamic Courts: Medicine and Crafts in the Service of Law*, 66–75.

[104] Aḥmad Bābā al-Tinbuktī, "*al-Lamaʿ fī-l-ishārah bi-ḥukm ṭābah*," 3–4.

preservation of which is a necessity in jurisprudence. To emphasize his point, and to prepare for his defense of tobacco, Aḥmad Bābā repeats al-Qarāfī's point that with the exception of wine where there is a clear Qur'anic prohibition, it is forbidden to ingest substances based on their ability to intoxicate, not on their essence. Fermented milk, therefore, is prohibited, while a small nonintoxicating amount of opium that does not cloud the mind is not. Since, Aḥmad Bābā argues, there is consensus that tobacco is not an intoxicant and, he continues, it is neither a narcotic nor an anesthetic according to the criteria established by al-Qarāfī, the most that can be said of it is that it produces languor (*mufattir*). There is, therefore, no doubt that it is permitted to smoke it and only the stupid or ignorant could disagree.[105] This last reference may have been to the above-mentioned Egyptian jurist al-Sanhūrī, or the judge of Timbuktu, Muḥammad ibn Aḥmad ibn ʿAbd al-Raḥmān. After some time shoring up his claim that tobacco produces only languor by considering views that it could function as a narcotic or anesthetic, he argues that even if it did cloud the mind, not everything that clouds the senses is forbidden – think only of mystics who lose awareness due to their deep love of God. And anyway, he notes, anesthetics are permitted when one has to amputate a limb.[106]

Toward the end of his *fatwā*, Aḥmad Bābā takes up the Prophetic Tradition that stated that every intoxicant (*muskir*) and languor-producing substance (*mufattir*) was forbidden.[107] Aḥmad Bābā argues that the languor producing substance forbidden here was that which leads to intoxication, and this was the reason the second term was included in the tradition. He then cites al-Khaṭṭābī's (d. 338/998) commentary for the type of languor produced in the limbs being a prelude to intoxication and therefore prohibited, for it leads to (*dharīʿa*) intoxication. Yet that which does not lead to intoxication at all but only to languor of the limbs does not fall under this prohibition: neither high quality refined butter (*al-samn al-ʿatīq*), nor warming medicines, much less entering a bathhouse (*dukhūl al-ḥamām*). In closing his argument, Aḥmad Bābā makes an argument for the beneficial nature of smoking, listing its medicinal benefits: heating the head and body, causing phlegm and food to descend to the stomach, quickening digestion, lessening colds, strengthening sight, exciting sexual desire, working against poisons such as snakebite, and so on. All of this, he claims, has been widely

[105] Ibid., 14. [106] Ibid., 16–19.

[107] Ibid., 30. The tradition can be found in Abū Dāwūd's *Sunan* in the chapter on drinks. On the debate surrounding intoxication in the first centuries of Islam and specifically the term *khamr* see Najam Haider, "Contesting Intoxication: Early Juristic Debates over the Lawfulness of Alcoholic Beverages."

TABLE 3.1: Representative (but not comprehensive) list of the scholars of the Great Tobacco Debate

Scholars in the Mashriq

Sālim b. Muḥammad al-Sanhūrī (d. 1015/1606)
 – forbade smoking.
Nūr al-Dīn Alī b. Yaḥyā al-Zayādī (d. 1024/1615)
 – defended smoking; provided it didn't cloud the mind.
Mar'i al-Karmī (d. d. 1032/1623-24)
 – disliked smoking, but didn't see enough evidence to forbid it.
Ibrāhīm al-Laqānī (d. 1041/1631)
 – forbade smoking.
Aḥmad al-Aqḥiṣārī (fl. Early 17th century)
 – forbade smoking.
'Alī al-Ujhūrī (d. 1066/1657)
 – defended smoking.
'Abd al-Ghānī al-Nābulusī (d. 1143/1731) [written in 1093/1682]
 – defended smoking.

Scholars in the Maghrib

Muḥammad b. Qāsim al-Qaṣṣār (d. 1011/1603)
 – jurist of Fes, advised al-Manṣūr to ban smoking.
'Abdallāh b. Ḥassūn (d. 1012/1604)
 – jurist of Salé, also advised al-Manṣūr to ban smoking.
Muḥammad b. Aḥmad b. 'Abd al-Raḥmān (d. 1016/1608)
 – judge of Timbuktu; forbade smoking.
Ibn Abī Maḥallī (d. 1022/1613)
 – defended smoking.
Aḥmad b. Muḥammad al-Būsa'īdī (d. 1046/1636-37)
 – believed tobacco had medical benefits.
Abū Qāsim b. al-Nu'aym al-Ghassānī (d. 1032/1623)
 – compiled collective *fatwā* against smoking.
'Abd al-Raḥmān b. Muḥammad al-Fāsī (d. 1626)
 – forbade smoking.
Aḥmad Bābā al-Tinbuktī (d. 1036/1627)
 – defended smoking.
Abū Zayd 'Abd al-Raḥmān b. Muḥammad al-Fāsī (d. 1036/1626-27)
 – forbade smoking.
Aḥmad b. Muḥammad al-Maqqarī (d. 1041/1632)
 – forbade smoking.

TABLE 3.1: Continued

Scholars in the Maghrib
Abū ʿAbdallāh Muḥammad al-ʿArabī b. Yūsuf al-Fāsī (d. 1052/1642) – defended smoking.
Abū Bakr al-Suktānī (d. 1063/1652) – believed the evidence was inconclusive either way.
ʿAbd al-Qādir al-Fāsī (d. 1091/1680) – concerned with the ritual purity of tobacco, forbade buying from foreigners.
Abū ʿAlī al-Yūsī (d. 1102/1691) – gave conflicting rulings on whether smoking was allowed.
Muḥammad ibn Jaʿfar al-Kattānī (d. 1927) – forbade smoking.

witnessed to be true. In addition, Aḥmad Bābā has testimony to this effect written in the handwriting of an unknown doctor of the Maghrib who, he notes, died before people began smoking tobacco. He was known for his knowledge of plants, herbs, and their properties, and held that tobacco was a beneficial medicine against ailments of the chest and, like honey, against shortness of breath; it drew out the corrupt humors in the chest and was beneficial against lung disease.[108]

Ibn Abī Maḥallī welcomes Aḥmad Bābā's support of his own position, and leveled a harsh critique of Muḥammad ibn Aḥmad ibn ʿAbd al-Raḥmān, the judge of Timbuktu who had prohibited smoking:

> Consider this! If he permits the consumption of small quantities of narcotics, though not intoxicants, that do not cloud the mind, then how can he prohibit substances that produce languor (*mufattir*), when these don't affect the mind at all?[109]

Ibn Abī Maḥallī notes here that it is legally permissible to anesthetize the body – to use soporifics – in preparation of an amputation and cites a number of scholars, including the prominent Medinese Mālikī jurist Ibn Farḥūn (d. 719/1319), as support for this position. For Ibn Abī Maḥallī, it is permissible to consume substances that change one's awareness of one's body, including narcotics (*mufsid*) as long as they do not influence one's mental abilities (*mā lam yablugh ḥadd al-taʾthīr fī l-ʿaql*). Only substances that fog the mind (*illā mā ghayyaba al-ʿaql*) are prohibited, he explained.[110] In addition, he maintains that both jurists and hadith scholars in the

[108] Ibid., 35–36. [109] Al-Qaddūrī, *Ibn Abī Mahallī*, 148. [110] See ibid., 148.

Mashriq and Maghrib hold that tobacco has proven beneficial medicinal properties against poison.[111] Indeed, a series of scholars, such as the *qāḍī* of Darʿa, Aḥmad b. Muḥammad al-Būsaʿīdī (d. 1046/1636–37), who had studied with Aḥmad Bābā and with ʿAbd al-Raḥmān al-Fāsī, argued that tobacco was beneficial for the health.[112] Al-Būsaʿīdī's assertion, in which he was joined by scholars from the Mashriq such as ʿAlī al-Ujhūrī (d. 1066/1657), was made on the basis of the author's legal authority without explicit reference to specific doctors or to the profession of medicine in general.

Following this initial discussion, Ibn Abī Maḥallī continues by taking aim at the legal opinion of the Egyptian scholar al-Sanhūrī, who had forbidden smoking, which he quickly counters with the authority of the permissive views of ʿAli al-Ujhūrī. He then summarizes a *fatwā* by the head of Egypt's Shāfiʿī legal community, ʿAlī b. Yaḥyā al-Zayādī (d. 1024/1615), to the effect that tobacco could not be forbidden as long as it did not harm the mind. Neither a languor-inducing substance (*mufattir*) nor an anesthetic (*murqid*) or narcotic (*mufsid*) that failed to cloud the brain, fulfilled this condition. As Ibn Abī Maḥallī explained, this position was in accordance with the judgment (*qā'ida*) of al-Qarāfī, whose views on this, Ibn Abī Maḥallī argued, had been followed by all prominent jurists after him, as al-Zayādī's *fatwā* demonstrated.

There was of course the question of what intoxication consisted of, and the possibility that not all humans became equally drunk. In exploring this possibility, Ibn Abī Maḥallī quotes a letter to him from the Egyptian scholar ʿAli al-Ujhūrī, who raised the hypothetical case of a vessel of wine from which seventy people drank, one of whom did not become intoxicated.[113] Since laws are not made on the basis of exceptions, it would follow that this individual would not be allowed to drink wine, even if the legal rationale (*ʿilla*) for the prohibition was inebriation. After all, in the opposite case, where if a large number of people drank lightly fermented date juice (*nabīdh*), fermented milk, or vinegar, and one of them became intoxicated, would all of them be prohibited from drinking these substances? The implication is that this would certainly not be the case, and the relevance

[111] See ibid., 149.

[112] On al-Būsaʿīdī, see Muhammad Makhlūf, *Shajrat al-nur al-zakiyya* (1930–31), 301. It is unclear here why Ibn Abī Maḥallī is referring to ʿAbd al-Raḥmān al-Fāsī as a supporter of smoking when, as discussed below, this scholar wrote a treatise against smoking. For an overview of the influential Fāsī family of scholars, which shows there doesn't seem to have been a different ʿAbd al-Raḥmān that Ibn Abī Maḥallī could have been referring to who would have been alive at this time see Lévi-Provençal, *Les historiens des Chorfas*, 242.

[113] See al-Qaddūrī, *Ibn Abī Mahallī*, 156.

of this discussion to tobacco is that, as Ibn Abī Maḥallī later makes clear, even if intoxication did result in a rare case from smoking, as it would be the exception, this would not be grounds for its prohibition.[114]

Ibn Abī Maḥallī's discussion of the issue is structured closely in conversation with the work of previous authors, especially his Egyptian contemporary al-Sanhūrī, to whom he wrote directly, seeking to change his position that smoking should be forbidden. Possibly in light of his earlier refutation of Muḥammad ibn Aḥmad ibn ʿAbd al-Raḥmān's citing of the Prophetic tradition prohibiting every languor-producing substance (*mufattir*), Ibn Abī Maḥallī stresses again that he does not understand how languor, itself, can be the cause for the prohibition of a substance if the senses and the mind have not been affected (the irony being, of course, that Muḥammad ibn Aḥmad ibn ʿAbd al-Raḥmān had made the same argument). Although he does not phrase it in this fashion, it is reasonable to argue that he held the legal rationale (*ʿilla*) for the prohibition of intoxicants to be the clouding of the mind, and in the absence of this condition, there could be no reason for tobacco to be forbidden and a single Prophetic Tradition to the contrary was insufficient in light of the available evidence. After all, he argued, a feeling of languor can also be induced through eating fatty foods, or can follow the taking of laxatives for medical reasons, or can be felt during sex after ejaculation, or after exiting the baths. Surely, the argument implies, none of these activities was forbidden because it produced languor?

After Ibn Abī Maḥallī finishes quoting his letter to al-Sanhūrī, he goes on to finish his own discussion by citing cases he had either heard of or witnessed himself in which tobacco had had clear medical benefits:[115] it helps one digest (*yahḍam)* food, counteracts poison (he had witnessed over forty individuals who were bitten by scorpions being treated with it), and was effective against snakebite. He recalls that when he himself was a boy he suffered from a head cold (*al-tarwīha*) to the point where he was bedridden, until he smoked and recovered. He had also seen it cause flies (*al-aḥnāsh*) to come up from the stomachs of many people and, finally, on the authority of what he calls a reliable source he gives the example of

[114] See ibid., 163.

[115] Ibn Abī Maḥallī observes that al-Sanhūrī was never able to answer as he died shortly after receiving his letter (as Batran noted, Ibn Abī Maḥallī had met al-Sanhūrī twice in Egypt (Batran, *Tobacco Smoking*, 44)), he notes that he received a very warm letter from al-Ujhūrī, extracts of which he goes on to quote. This acknowledgment from a prominent Egyptian scholar was important to Ibn Abī Maḥallī as it spoke to his own juridical authority and to the soundness of his arguments (see ibid., 178–79).

a person who suffered from head pain until he smoked it and a worm came out of his nose, leaving his condition much improved.[116]

These empirical testimonies of tobacco's medical efficacy, as well as the intermittent appeals to the authority of doctors, are, however, noticeably less important for Ibn Abī Maḥallī than his extensive discussion on how to understand tobacco's effects on the body – that is to say, his self-fashioning as an expert on states of human consciousness. It is on this basis that some subsequent opponents of smoking in the Maghrib formulated their arguments, such as the collective legal opinion of Fassi scholars in 1026/1617 and the treatise of Abū Zayd ʿAbd al-Raḥmān b. Muḥammad al-Fāsī (d. 1036/1626–27).[117]

3.10 STRATEGIC SCRIPTURALISM: MATCHING DIFFERENT EMPIRICAL EVIDENCE WITH VARIOUS AUTHORITATIVE TEXTS

At the beginning of Ramadan in 1026/1617, a Fassi jurist, Abū al-Ḥasan Sīdī al-Ḥājj ʿAlī b. Aḥmad (d. 1032/1622), famed as al-Shāmī al-Khazrajī al-Anṣārī,[118] wrote regarding smoking to the chief judge (*qāḍī al-jamāʿa*) of Fez, Abū Qāsim b. Muḥammad b. Abī al-Naʿīm al-Ghassānī (d. 1032/1623). Al-Ghassānī, who was known for energetically pursuing corruption, and who would be dramatically shot down in the streets of Fez six years after he answered this query, had studied with Aḥmad Bābā and received multiple diplomas (*ijāza*) from him.[119] In formulating his response to al-Shāmī al-Khazrajī al-Anṣārī, al-Ghassānī took the unusual step of assembling a collective legal opinion that bore the names of nineteen Fassi jurists, although the two most important were doubtlessly al-Ghassānī himself and the *mufti* of Fez's most important religious institution, the Qarawiyyīn, Abū al-ʿAbbās b. Sīdī Aḥmad b. Muḥammad al-Maqqarī (d. 1041/1632).[120] The latter, who achieved fame as the author of

[116] See ibid., 186.

[117] See ʿAbd al-Raḥmān b. Muḥammad al-Fāsī, *Jawāb ʿan ḥukm ṭāba*, 219. As Batran noted, al-Fāsī's treatise was in response to that of his nephew, Muḥammad al-ʿArabī al-Fāsī (d. 1052/1642) al-Fāsī (Batran, *Tobacco Smoking*, 46).

[118] I have found little information on his scholar. Al-Kattānī gives only his death date and that he was one of the students of al-ʿĀrif al-Fāsī (*Salwat al-anfās*, vol. 3, 399).

[119] On al-Ghassānī's life, see al-Kattānī, *Salwat al-anfās*, vol. 2, 116–17, and for his having studied with Aḥmad Bābā, see Zouber, *Aḥmad Bābā*, 67.

[120] On the life of al-Maqqarī see Sabahat Adil, "Memorializing al-Maqqarī: The Life, Work, And Worlds of a Muslim Scholar." Adil addresses this phase of al-Maqqarī's life in ibid., 112–16.

the *Nafḥ al-ṭīb*, the unequaled anthology of history and literature of al-Andalus and a study of the life and works of the fourteenth-century Granadan vizier Ibn al-Khaṭīb, left his position and Fez soon after these events and traveled to the East, where he settled in Cairo (he figures in the epigraph for Chapter 2). Like al-Ghassānī, al-Maqqarī had studied with Aḥmad Bābā, and like him, he fiercely opposed the position taken by his teacher on the subject of smoking.

The legal opinion is a compilation, with each scholar adding his own opinion regarding the prohibited nature of tobacco, although after al-Ghassānī and al-Maqqarī, few of the other seventeen scholars do more than add their assent. Al-Ghassānī begins the first opinion with a laundry list of the evils caused by tobacco, which has weakened Muslims and helped the nonbelievers (*al-mulḥidīn*): smoking corrupts the heart, and extinguishes its light; it drives away the beneficial angels who stand over man's nose and mouth, as it does those angels who dwell in mosques; it leads to financial loss and the corruption of the character of the one who uses it; those who do it are similar to the people of the fire; smoking leads to the use of other forbidden things *(munkarāt)*, and is an excuse to gather with people who commit horrendous acts (*ahl al-fawāḥish*), where they ingest it along with hashish, opium, wine, and other intoxicants; and it produces lassitude (*mufattir*) in the bodies of those who ingest it, with the Prophet having said that every such substance was forbidden. He summed up his opening statement, saying that anyone who had the least understanding of and experience with the principles of jurisprudence (*man la-hu fī uṣūl al-sharī'a adnā dirāya wa khibra*) would see fit to forbid it.[121] While most of al-Ghassānī's list of objections to smoking were framed as ethical, he also stressed the Prophetic tradition prohibiting substances that produced lassitude, a point that al-Maqqarī made as well.[122] Al-Ghassānī proceeded to reject tobacco's possible medicinal properties, arguing both that such claims were nonsense and that it was evident that it blackened the heart and made the mouth stink. He finished his invective with an ad hominem, stating that the low intelligence and poor character of those who smoke is confirmed by tobacco being used by lesbians (*al-saḥḥāqāt min al-nisā'*) and men who have sex with boys (*al-fi'l bi-l-murdān min al-rijāl*), and by its being used publicly.[123] The subsequent scholars add little

[121] Al-Ghassānī, *Jawāb 'ulamā' Fās 'an al-'ushba al-khabītha*, 76r.

[122] Ibid., 80v. This Prophetic Tradition is also a central argument in the twentieth century work of Muḥammad Ja'far al-Kattānī (see *Ḥukm al-tadkhīn 'inda al-a'imma al-arba' wa ghayri-him*, 74–79).

[123] Al-Ghassānī, *Jawāb 'ulamā' Fās*, 79v.

to the substance of al-Ghassānī's argument, although their tone is less harsh, and is then followed by al-Ghassānī's judgment (*ḥukm*): tobacco is absolutely forbidden under all conditions (*bi-l-taḥrīm al-muṭlaq alladhī lā rukhṣa fīhi*).[124]

It is possible that the legal opinions of those jurists following al-Ghassānī and al-Maqqarī were considerably shortened for the purposes of this collective document, as the opinion of ʿAbd al-Raḥmān b. Muḥammad al-Fāsī (d. 1036/1626), third of the nineteen, is extant separately in greater length. Unlike al-Ghassānī who assumed the validity of the tradition in which the Prophet had prohibited both languor-inducing and intoxicating substances without adducing empirical evidence, ʿAbd al-Raḥmān al-Fāsī begins his *fatwā* by claiming that smoking shares with wine, hashish, and opium the characteristic of confusing the mind (*tashwīsh al-ʿaql)* thereby implicitly questioning Aḥmad Bābā and Ibn Abī Maḥallī's claim that the languor produced by smoking did not lead to a clouding of the mental facilities. Citing experience (*tajriba*), 'Abd al-Raḥmān argues that smoking brings about a state akin to the intoxication produced by drinking a small amount of wine, and that like wine, it is addictive (*yushhad li-dhālik ayḍ^{an} al-idmān la-hā*). Against the claim that tobacco can be medicinal, he observes that no medicine produces the type of desire (*al-wulūʿ*) that tobacco does.[125] The claim that tobacco possesses medicinal benefits seems to require additional refutations, for ʿAbd al-Raḥmān proceeds not only to quote the Companion Ibn Masʿūd as to God not placing remedies in forbidden substances, but also to quote "one of the most excellent and skilled doctors" as to tobacco having no medicinal properties.[126] A certain Tunisian had written a poem as to it inducing languor (*mufattir*), and it was certain that such substances, like intoxicants, were forbidden. Here ʿAbd al-Raḥmān pauses to address the issue of a smoker who denies feeling any languor when smoking:

> If you say that we see one who consumes and does not feel languor (*la yashʿur bi taftīr-hā*) or you don't acknowledge this, then we say: the lack of feeling it does not

[124] Ibid., 88r. The ruling is followed by an anecdote reminiscent of the one given by al-Laqānī: there was a Muslim who was imprisoned in the land of the enemy. God freed him and brought him out of this land to the lands of the Muslims, whom he told of what he had seen: he had been the slave of an unbeliever and had grown a great deal of tobacco. In doing so, he had been responsible for collecting the urine of humans and beasts and using it as a base to grow tobacco, for it was the custom of the unbelievers to grow tobacco in their country and then to sell it to Muslims (ibid., 88v).

[125] ʿAbd al-Raḥmān al-Fāsī, *Jawāb ʿan ḥukm ṭāba*, 218.

[126] Ibid., 219.

refute this, due to what its nature has imbued it with, as we have indicated. Many of those who have experience of it or of wine have acknowledged and witnessed that it induces languor (*iftār-hā*).[127]

Following a similar logic to Ibn Abī Maḥallī above, albeit to opposite effect, ʿAbd al-Raḥmān argues that even if someone were to smoke and not experience a feeling of languor, this was like someone who drank alcohol without growing drunk: it did not change the fact that the substance was forbidden in and of itself. It is only at this point, after having established the effects of tobacco upon the body that ʿAbd al-Raḥmān turns to citing scriptural evidence as well as the opinions of authoritative scholars such as al-Suyūṭī and Ibn Ḥajar that all intoxicants are forbidden.[128] Ibn Ḥajar, for his part, had stressed in *Fatḥ al-Bārī* – his commentary on Bukhārī's hadith collection – that the reason for the prohibition of wine lay in its impairment of the mind – man's tool of distinguishing and perception that he required in order to fulfill his responsibilities to God.[129]

The issue remained unsettled – al-Ghassānī's efforts to use a collective legal opinion to persuade his countrymen does not seem to have been effective, and later efforts such as that of the prominent scholar Abū Zayd ʿAbd al-Raḥmān al-Tamanārtī, whose *fahrasa* we encountered in Chapter 2, to refute Aḥmad Bābā and condemn the practice of smoking, added little to the debate beyond the author's authority.[130]

A different approach to the question of smoking was taken by Abū ʿAbdallāh Muḥammad al-ʿArabī b. Yūsuf al-Fāsī (d. 1052/1641–42) in his *Sahm al-iṣāba fī ḥukm al-ṭāba*, nephew of the aforementioned ʿAbd al-Raḥmān b. Muḥammad al-Fāsī. Arguing that, overall, the issue is inconclusive and therefore smoking is permitted, Abū ʿAbdallāh al-Fāsī does not mention the tradition regarding every *muskir* and *mufattir* being forbidden at all, observing that no harm is witnessed coming from smoking:

What is being asked about has benefits, and in it is a cure for many illnesses, as has been witnessed. A large number of people smoke it and are not harmed. If they would be harmed, then most of the people would die or suffer harm. The default for it is permission, and we should stay with that basis (*al-baqāʾ maʿa al-aṣl*).[131]

[127] Ibid. [128] Ibid. [129] Quoted in al-Kattānī, *Hukm al-tadkhīn*, 90.

[130] al-Tamanārtī's *fatwā* on the subject, included in his *fahrasa*, is extensive, but discusses the subject largely on formal legal and moral grounds (al-Tamanārtī, *Fawāʾid al-jamma*, 476–502), although he did note that one reason for prohibiting tobacco was that it harmed the health and that he had witnessed it causing intoxication (ibid., 481).

[131] Ibn Yusūf al-Fāsī, *Sahm al-iṣāba*, 210.

In what may be a reference to al-Laqānī's description of Europeans dissecting a dead smoker, Abū ʿAbdallāh al-Fāsī argues in part that simply because smoking may have a negative effect on bad people it should not be prohibited:

As for what is mentioned about inhaling smoke (*siff al-dukhān*) burning the chests of the people of the left (*ahl al-shimāl*), there is no proof in this, and if it were true it would also not indicate prohibiting it, for what is a sign of a bad people does not indicate that it should be prohibited.[132]

Yet the broader issue is, again, that there was plenty of evidence that tobacco did not harm the health and that therefore, there was no reason to ban it:

I answer that if it were harmful then death or sicknesses would spread among the people to the degree that smoking has spread among them. Nothing of this has happened, so it isn't harmful. Yes, if it were agreed that there was an association of sickness with consuming it in some people, and a judgment of experience or if the opinion of a doctor were that it [damage] was from it, then it would be prohibited in that person's case, and the same would be the case if medicine made a judgment of this regarding its consumption. That is a general ruling regarding all things that are consumed, and does not single out that which has been asked about. The scholars have said that no plant is forbidden save that which ends life such as poison, or which puts an end to one's reason, such as wine, or through which harm occurs. To the point that they said that if bread, for example, would cause harm at a given time, it would be forbidden in the case of it being harmful, and so on. Similar to this opinion is found in al-Ghazālī's *Iḥyāʾ*, and the jurist here judges according to the testimony of the doctor. Save for wine, which is always forbidden in all cases.[133]

Abū ʿAbdallāh al-Fāsī disagrees with opponents of smoking like his uncle ʿAbd al-Raḥmān al-Fāsī not so much in the relevant scriptural texts – they both believe that inebriants are prohibited, although Abū ʿAbdallāh does strikingly omit the Prophetic tradition prohibiting languor-inducing substances as well – but on the effects of smoking on the body, both in terms of its potential harmfulness but also in the crucial and vexing question of how to evaluate the subjective experience of smoking. Whereas Aḥmad Bābā and Ibn Abī Maḥallī had argued strenuously against the feeling of languor being prohibited, and ʿAbd al-Raḥmān al-Fāsī had held that smoking was tantamount to drinking alcohol, Abū ʿAbdallāh addressed the effects of smoking on the mind as follows:

What happens rarely to some of those who smoke is not a confusion of the mind in any way, but is akin to one who inhales fire (*nāfikh al-nār*), and is similar to being dizzy

[132] Ibid. The reference to the people of the left is to the Qur'an 56 (*Sūrat al-Wāqiʿa*): 43–47.
[133] Ibid., 212.

while the mind remains sound, not influenced in any way. That is what a smoker claims, and whoever claims something else, let him prove it and clarify the type of confusion. Is it inebriation, being under the effect of narcotics, anesthesia (*irqād*), or loss of reason (*tajnīn*)? Shihāb al-Dīn al-Qarāfī has stated in his *Qawā'id*, and Shaykh Khalīl in his *Tawḍīḥ*, and the preacher Ibn Juzayy in his *Qawānīn* – all three of them defined it as inebriation.[134]

The effects of smoking are thus in no way analogous to drinking, and this observation is based on what could be loosely referred to as a thought experiment that suggests that inhaling smoke from a fire produced dizziness.[135] This empirical claim of the state of the body is then bolstered with a citation of three important Mālikī jurists.

What is evident from even this brief and cursory reading of texts from the Great Tobacco Debate is that disagreement when it came to the effects of tobacco on the body existed regarding both the nature of tobacco's influence on the body and on which categories of influence were prohibited. Yet, the importance of the embodied experience of smoking tobacco and of the body itself was not disputed, and was in fact the basis for much of the discussion.

I have attempted here not to so much offer a comprehensive overview of Muslim responses to the introduction of smoking as to draw attention to how important the body and its various states of consciousness were to these responses. The prominence of the body in these debates, where it was admittedly but one of the major themes addressed, and the way both it and the state of the human mind were implicated in arguments both for and against smoking speak to the importance of medicine in the articulation of human understanding of God's revelation. As I have had occasion to explore in my work on Muslim responses to epidemic disease, it is important that we not read the different attitudes of, for example, Ibn Abī Maḥallī and ʿAbd al-Raḥmān al-Fāsī, toward the hadith from the Prophet that prohibited languor-inducing substances as one of religious traditionalism verses rational empiricism.[136] All participants in this debate availed themselves of scriptural, legal, and empirical evidence to prove their respective points. Their interpretations of these types of evidence differed and were

[134] Ibid., 214. For the reference to Ibn Juzayy (d. 785/1383), see Ibn Juzayy, *al-Qawānīn al-aḥkām al-sharʿiyyah*, 170, where he quotes al-Qarāfī. I have not been able to consult Khalīl's *Tawḍīḥ*, but an early twentieth commentary on his famous *Mukhtaṣar* defines inebriation in language taken wholly from al-Qarāfī. See al-Azharī (d. 1335/1916–17), *Jawāhir al-iklīl*, vol. 2, 295.

[135] For a brief discussion of "thought experiments" in the presentation of how nature works in the writings of seventeenth century European natural philosophers, see Steven Shapin, *The Scientific Revolution*, 84.

[136] See Stearns, "Contagion in Theology and Law."

doubtlessly affected, as jurists and moralists continue to be, by their own personal proclivities, experiences, and scholarly networks. What I have suggested here instead is that knowledge of the natural sciences and the world of natural phenomena more generally played an important role in these debates and while some of the participants had specialized knowledge of these sciences, far more of them drew on general understandings of these sciences' contents and significance. These general understandings speak to the degree that the natural sciences had spilled beyond their discipline specific conversations and debates into the broader intellectual and social debates that preoccupied Moroccan society.

3.11 CONCLUSION

The scholarship of the past few decades has demonstrated that the Islamic legal tradition was characterized by both stability and flexibility after the formation of law schools in the fifth/eleventh century.[137] This flexibility is well illustrated in the legal texts examined here, with individual jurists drawing creatively on both legal and scientific precedents in order to craft authoritative opinions that would reflect both the jurists' interpretation of the intention of the scriptural sources as well as the exigencies of the Muslim community. Yet, while the old stereotypes of Islamic law's static or arbitrary nature can now be safely discarded, all too often scholars continue to treat developments in legal discourses in isolation, not considering possible relationships of mutual influence with developments in, for example, theology or Sufism, or the natural sciences, for that matter.

The types of legal materials examined here that circulated in the Muslim West during the post-formative period speak to the presence in Islamic jurisprudence of the natural sciences and their authority concerning a narrow but significant group of topics. They also offer us the chance to move past previous theories that had viewed the natural sciences in the Muslim world as having entered into decline either because their study had been appropriated by religious scholars for the purposes of ritual law, or because their epistemological authority was relegated by most scholars to a different sphere and with there being no attempt to reconcile their claims with those of the religious sciences.[138] While there is truth to both of these

137 See the sources on the *ijtihād/taqlīd* debate cited at the beginning of this chapter in Section 3.1.

138 I'm referring to A. I. Sabra's appropriation thesis and Ahmed Dallal's compartmentalization thesis (Dallal, *Islam, Science, and the Challenge of History*).

explanations concerning how Muslim scholars writing in genres outside of the natural sciences themselves framed them – some of the legal opinions discussed here offer support to the appropriation thesis – both theories are limited by their implicit acceptance of the teleology underlying the narrative of the rise of modern science. Both theories, are, in other words, predicated on the notion of a decline that they attempt to explain. This chapter, and *Revealed Sciences* as a whole, has aimed to show the extent and nature of the presence of the natural sciences in Muslim intellectual life as a whole without assuming that these sciences were declining, or that their diffusion and influence on contemporary social issues was a symptom of a lessening of their development. Exploring this last point in greater depth, however, requires turning to texts written in the natural sciences themselves during the long seventeenth century, the subject of Chapter 4.

EXCURSUS

Kuhn and the History of Science in Islamicate Societies

The taproot and trunk of the tree of the astral sciences are buried in the Mesopotamian desert, with subsidiary roots in Egypt and China (I have lopped the Mayas off this arboreal image, as they are self-rooted). From Babylonia the tree branched out to Egypt, to Greece, to Syria, to Iran, to India, and to China; grafted onto different cultural stocks in each of these civilizations, it developed variant leaves, shoots, and flowers. The process of the intertwining of these diverse varieties of astronomies throughout Eurasia and North Africa was amazingly complex, as ideas, mathematical models, parameters, and instruments circulated rapidly over the vast expanse of divergent traditions. Out of this process modern Western astronomy sprang from a rather late branch that grew from and was fed by an incredibly complicated undergrowth. For very complex reasons this modern Western astronomy has choked off all of its rivals and destroyed the intellectual diversity that mankind enjoyed before it moved from simple communication to Western domination. We cannot know what the Islamic, Indian, or Chinese astral sciences might have become had this not happened, except that they would not have become what our culture has produced. But unraveling the intertwined webbing of these sciences is a fascinating and a rewarding task for a historian, and one in which much remains to be done. I strongly recommend to those of you who have the opportunity thus to broaden your perspectives to grasp it.

David Pingree, *"Hellenophilia versus the History of Science"*

Imagine an evolutionary tree representing the development of the modern scientific specialties from their common origins in, say, primitive natural philosophy and the crafts. A line drawn up that tree, never doubling back, from the trunk to the tip of the same branch would trace a succession of theories related by descent. Considering any two such theories, chosen from points not too near their origin, it should be easy to design a list of criteria that would enable an uncommitted observer to distinguish the earlier from the more recent theory time after time.

Among the most useful would be: accuracy of prediction, particularly of quantitative prediction; the balance between esoteric and everyday subject matter; and the number of different problems solved. Less useful for this purpose, though also important determinants of scientific life, would be such values as simplicity, scope, and compatibility with other specialties. Those lists are not yet the ones required, but I have no doubt that they can be completed. If they can, then scientific development is, like biological, a unidirectional and irreversible process. Later scientific theories are better than earlier ones for solving puzzles in the often quite different environments to which they are applied. This is not a relativist's position, and it displays the sense in which I am a convinced believer in scientific process.

Compared with the notion of progress most prevalent among both philosophers of science and laymen, however, this position lacks an essential element. A scientific theory is usually felt to be better than its predecessors not only in the sense that it is a better instrument for discovering and solving puzzles but also because it is somehow a better representation of what nature is really like. One often hears that successive theories grow ever closer to, or approximate more and more closely to, the truth. Apparently, generalizations like that refer not to the puzzle-solutions and the concrete predictions derived from a theory but rather to its ontology, to the match, that is, between the entities with which the theory populates nature and what is "really there."

Perhaps there is some other way of salvaging the notion of "truth" for application to whole theories, but this one will not do. There is, I think, no theory-independent way to reconstruct phrases like "really there"; the notion of a match between the ontology of a theory and its "real" counterpart in nature now seems to me illusive in principle. Besides, as a historian, I am impressed with the implausibility of the view.

Thomas Kuhn, *Postscript to* The Structure

The metaphor of the tree, used by both Pingree and Kuhn, is an attractive one when applied to the subject at hand. By conceiving of seventeenth-century Moroccan scholars of the natural sciences as yet another branch of the tree of knowledge growing from a common human desire for learning, one glimpses a vision of an alternate world of scholarship, one with its own validity that disappeared following the expansion of one strand of European scientific thought. Linking such a vision with Kuhn, however, involves its own violence, as the above quote clarifies, for while Kuhn is often linked to the much-debated notion of incommensurability, he did believe in the progressive ability of scientific theories to provide greater insights into the natural worlds (to solve more puzzles). He is therefore an unlikely thinker to advance to argue for the prevalence of the natural sciences in the post-formative Islamicate world – in part because, arguably, he would not have recognized many of the sciences discussed in this book as such. Yet a closer consideration of the implications of Kuhn's broader project provides suggestions for how to take up Pingree's call for a history of those branches of human inquiry that flourished before what he called Western domination.

One of the chief explanatory attractions of Kuhn's choice to characterize scientific communities as characterized by paradigms that define the questions that they collectively pursue was its neat structure. Within these paradigms, he described scholars slowly and steadily making progress, a process he characterized both as puzzle-solving and normal science. For Kuhn, consistent with many historians of an earlier generation, the quintessential case of normal science and its contrasting category of revolutionary science was astronomy and the Copernican Revolution and his evocative description of how the writings of Copernicus, Kepler, Brahe, Galileo, and subsequently Newton put an end to the Ptolemaic understanding of the heavens has been profoundly satisfying for many generations of students of the history of science. At the heart of this description lay the moment when a believer in one paradigm – in this case the Ptolemaic understanding of the heavens revolving around the earth – having grown frustrated with the increasing number of observable anomalies in this paradigm, changes to adopting a radically different way of interpreting the evidence at hand. The moon becomes a satellite, not a planet. The earth revolves around the sun, and the universe becomes unbounded. The paradigm shifts and a new set of questions emerge: a new research agenda leading to a distinct normal science. Kuhn emphasized that there was something inexplicable about the moment that the observer stops trying to fit new evidence into an old paradigm and instead rejects it for a new explanatory model. Critically, it was not an obvious or inherently rational decision as much as a moment of instinctive insight and conviction that subsequently drove the collection of new evidence that it was able to more satisfactorily explain according to the new paradigm's criteria. The difficulty to explain how and why a paradigm shift takes places was associated with each paradigm being only internally coherent and thus incommensurable with other paradigms, meaning that one could not understand the world from within two paradigms simultaneously, nor could one contemplate reality from outside a paradigm.

In the over half a century since Kuhn wrote *Structure*, the work has been much debated, criticized, and defended. I am not so much interested here in reviewing the methodological coherence of terms like paradigm and incommensurability as I am in returning to the broader implication of Kuhn's work that science, like the evolutionary process itself, has no telos, no truth toward which it proceeds, but that it advances through long periods of puzzle-solving and dramatic revolutionary ruptures that set up a new set of research questions and refashion the observer's world. Kuhn paired the absence of an absolute progress in science with a keen awareness of the importance of institutions and social factors in perpetuating and reinforcing the veracity of a given paradigm. This contingency of our understanding of nature and its dependence on social reproduction has been appealing to historians attuned to the importance of political and cultural contexts of the study and transmission of the natural sciences, and while Kuhn was certainly not the first to observe it, it acquired additional force through association with his discussions of scientific communities being structured according to adherence to paradigms. Awareness of this contingency is additionally useful when it comes to questioning the master narratives that have characterized our historical narratives of both science and the Islamic world and which emerged with force in the nineteenth century: the link of Europe and its colonies to modernity, progress, and enlightenment and their putative absences in the rest of the world. As sketched out in the Introduction, the nineteenth century was also when debates among both Middle Eastern scholars and Orientalists produced a description and chronology of Muslim intellectual history that has maintained salience until today.

Historians have subjected these very narratives to an incisive critique in the past decades, with the problematic nature of modernity attracting arguably the greatest number of pages (and the most vitriol). The boundaries between categories such as medieval and modern, for example, have productively been shown to be constructed around specific social and political processes linked to nineteenth-century European colonial expansion, and the uniform nature of time itself as a measure of development and change has been shown to be insufficient to explain the differentiated social and intellectual processes taking place in Europe over the past five centuries. Further, the very opposition of human society with a nature that can be described, controlled, and exploited has been revealed to be culturally and historically contingent and that, as has been noted by proponents of the ontological turn, there is no one defining way to experience the world. These interventions and others – many others – have been launched from within the Western academy as a critique of the explanatory power of the concept of Western modernity, and they also have implications for how we understand the intellectual pursuits of communities of Moroccan scholars in the seventeenth century. Here we return to Kuhn, albeit somewhat loosely (which may indeed by the most useful way to read him).

During the long seventeenth century, Moroccan scholars pursued extensively the study of the natural sciences, and the way that they thought about these sciences permeated their thinking about Islamic law and the workings of the natural world in general. The study of the natural sciences took place in urban and rural institutions as well as privately and points to a broad social acceptance and interest in their study and transmission. In Kuhnian terms, the paradigms that circumscribed their study of medicine, astronomy, astrology, and alchemy remained stable: these scholars pursued what Kuhn would have called normal science. To pick up the metaphor introduced above, any new branches that appeared on the tree of our understanding of the natural world did not move far from where they left their stems. Yet, what Kuhn's analysis lends us is a greater appreciation of the intellectual legitimacy of their collective enterprise within human history. These Moroccan scholars did not experience paradigm shifts in the bodies of knowledge that they pursued that led to their endeavors being legible to us today as modern science, but the epistemological humility that comes with Kuhn's claim that there is no extra- or trans-paradigmatic truth, allows us to preserve them from representing a decline narrative without forcing them into a competing narrative of scientific progress. Intellectual legitimacy alone could be considered a small thing, and yet it shifts the meaning of the story this book tells. It goes from being one characterized by paternalism – "Commendable that Muslims continued to study the natural sciences even if these were largely pseudo-sciences" – to one where the ontology (and not just epistemology) of these scholars is respected as paradigmatically coherent. The chief difficulty we have in accepting this shift may lie in the paradigms these scholars worked within being so similar to the ones that proponents of modern science in Europe argued that they had superseded: Galenism, Ptolemaic astronomy, or the alchemical belief in a common material substrate of all matter. The value of emphasizing this legitimacy, however, lies not only in practicing the kind of historical humanism advocated by Marshall Hodgson – laudable and difficult though this is – but more profoundly in reflecting on the social and historical contingency of our own ontologies.

4

Writing the Mathematical and Natural Sciences

Know that God, may His name be exalted through the subtlety of His wisdom and the brilliance of His handiwork, created the mind and made the sciences and bodies of knowledge its sustenance. It was easy for Him to arrange this from what was in existence. The Exalted said: "Surely in the creation of the heavens and earth and in the alternation of night and day there are signs for men possessed of minds."[1] He also said: "In the earth and in yourselves there are signs for those having sure faith."[2] There is nothing that comes into existence on heaven or earth or between the two without it being a possible source of sustenance for the mind through acquiring one or more sciences. The ability to do so differs according to minds' differing levels of intelligence and lethargy, and the differing gifts and insights from Exalted God. This acquisition of knowledge takes place with regard to the quantity, quality, and nature of all essences and their attributes. He whom Exalted God grants perspicacity benefits from these matters, while the sluggish considers them obscure.

This is what happened with the philosophers in the matter of establishing proof, in philosophy, in geometry, and in all kinds of crafts and professions, types of devices, various marvelous sayings and acts. Whomever Exalted God grants understanding and light through mystical experience possesses power and superiority to the point that a bird could not fly without benefitting from his flight, nor a door creak without benefitting from his creaking, nor a speaker speak without benefitting from his speech, to which the speaker would not be able to answer, nor would anything come to his mind. This state is well known among the gnostics, the lovers of God, and

[1] Q 3: 190.
[2] Q 51: 20–21. I have adapted the syntax of Arberry's translation to the quoted passage.

> the virtuous disciplines of the people of the path. Everything is God's to command.
>
> Al-Yūsī, *The Discourses*

> Know, O brother, that people differ on its essence and nature and what it is. Know that I exerted myself for a long time in the science of the art without discovering anything, that is, of the truth. Then, Exalted God opened its door for me, and taught me what kept it veiled (*mā inghalaqa ʿalayhā ḥijāb-hā*). I saw it then in truth and saw that what the scholars had said regarding it was true. I contemplated what was in it with the effort of my insight (*bi-l-naẓr jahdī*), exerting myself and my thought in the matter. Morning and evening I devoted my study to the books of the alchemists (*al-ḥukamā'*) and to experiments, and its wonders became evident to me.
>
> Al-Mirghitī, *On Alchemy*

4.1 PRELUDE: MATERIAL FACTORS

The vast majority of works in the natural sciences that were written in Morocco in the long seventeenth century remain in manuscript, scattered among libraries throughout the country and beyond in collections in the Middle East, Europe, and the United States. This simple fact has influenced our understanding of Moroccan contributions to these fields as much if not more than any of the historiographical considerations laid out in this book's Introduction. Those historiographical considerations, as with recent writing on the commentary traditions of the post-formative period, suggest that historians have largely unwittingly ignored a substantial body of scholarship written by Muslims in the Early Modern period because it was written in genres they held to be derivative.[3] Yet there is something of the chicken and the egg about this observation, for while historians have certainly paid less attention to the production of Muslim scholars writing on the natural sciences in the early modern period because they assumed there was little of interest there compared to earlier centuries, they also have simply not had access to materials that would have encouraged them to revisit their presuppositions. One recent publication that speaks to this claim is Kacem Aït Salah Semlali's groundbreaking if uneven history of alchemy in Morocco, which I will draw on below in this chapter: when laying out his argument of the pervasive study of alchemy in Morocco during the late Medieval and Early Modern periods, Semlali refers repeatedly to manuscripts he accessed

[3] For a general overview of the project of recording Morocco's manuscript heritage, see Aḥmed Chouqui al-Binebine, "Fahrasat al-makhṭūṭ al-ʿarabī: al-tajriba al-maghribiyya."

in private Moroccan collections, inaccessible when not unknown to scholars in Middle Eastern and Western academies.[4] What is true of alchemy also has bearing on astronomy, astrology, and medicine, the other natural sciences discussed in this chapter. This situation is changing as more and more manuscripts are digitized and made accessible online, but the slow pace at which these works are being edited (not to mention translated) will continue to impede the study of these fields. Much of the work that has been done, as is the case with that of Semlali, has been done by Moroccan scholars whose work has failed to attract sufficient attention in Western scholarship.[5]

It is worth stressing, if only in passing, the material factors that have contributed to this lack of awareness, in itself part of the larger phenomenon of there being a striking rift between Moroccan and Western academic work that is only confirmed by the admitted partial exception of Francophone scholarship: most academic works printed in Morocco enjoy only a small print run and are rarely reprinted; Western scholars often ignore (or cannot access) secondary scholarship published in Arabic (as opposed to French) by Moroccan scholars; Moroccan academics work under difficult financial constraints with publishing often not leading to the kind of professional advancement in does in Europe or America. While these factors are clearly similarly valid more broadly for the uneven relationship between Moroccan and Western academics – a relationship which at its most problematic could be characterized as neocolonial – for the premodern period they are additionally exacerbated by a rough division of labor where much of the cataloging and editing work that is necessary for works to be made available is done by Arab academics in Arabic in small print runs, while works of analysis and synthesis largely appear in European languages.

The factors I have listed here, part of what might be called clumsily the political economy of textual availability, are all especially salient in the case of the natural sciences and go some way toward explaining the lack of sophisticated scholarship on the natural sciences in Morocco in the Early Modern period. They also function to partially explain the limits of this chapter, and of how a detailed depiction of the landscape of scholarship on the natural sciences in Morocco during this period is currently beyond our

[4] Kacem Aït Salah Semlali, *Histoire de l'alchemie et des alchemists au Maroc.*

[5] Two of the most important Moroccan scholars who have produced numerous studies on Morocco's manuscript heritage and whose work has been widely cited in European languages are Muḥammad al-Mannūnī (d. 1999) – especially those studies gathered in the four volume *Qabs min ʿaṭāʾ al-makhṭūṭ al-maghribī* – and Aḥmed Chouqui al-Binebine, author of, among many other works, *Histoire des bibliothèques en Maroc.* For an evocative picture of the current politics of the hunt for manuscripts in southern Morocco, see Roman Simenel, "Le livre comme trésor."

reach. In its place, this chapter offers an impressionistic account of the texts available alongside a close reading of selected medical, astronomical, and alchemical texts.

4.2 INTRODUCTION

Who wrote texts in the mathematical and natural sciences in Morocco during the long seventeenth century, what did they write, and how many of their works are extant? While precise answers to these questions are not available to us for the reasons sketched out above, a survey of manuscript catalogs is a necessary first step to providing an overview considering how few seventeenth century works have been edited and printed in general. The libraries, both public and private, that have manuscript holdings from the period in question that are accessible, include the following:[6]

1. The Moroccan National Library (Rabat): approximately 34,000 titles.
2. The Hasaniyah (Royal) Library (Rabat): approximately 40,000 titles.
3. The ʿAllāl al-Fāsī Institute (Rabat): 2,403 titles.
4. The Ṣbīḥī Library (Salé): over 2,000 titles.
5. The Qarawiyyīn Library (Fez): approximately 5,600 titles.
6. The Ibn Sūda Library (Fez): 467 volumes.
7. The Library of the Great Mosque of Meknes: 663 titles.
8. The General Library and Archives (Tetouan): 3,500 titles.
9. The Dāwūdiyah Library (Tetouan): 752 titles.
10. The King ʿAbd al-ʿAzīz Āl Saʿūd Foundation (Casablanca): approximately 1,980 titles.
11. The Ibn Yusūf Library (Marrakesh): approximately 1,840 titles.
12. The Library of the Nasiriyah Zawiya (Tamgrout): 4777 titles.[7]
13. The ʿAbdallāh Gannūn Foundation (Tangier): 512 volumes.

[6] I have drawn here with gratitude on Jocelyn Hendrickson, "A Guide to Arabic Manuscript Libraries in Morocco, with Notes on Tunisia, Algeria, Egypt, and Spain," and Jocelyn Hendrickson and Sabahat Adil, "A Guide to Arabic Manuscript Libraries in Morocco: Further Developments." In the first of these articles Hendrickson gives a breakdown of the collections that compose holdings of the National and Royal libraries, including remarks on when manuscripts from other libraries were incorporated into them. For an overview of the history of the libraries of Morocco – a subject that overlaps with the history of Morocco's manuscripts – see, along with the cited works of Binebine, Latifa Benjelloun-Laroui, *Les bibliothèques au Maroc.*

[7] For the Nasiriyah Zawiyah, unlike for the other libraries where I have relied on the numbers given by Hendrickson (and Adil), see the new catalog of Ḥamīd Laḥmar, *Al-fihris al-waṣfī li-l makhṭūṭāt khizānāt al-zāwiya al-nāṣirīya.*

14. The Waqf Library of the Mosque of Mawlay ʿAbdallāh Sharīf (Wazzān): 651 titles.
15. The Waqf Library in Asfi: 419 titles.
16. The Waqf Library of the Islamic Institute (Tetouan): 183 titles.
17. The Waqf Library of Zarhūn: 169 titles.
18. The Waqf Library of the Islamic Institute (Salé): 149 titles.
19. The Library of the Ḥamziyya-ʿAyyāshiyya Zawiya (south of Midelt): 1,908 titles.[8]
20. The Waqf Library of the Mosque of Essaouira: 126 titles.
21. The Waqf Library of the Great Mosque of Tangier: 115 titles.
22. The Library of the Madrasa of the Qā'id al-ʿAyyadī (Kelaa Sraghna): 85 titles.
23. The Waqf Library of al-Qasr al-Kabir: 22 titles.
24. The Library of the Mausoleum of Sidi Awsidi (Taroudant): 15 titles.
25. The Library of the Regional Council of Scholars of Casablanca: 142 titles.
26. The Library of the Great Mosque of Taza: 840 titles.
27. The Library of Muhammad V University: 395 volumes.
28. The Library of Mohamed Sekkat, University Hassan II, Casablanca: approximately 550 manuscripts.

This list, while long, is far from comprehensive and does not include a range of smaller private libraries that fall outside of the control of the Ministry of Pious Endowments, educational institutions or private establishments, much less those Moroccan manuscripts held by foreign libraries.[9] In addition, the challenges of cataloging these collections, especially those of the National and Royal libraries, and the number of new titles that have been uncovered by newer cataloging efforts, speaks to the need for us to understand our knowledge of these libraries' holdings as tentative. Still, the list offers an impression of the sheer amount of intellectual production that is only partially accessible to researchers, and which, as scholars work through it in the coming decades will continue to improve our understanding of Morocco's intellectual history. It also explains some of the difficulty in establishing a clear overview of the

[8] Here I have relied on Ḥamīd Laḥmar, *Al-fihris al-waṣfī li-l makhṭūṭāt khizānāt al-zāwiya al-ḥamzīya al-ʿayyāshīya*.

[9] See for example Najīb, *Fihris ma lam yufahras min al-makhtutat al-ʿarabiyya fi al-khizanat al-khaṣṣa bi-l-mamlaka al-maghribiyya*, which covers the holdings of eleven private libraries in the Sūs. On the libraries of Taroudant specifically, see Abū al-Ṣawāb, "Dirāsa biblīūghrāfiyya li-baʿḍ al-makhṭūṭāt al-mawjūda bi-l-maktabāt al-ʿāmma wa-l-khāṣṣa bi-Tārūdānt." The most important single holding of Moroccan manuscripts outside of Morocco is quite possibly that of the Escorial. On the fascinating history of the Escorial's manuscripts, and how the personal library of Moulay Zaydān ended up in Spain, see Hershenzon, "Traveling Libraries."

number of manuscripts written in the sixteenth–eighteenth centuries in the natural sciences. Not only does one need to consult the various catalogs for the mentioned libraries, but often manuscript titles that clearly relate to the mathematical and natural sciences are anonymous or their author is unknown, making it unclear when they were written. With these caveats in mind, an initial survey, based on the library catalogs listed above, gives an impressionistic overview of works in these sciences written in Morocco between the sixteenth and eighteenth centuries, the vast majority of which remain in manuscript, demonstrating that roughly 120 extant works were written in Morocco on the mathematical and natural sciences during this period (see Appendix 2).[10] The majority of these texts dealt with matters related to astronomy and astrology (72), followed by medicine (27), with much smaller numbers of texts being devoted to lettrism (10), and alchemy (5) – and the nature of other works being difficult to ascertain from catalog entries alone. Broadly speaking, these texts speak to the larger themes explored in this book so far: a serious interest in the mathematical and natural sciences among Moroccan scholars that was reflected both in summaries and commentaries on earlier works as well as in original works. As discussed in the Introduction, the content of these works is unlikely to strike us as highly innovative, although it bears repeating that our current level of knowledge of scholarship for this period makes it difficult to chart developments in a nuanced fashion. Still, their scope and nature reveal a broad interest in and fascination with the natural world that we can productively take up with what Sonja Brentjes has called the "horizontal perspective" that is more common among Europeanist historians of science of the same period.[11] Thus, we will approach the texts in this chapter for what they reveal of the practice and role of the natural sciences in the society their authors lived in – bluntly: as a window on social history as much as a series of microstudies in intellectual history – and not as a search for that problematic beast, progress. This is not to reject the presence of difference and development across the intellectual traditions charted here so much as it is an aversion to placing the primary significance of these texts in their relation to a macro-intellectual history that has persistently downplayed the significance of the Early Modern

[10] A proper evaluation of this number would require a clear picture of how many manuscripts of Moroccan works overall from the sixteenth to eighteenth centuries are extant. Unfortunately, considering the partial information we have about the manuscript holdings of local and private libraries in Morocco, such an estimate is not feasible at the moment.

[11] See Sonja Brentjes, "The Mathematical Sciences in the Safavid Empire: Questions and Perspectives," 328–29.

period. It is difficult, admittedly, to reconstruct these texts' social significance much beyond their intellectual context, and here this chapter will build on previous ones and use individual texts to discuss specific sciences.

4.2.1 Teleological Categories: Science, Religion, and Magic

Historians of medicine who work on premodern Europe will be familiar with the danger of projecting modern categories, theories, and understandings back onto their premodern subjects and texts, and are sensitive to the need to carefully parse the diversity and richness in premodern sources. This is, in fact, often the central attraction of many scholars to the field. Whereas the history of science, including the history of medicine, largely began as a discipline that sought to describe the ways in which the efforts of past scholars contributed to a progressive development that led to modern understandings of science, the last few generations of scholarship on the European Middle Ages and Early Modern period has made great contributions to our understandings of premodern medicine as a set of fields worthy of study in their own right and not primarily as antecedents of forms of modern knowledge. The same cannot be said of the study of the natural sciences in the Muslim world. The bulk of the work over the last century on scientific production been focused on three moments: 1) the initial translation and appropriation of largely Greek and Syriac texts in the ninth–tenth centuries into Arabic; 2) then the dynamic development of the natural sciences in the Islamicate world, specifically with regard to Ptolemaic theory in astronomy, and Galenic humoral theory in medicine; and 3) the translation of Arabic texts into Latin in the Middle Ages and the Renaissance. When our field paid attention to the substantial body of texts in the natural sciences that were produced in the Muslim world after the end of translations into Latin and the European consciousness, it evaluated these materials almost entirely in terms of how they conformed to or departed from our understandings of what was proper or real science. To be sure, in the case of medicine, for example, the differentiation between Galenism and religious and occult texts also rests on an implied differentiation between an elite medical tradition represented by Graeco-Arabic medicine and a popular medical tradition represented by a genre called Prophetic medicine that came into its own in the fourteenth century and which included frequent references to Prophetic Tradition, the Qur'an, and which routinely included prayers as being efficacious against disease. When scholars of an earlier generation – upon whose formidable achievements in the realm of translating, editing, and commenting on manuscripts much of our current scholarship rests – read postclassical Islamicate medical texts through the lens of elite humoral medical versus popular prayers and magic, they distorted not only the texts but the social contexts that had produced them. In this, they reflected as a much a general modern bias against early modern understandings of nature as they did a particular understanding of medicine.[12] What is true of medicine is also valid for the other sciences, and with its readings of individual texts from the sciences of astronomy, medicine, and alchemy, this chapter charts ongoing developments in these fields in the post-translation period.

4.3 ASTRONOMY: KEEPING TIME AND FIXING PLACE

This chapter's central focus is on the natural sciences as defined by al-Yūsī in Chapter 2, yet it addresses astronomy as well, even though in al-Yūsī's enumeration and division of the sciences astronomy falls under the mathematical sciences. The principal reason for this inconsistency lies in the at times fluid boundary between astronomy, timekeeping, astrology, cosmology, and geography. This fluidity is reflected both in the taxonomy of the sciences of ʿAbd al-Raḥmān al-Fāsī also discussed in Chapter 2, where astronomical and astrological sciences are intermixed, and in the treatises of al-Mirghitī and al-Rūdānī on timekeeping discussed here. I have therefore included the discussion of astronomical and timekeeping texts while not discussing other mathematical materials.

It is clear in the appendix to this chapter (Appendix 2), but also in Chapters 2 and 3 from the accounts of scholars' studies, that astronomy and timekeeping were comparatively popular subjects of study in seventeenth-century Morocco. This was not a new development in the Muslim world. As discussed in passing in Chapter 3, David King and A. I. Sabra, among others, demonstrated some decades ago, that astronomical timekeeping, as opposed to cosmology or astrology, was especially popular in Muslim scholarship from the first centuries of the Islamic era onwards due to the important role it played in supporting the observation of ritual obligations such as prayer and fasting by providing precise times. This clear utility of such observations gave rise to the office of the timekeeper (*muwaqqit*), which first appeared in the Mashriq in the seventh/thirteenth century and spread by the

[12] Compare here with the astute comments in chapter 7 ("The Wider Worlds of Chymistry") of Lawrence Principe, *The Secrets of Alchemy.*

end of that century to al-Andalus.[13] In Morocco, the treatises on timekeeping (*mīqāt*) by the renowned mathematician Ibn al-Bannā' (d. 721/1321) and al-Jādirī (d. 839/1435) built on this earlier Andalusi tradition and al-Jādirī's treatise proved especially influential in the Moroccan scholarship under consideration here. The main subjects taken up in these two treatises, as summarized by Emilia Calvo, were: 1) the conversion between solar and lunar calendars; 2) spherical astronomy; 3) the use of shadows; 4) time reckoning; 5) topics in trigonometry; and 6) the azimuth of the direction of prayer.[14] The eleventh/seventeenth century astronomical treatises of al-Mirghitī and al-Rūdānī largely followed the framework established by these earlier works.

Of the two, al-Rūdānī achieved greater fame during his own lifetime and he has attracted more attention in recent decades.[15] This status is reflected in and to some extent the result of the glowing description al-Rūdānī was given by his contemporary ʿAbdallāh b. Muḥammad al-ʿAyyāshī (d. 1090/1679), author of a famous account of his second pilgrimage to Mecca, entitled *Water of the Meeting-Places (māʾ al-mawāʿid)*. Al-ʿAyyāshī was well connected among the Moroccan scholars and Sufis of his day. His father had been active at the Dilā' lodge and, under the direction of the head of the Dilā' lodge, had founded his own lodge in the High Atlas in 1044/1634–35.[16] Al-ʿAyyāshī himself had studied with the founder of the Nasiriyya order, Maḥammad b. Muḥammad b. Aḥmad b, Nāṣir al-Darʿī (d. 1085/1674–75) in Tamgrout, was friends with al-Yūsī, and had written on theological matters, including the controversy that broke out at Sijilmasah on the question of the belief of common folk.[17] Yet his most famous work was his account of his 1661–63 pilgrimage to Mecca, a work that contained numerous vignettes of the scholars whom he met traveling between Morocco and Mecca. Through the lives and careers of these scholars al-ʿAyyāshī's book offers a clear demonstration of the many and strong connections between scholarly networks in Morocco, North Africa, and the Hijaz, a connection of which al-ʿAyyāshī's own trip was a clear example. Al-Rūdānī was himself also an

[13] See the foundational article of David King, "On the Role of the Muezzin and the *Muwaqqit* in Medieval Islamic Societies," at 298–300, and Emilia Calvo, "Two Treatises on Mīqāt from the Maghrib (14th and 15th Centuries A.D.)."

[14] Calvo, "Two Treatises on Mīqāt," 163–64.

[15] On al-Rūdānī's life and work, see al-ʿAyyāshī, *Riḥlat al-ʿAyyāshī*, vol. 2, 43–61; El-Rouayheb, *Islamic Intellectual History*, 160–70; al-Maslūtī, *Muḥammad b. Sulaymān al-Rūdānī*; Mohammed Lakhdar, *La vie littéraire au Maroc sous la dynastie ʿalawide*, 81–88.

[16] Al-ʿAyyāshī, *Iqtifā' al-athar baʿd dhahāb ahl al-athar: fahras Abī Sālim al-ʿAyyāshī*, 21.

[17] See ibid., 116–18; al-Yūsī, *The Discourses*, 233–41; al-ʿAyyāshī, *al-Ḥukm bi-l-ʿadl wa-l-inṣāf al-rāfiʿ li-l-khilāf.*

excellent example of these connections, especially with regard to the natural sciences. Al-ʿAyyāshī met him in Medina for the first time, not having known him in Morocco, and gave an extended depiction of al-Rūdānī's life, character, and intellectual achievements. Born in Taroudant in the south of Morocco, al-Rūdānī studied first with Muḥammad b. Nāṣir al-Darʿī (d. 1052/1642), with whose son al-ʿAyyāshī had himself studied. He then traveled throughout the south, studying in Sijilmāsah, and then moving on to Marrakesh, Tadlā, and then Fez, studying with prominent Sufis in each of these places. Yet the networks that connected those scholars linked by their affiliation to Sufi brotherhoods and their focus on spiritual exercises overlapped with those that connected specialists in the natural sciences.

Al-Rūdānī came to Fez to study astronomy, astrology, mathematics, logic, and related sciences and became known for his talents in these fields. His studies in these fields were interrupted, however, when Muḥammad b. ʿAbdallāh Maʿn (d. 1062/1652), the head of the Makhfiyya lodge in Fez, told him to stop his studies of these sciences, to return to his parents and to listen to their counsel. Al-Rūdānī submitted to this advice, until his parents gave him permission to travel again and he went to Marrakesh, where he studied with various teachers including the very Muḥammad b. Saʿīd al-Mighrītī whose own treatise on timekeeping and astrology we will come to in Section 4.5, and whom we previously met in Chapter 2. Subsequently, he began his travels east through Algeria, Egypt, the Hijaz, and Istanbul, before settling in Medina, where he served for a time a mufti of the holy cities of Mecca and Medina. He never returned to Morocco, and died in Damascus.[18]

I have given this much detail on al-Rūdānī's early studies to demonstrate the extent of his early studies in Morocco and how they formed his subsequent career – lest it be thought that because he wrote his works in the East that he is unrepresentative of his homeland. Al-Rūdānī's connections to Morocco are further emphasized by al-ʿAyyāshī having received from him as a gift a fascinating astrolabe – al-ʿAyyāshī describes this object in detail as he does al-Rūdānī's skill in making astronomical devices, and by one of the few extant copies of al-Rūdānī's *Maqāsid al-ʿawālī bi-l-qalā'id al-laʾālī* having been preserved in the Hamziyya-ʿAyyashiyya lodge – founded by al-ʿAyyāshī's father – to which he donated a number of manuscripts following his return from his travels.[19]

[18] Al-ʿAyyāshī, *Riḥlat al-ʿAyyāshī*, vol. 2, 43–44.

[19] I am only aware of two copies of this manuscript: Hamziyya 1787 (pages 237–420) and that of the Hathi Trust, which is missing the section containing the last forty pages of the Hamziyya manuscript. The colophon at the end of the Hamziyya copy says that this copy

Striking in the remainder of al-ʿAyyāshī's account of al-Rūdānī is his depiction of a man of deep piety who lived largely withdrawn from society, making his living by making sandals and astronomical instruments, leaving his house only at night to do necessary errands and meet other Sufis in their homes. In his narrative, al-Rūdānī's piety is reflected in a gathering with a number of jurists in Istanbul, where he was offered both coffee and tobacco. Asked if he refused out of asceticism, he responded that he believed that these substances were forbidden, and proceeded to defeat his opponents in a debate on the question, although, unfortunately, considering the discussion of smoking in Chapter 3, al-Rūdānī does not mention why he believed tobacco was forbidden.[20] This account of his subject's piety blends seamlessly into al-ʿAyyāshī's laudatory account of al-Rūdānī as both an astronomer and a maker of astronomical instruments:

He was skilled in using astrolabes and other timekeeping devices such as quadrants (*al-arbāʿ*), dials (*al-dawāʾir*), *al-anṣāf*,[21] and clocks (*al-makānāt*). One of the most wondrous things I saw of his making was that, through effective art and fine reasoning he made it so you could hardly see the cracks in cracked glass vials. Of the most subtle of his creations, the most exquisite of what he fashioned, the greatest of his inventions, was a beneficial device that joined the sciences of timekeeping and astronomy. Nothing like it had been made, and its shape had no parallel – indeed he invented it with the brilliance of his thought and his extraordinary skill. It was a globe of round shape, smooth polish, covered with white, mixed with black paint. The one looking at it would consider it to be a golden egg[22] due to its brilliance, covered with circles and inscriptions. Another hollowed sphere was attached to it, split into two halves, on which were perforations and indentations for, among other things, the circles of the astrological houses. It was spherical like the one below it, polished and painted green, with a latch (*walm*[an]) revealing what was beneath it (*yabdū min allattī taḥta-hā*) – a beautiful appearance and

of the work was finished on the 18th of Dhu al-Qaʿdah, 1080/April 4, 1670 and a note on the side says that it was copied by an Aḥhmad b. Ḥasan al-Imlīṭī al-Mālikī al-Wafāʾī al-Azharī, in Mecca. If the date of the completion of the copy is accurate, al-ʿAyyāshī could not have brought this particular copy back from the trip in which he met al-Rūdānī and the work must have arrived at the lodge later.

[20] Al-ʿAyyāshī, *Riḥlat al-ʿAyyāshī*, vol. 2, 45–46. For his withdrawing from the general population in al-Madina, see ibid., 49–51, and living from his work, see ibid., 52–53. Al-ʿAyyāshī's other principal example of al-Rūdānī's demanding religious views is that the latter refused to wear wool bought from Christian merchants, which required some explanation on his part as the majority of the Egyptian scholars of his day had their clothes made from such wool (ibid., 46–49).

[21] I am unsure what this word refers to.

[22] Reading *ʿasjad* for *ʿajsad*.

striking significance. This was the device that rendered all others unnecessary when it came to timekeeping and astronomy. It was easy to use (*suhūlat al-mudrik*) due to the things on it being clear, the imagined circles related to astronomy, and the spaces between them visible (*mashāhid fī-hā*). It could be used on all countries, regardless of its longitude and latitude. In short, it can hardly be described, and only one who had witnessed it could know its power and characteristics.[23]

It was a copy of this sphere that al-Rūdānī would give to al-ʿAyyāshī as a gift and to which he devoted a treatise, *Quenching One's Thirst for the Universal Device* (*al-Nāqiʿah ʿalā al-ālah al-jāmiʿah*).[24] Al-ʿAyyāshī describes how he studied this treatise with al-Rūdānī with the spherical astrolabe itself in order to understand how to use it, and how as soon as people heard of the device, they clamored to get access to it, although al-Rūdānī would only sell it for a high price. Most strikingly – al-ʿAyyāshī expresses his wonder here as well – is that al-Rūdānī fashioned the astrolabe out of what appears to have been paper-mache:[25]

I was told that he took paper (*al-kāghid*), placed it into water until it dissolved and became like paste. He then placed gum Arabic in water until it dissolved and mixed it well with the dissolved paper. One then takes a sphere and endeavors to wrap it so that all parts are equally covered in relation to the center so that if you placed it on a smooth surface it would stop on a single point. I was told that it kept breaking for him until he took a nail and placed it in its center and then took half of a brass circle both sides of which were pierced, and placed both sides on the ends of the nail that extended beyond the sphere's two sides. He then began to go around the half of the mentioned circle with the paste (*al-ʿajīn al-mukawwar*) until there were no

[23] Al-ʿAyyāshī, *Riḥlat al-ʿAyyāshī*, vol. 2, 53. Compare with the translation of this passage in M. Lakhdar, *La vie littéraire au Maroc*, 83, and that in El-Rouayheb, *Islamic Intellectual History*, 163.

[24] This treatise, the title of which has several variants, has been edited and translated into French by Charles Pellat in "Muhammad al-Rudani: al-Naqiʿa ʿala al-ʿala al-nafiʿah," and "L'astrolabe sphérique," 83, 165. Pellat recounts in the Arabic introduction to his edition of the text how he based it on a manuscript from the Ḥamziyya-ʿAyyāshiyya zawiya that he had been given by Muhammad Lakhdar. In the French introduction to his translation of the text, Pellat notes that al-ʿAyyāshī's description of the sphere differs from al-Rūdānī's own description. See, nonetheless, Pellat's illustrations in his French translation of the two spheres that comprise the spherical astrolabe as a whole (Pellat, "L'astrolabe sphérique," 86, 87, 89).

[25] Al-ʿAyyāshī, *Riḥlat al-ʿAyyāshī*, vol. 2, 57. Compare with what seems to have been an actual example of such an astrolabe, ca. 10cm in diameter, that was signed by al-Rūdānī and auctioned off at Christies in 2015 for GBP 722,500 (www.christies.com/lotfinder/Lot/a-rare-and-important-spherical-astrolabe-signed-5930901-details.aspx). My gratitude to David King for this reference, who also noted to me that the auctioned astrolabe was universal, not spherical, and that the astrolabe described in al-Rūdānī's treatise was superior to it.

more protrusions, the indentations were filled and it became a perfect circle. He then painted its surface white, wrote on it what needed to be written and brushed over the writing with lacquer (*dahn al-kattān*). Because of this the writing would not be erased, even if it became wet from the hand's sweat or another source. The sphere above it is made in the same fashion, save that it is pierced while still wet.

After perfecting this technique, al-Rūdānī was able to produce such spheres relatively quickly, and was widely recognized as gifted in this art. Al-ʿAyyāshī stresses that, while skilled in all aspects of astrology, al-Rūdānī refused to predict future events, believing this to be religiously forbidden.[26] His independence in his thinking on astronomical matters was laid out in a poem on timekeeping, the title of which al-ʿAyyāshī does not mention, and in a gloss on this poem, in which he produced the "best astronomical table (*zīj*) of our time according to the leaders in this art." Al-Rūdānī drew on the solid observations made by his predecessors, such as those found in the tables of Ulugh Bey, and carried out his own observations in all areas needed to produce his table.[27] I have not found reference to extant copies of the poem itself – *The Necklace of Pearls for Establishing Days and Nights* (*Qalāʾid al-laʾālī fī ʿamal al-ayyām wa-l-layālī*) – but it is included in al-Rūdānī's own commentary on it, entitled *Higher Goals Regarding the Necklace of Pearls* (*Maqāsid al-ʿawālī bi-l-qalāʾid al-laʾālī*).[28]

Al-Rūdānī begins his extensive commentary – it runs to over 180 pages in the Ḥamziyya manuscript – with the trope of humility. The text is, he writes," a commentary on a poem that I cobbled together on the science of timekeeping (*al-tawqīt*), even though in both this science and others I am, as is said:

Like one who urges on camels while possessing none
And who grazes his flocks without having any."[29]

In the first lines of the quoted poem, he states that knowing the times is an important legal obligation, and then directly takes up the question of the Arab calendar. In the following he structures the commentary with the citing of one or two lines from the poem and then elaborating, sometimes at length, on the subjects raised in the line. In this first section, which deals with

[26] Al-ʿAyyāshī, *Riḥlat al-ʿAyyāshī*, vol. 2, 57–58.

[27] For a general overview of the genre of *zīj* and especially its fate in the Maghrib, see King and Samsó, "Zīdj," EI2. For a detailed overview of the *zīj* in Morocco from the seventh/thirteenth century onwards, see Samsó, "An Outline of the History of Maghribī Zijes from the End of the Thirteenth Century."

[28] Lakhdar notes that *Qalāʾid al-laʾālī* was printed in India, but I have not found a full citation. There is some doubt regarding my translation of *ʿawālī*.

[29] Al-Rūdānī, *Maqāsid al-ʿawālī*, 237. The line is famous, but I have not found it attributed to an author.

the nature of different calendars and the calculation of the beginnings of years and months, al-Rūdānī compares the observations of earlier astronomers, noting, for example, how Ibn Yūnus al-Ḥākimī al-Miṣrī – possibly Abū al-Ḥasan ʿAlī b. Abī Saʿīd ʿAbd al-Raḥmān b. Aḥmad b. Yūnus al-Ṣadafī (d. 399/1009), a famed astronomer who lived in Cairo – held that the first day of the *hijrī* calendar had been a Thursday, whereas the prominent Syrian timekeeper Ibn al-Shāṭir (d. 777/1375) wrote that it was a Friday.[30] Similarly, in the subsequent section "on the beginning of the year, its months and intercalations (*kabs-hā*)," he juxtaposes the understandings of intercalary years presented by a first group that included the celebrated Moroccan astronomer ʿAbd al-Raḥmān b. Abī Ghālib al-Jādirī (d. 839/1435) in his influential *Garden of Flowers Regarding the Science of Nights and Days* (*Rawḍat al-azhār fī ʿilm al-layl wa-l-nahār*), and Ibn al-Mawwās, who had written a commentary on al-Jādirī's work, with those of a second group, including al-Ṣūfī (d. 376/986) in his *Tables* (*Zīj*) and Abū ʿAlī al-Ḥasan Ṣadr al-Marrākushī (d. 660/1262) in his *Collection of the Principles and Goals Related to the Science of Timekeeping* (*Jāmiʿ al-mabādiʾ wa-l-ghayāt fi- ʿilm al-mīqāt*).[31]

In the lunar *hijrī* calendar, for the purposes of calculation of astronomical tables, Muslim astronomers separated the lunar months into months of thirty and twenty-nine days, adding an intercalary day to the month Dhū al-Ḥijja every few years – eleven times every thirty years – to preserve the lunar month (which is slightly longer than 29.5 days). It is worth remembering that these added days were not actually used in terms of determining the beginning of the month, which was done by sighting the new moon, but in order to establish past dates. A necessary fiction, in other words, for the regularization of the lunar calendar. Al-Rūdānī notes that the authorities have come to disagree on the details of this issue, with the first group described above holding that the extra day should only be added if the number of days that the month had fallen behind in a year was more than half a month – it was eleven days a year[32] – whereas the second argued that it should also be added when the missing number of days reached exactly half a month. After reviewing the details of the mathematics involved in the disagreement, al-Rūdānī provides his reader with a table based on the

[30] Al-Rūdānī, *Maqāsid al-ʿawālī*, 238. The debate on the nature the first day of the Hijrī calendar is already mentioned in al-Birūnī's *al-Āthār*. See F. C. De Blois, "Tāʾrīkh," *Encyclopedia of Islam* (2nd ed.). In the following I have drawn on both De Blois and van Dalen's sections in this extensive entry in the EI2.

[31] Al-Rūdānī, *Maqāsid al-ʿawālī*, 245. Ibn al-Mawwās seems to be Muḥammad b. Aḥmad b. Muḥammad b. ʿĪsā b. Aḥmad al-Mawāsī (d. 911/1505).

[32] I don't follow the reasoning here, as that would seem to be in relation to the solar calendar, but the intercalary days are there to keep the hijrī calendar tagged to the moon.

view of the first group mapping out over a 210 year cycle which years have this extra day or not – noting in passing that his calculations are at odds with those given in the *Zīj al-Sulṭānī* of Ulugh Beg (d. 850/1445).[33] As he ends this section, al-Rūdānī states that here and throughout the work he has drawn attention to places where he has disagreed with others – he seems to be thinking especially of al-Jādirī – on other issues one assumes.

As al-Rūdānī proceeds in his treatise to discuss the Roman solar calendar, he expands his sources to refer to Ptolemy, Euclid, Abū al-Rayḥān al-Bīrūnī's (d. 440/1048) primer on mathematics, astronomy, geometry, and astrology, *Instruction in the Foundations of Practicing Astrology* (*al-Tafhīm li-awā'il ṣinā'at al-tanjīm*), the Cordoban Ibn Abī al-Shukr (d. 680/1280)'s *Crown of Astronomical Tables* (*Tāj al-azyāj*) – the last an earlier example of a scholar from the Muslim West who also journeyed east, in Ibn Abī al-Shukr's case to the observatory at Marāgha where he studied with the renowned Naṣīr al-Dīn al-Ṭūṣī (d. 672/1274).[34] After his discussion of Roman, Coptic, and Persian calendars, he turns to a section on establishing the Ptolemaic chord (*al-watar*), the sine (*jayb*), and the versine (*saham*), including an extensive set of tables.[35] He then takes up the use and measurement of shadows, followed by the establishing of solar declinations (*al-mayl al-awwal*), regarding which he approvingly praises the values listed by Abū ʿAlī al-Marrākushī in his *Principles and Goals*.[36] After the discussion of declinations, he turns the question of establishing longitudes and latitudes, regarding which he gives a three-page chart for locations from al-Andalus to Central Asia, measuring the longitude from the Canary Islands (*al-jazā'ir al-khālidāt*) and the latitude from the equator.[37] Al-Rūdānī introduces this chart observing that the values for north and west Africa locations provided by eastern authorities such as Sulṭān (Beg) and al-Ṭūsī had substantial errors in them, and that one should rely on Maghrebi sources for these locations.[38] It is therefore of interest to compare the values given by al-Rūdānī for Moroccan locations with those

[33] Al-Rūdānī, *Maqāsid al-ʿawālī*, 246. Al-Rūdānī's awareness of Ulugh Beg's *zīj* is worth noting, as Samsó has argued that Moroccan scholars first became aware of it at the end of the seventeenth century ("An Outline of the History of Maghribī Zijes," 98–99). While I have not being able to date the composition of the *Maqāsid al-ʿawālī*, the Hamziyya copy was made on the 18th of Dhu al-Qaʿdah, 1080/April 4, 1670. It remains unclear, of course, when this copy arrived in Morocco.

[34] Al-Rūdānī, *Maqāsid al-ʿawālī*, 246. [35] Ibid., 271–87, tables on 281–87.

[36] Ibid., 304.

[37] Ibid., 325–27. Since the Canary Islands lie sixteen degrees West of Greenwich, this should be taken into account when comparing al-Rūdānī's coordinates to current estimates.

[38] Ibid., 324.

TABLE 4.1 Comparison of al-Rūdānī's coordinates with previous coordinates given by Eastern and Western scholars

		Long I	Lat I	Place
	ZAH (same)	21 00	31 30	Marrakesh
21 20	31 30 (MAR)			
24 00	33 00 (BAN)	24 00	33 00	Salé
	ATH, BIR, SNB (same)	05 30	22 00	Sūs al-Aqsa
31 40	28 00 (ZAY)	05 40	28 00	Ifrane
15 17	28 00 (MAR)			
24 10	35 10 (MAR)	24 00	35 00	Tangier
25 00	33 00 (BAN)	25 00	33 40	Fez
24 30	34 08 (BAN)	24 30	34 00	Maknes
	MAR (same)	17 00	29 00	Massa
	MAR (same)	22 31	32 35	Azzemour

summarized in E. S. and M. H. Kennedy's *Geographical Coordinates of Localities from Islam*, which draws on largely Medieval and largely Eastern Islamic sources.[39]

Looking at the coordinates of ten Moroccan towns, it is clear that for almost all of them al-Rūdānī has relied on the astronomical tables of two Moroccan predecessors, Aḥmad b. Muḥammad Ibn al-Bannāʾ's (d. 721/1321) *Minhāj al-ṭālib li-taʿdīl al-kawākib* (BAN) and the already mentioned *Collection of the Principles and Goals* of Abū ʿAlī al-Ḥasan Ṣadr al-Marrākushī (d. 660/1262) (MAR).[40]

The juxtaposition of al-Rūdānī's comment regarding eastern sources and his reliance on Western tables for the coordinates of Moroccan towns suggests that he had access to a range of tables by eastern and western sources when he wrote this text in Medina.

Following this table, in a section on how to establish the longitude of one's location, al-Rūdānī takes up Ptolemy's theory of the division of the world into four quarters, two south of the equator and two north, with only one of the northern quarters known to be inhabited.[41] The question of

[39] E. S. and M. H. Kennedy's *Geographical Coordinates of Localities from Islam*. When considering the image from the al-Rūdānī manuscript, the comments by Kennedy and Kennedy on conventions for writing coordinates in the *abjad* system are useful, two relevant examples being that the dots on the yā' and the jīm are customarily omitted (with the tail of the latter abbreviated to differentiate it from the ḥā') (ibid., x).

[40] See E. S. and M. H. Kennedy's *Geographical Coordinates*, xvii–xviii and xxv for an explanation of the acronoyms in the table that refer to eastern sources.

[41] Al-Rūdānī, *Maqāsid al-ʿawālī*, 331. This was not actually what Ptolemy says, although if one read the *Almagest* in haste one could have that impression. See Book II where Ptolemy

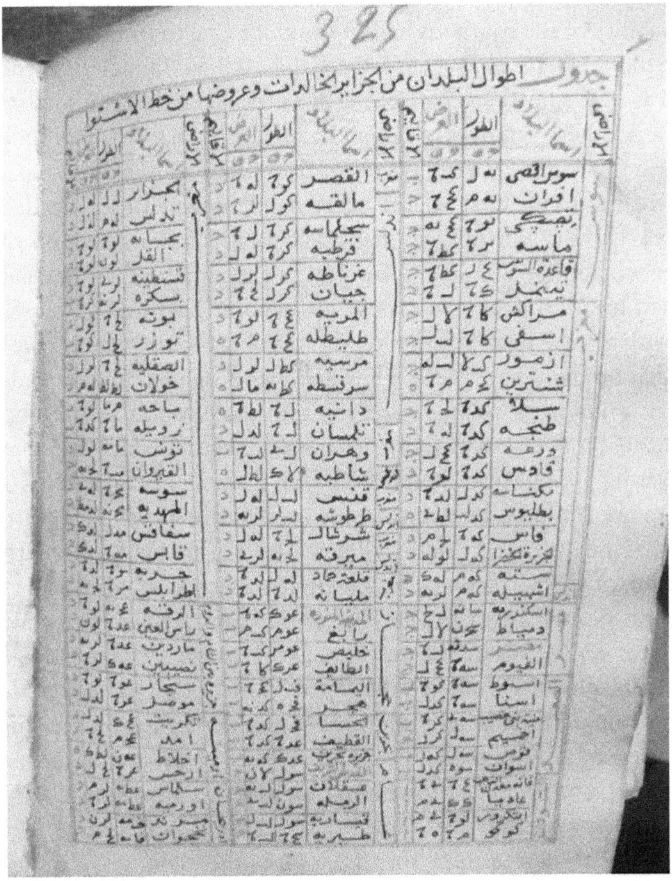
جدول اطوال البلدان من جزاير الخالدات وعروضها من خط الاستوا

FIGURE 4.1 Maqāṣid al-ʿawālī, Ḥamziyya 1787.

whether the southern quarters of the globe can be inhabited, it being hotter there and God having located larger bodies of water on that half of the

writes, "If one considers the earth to be divided into four quarters by the equator and a circle drawn through the poles of the equator, our part of the inhabited world is approximately bounded by one of the two northern quarters." (G. J. Toomer, *Ptolemy's Almagest*, 75). In his *Geography* Ptolemy was more explicit on this subject and noted that the inhabitable part of the world ran from 16 degrees south of the equator to 63 degrees north of it (Ptolemy, *Geography*, 160). In his popular seventh/thirteenth century textbook on Ptolemaic astronomy, al-Jaghmīnī juxtaposed Ptolemy's views from both the *Almagest* and the *Geography*, although he too seems to have understood Ptolemy to be saying in the *Almagest* that only one northern quarter of the world was inhabitable (Sally R. Ragep, *Jaghmīnī's Mulakhkhaṣ*, 148).

earth, has prompted some debate. Al-Rūdānī seems to be a firm believer in the uninhabitability of the southern half of the globe due to the heat there. He counters the view of al-Ṭūsī and Quṭb [al-Dīn] al-Shirāzī that the presence of seas in the northern half of the globe challenged the argument of water in the southern quarters of the globe by noting that seas had only been attested in the south-eastern quarter of the globe. Travelers who had traveled toward the south-west on the other hand, as when they headed toward the source of the Nile in Egypt had come to places 10 degrees north of the equator where they had seen snow-covered mountains, and beyond which they had been no ocean, just very hot regions.[42] Al-Rūdānī closes this section with a vehement critique of the astrologers (*al-aḥkāmiyyūn*) who believe that on the other side of a burning path – defined by the lands falling between a line that connects the sun and the earth at sunset and which runs through the house of Libra at nineteen degrees and a line that connects the setting moon with the earth through the house of Scorpio at 3 degrees – there are could be human settlements. This, he believes, is pure fantasy.[43] In the subsequent section on climes, we learn that in the seventh clime, far north of 10–50 degrees of longitude, where the majority of inhabitable lands lie, the Russians, Slavs, and Bulghars are forced to live in bath houses for six months of the year due to the intensity of the cold.[44]

In the following pages, al-Rūdānī takes up a number of subjects, including establishing the greatest height and lowest points reached by the sun and the planets, establishing the end of dusk and the beginning of dawn, knowing the times of the rising of the lunar mansions, converting terrestrial ascendants (*al-maṭāli'a al-baladiyya*) to celestial ascendants (*al-maṭāli'a al-falakiyya*), determining the direction of the qibla, establishing the four cardinal directions as well as the azimuth of a room with a skylight (*bādhāhinj*) with the help of a tripod (*kūnyā*); an unfinished table of the celestial coordinates of planets, and establishing the rising and setting of planets, among others.[45] He felt that his reader might not be familiar with what was meant by a *kūnyā*, and supplied illustrations to clarify two examples.

[42] Ibid., 332. [43] Ibid., 332–33.

[44] Ibid., 334, and compare with ibid., 335, where al-Rūdānī describes the inhabitants of the Thule islands as living in bath houses. A similar claim in remarkably similar language is found in al-Jaghmīnī as well, although it concerns there the unnamed inhabitants of an island at latitude 63 North (Ragep, *Jaghmīnī's Mulakhkhaṣ*, 152). For many geographers in the Ptolemaic tradition, "Furthest Thule" was an evocative allusion to the furthest north humans lived (though note that al-Jaghmīnī claims that humans resembling beasts lived as far as 66 degrees north (ibid.)).

[45] Ibid., 397–98.

FIGURE 4.2 Maqāsid al-ʿawālī, Ḥamziyya 1787.

Overall, al-Rūdānī compares the views of earlier scholars, and at times does not hesitate from noting where they erred, as he does notably regarding a medical matter: he notes that Democritus and Galen had erred in prescribing hot remedies to the inhabitants of warm climes; they should have instead of prescribed cold remedies. This is why the people of India counseled eating all types of roots (*al-afāwiya*) produced by the earth and ascribe their health to this, and the people of Azerbaijan are used to cow meat and green melons during winter.[46]

In his *Maqāsid al-ʿawālī bi-l-qalāʾid al-laʾālī* al-Rūdānī covered ground that was broadly speaking not that different from previous Moroccan authors of works on timekeeping. His engagement with the genre, however, was active and critical, showing the ongoing vitality of scholarship in this field. In this, his work was similar to that of his one-time teacher al-Mirghitī's *The Enjoyable Commentary on "What is Sufficient"* (*al-Mumtiʿ fī sharḥ al-muqniʿ*), another commentary on a poem on timekeeping, although this one includes greater discussion of astrology. We discussed al-Mirghitī's life in the context of his scholarly autobiography in Chapter 2. Unlike al-Rūdānī, whom he taught at one point in Marrakesh, al-Mirghitī never left Morocco, and as can be seen in Appendix 2, his numerous works are well represented in Moroccan manuscript collections today. *The Enjoyable Commentary*, composed in 1040/1630–31, is a gloss on al-Mirghitī's own poem entitled *What is Sufficient of the Knowledge of Abū Miqraʿ*, referring to a poem on timekeeping by the eighth/fourteenth century Moroccan scholar Abū ʿAbdallāh Muḥammad b. ʿAbd al-Ḥaqq b. ʿAlī al-Baṭṭīwī (fl. 730/1330), whose nickname – the one with the staff – al-Mirghitī explained as stemming from his never being separated from his

[46] Al-Rūdānī, *Maqāsid al-ʿawālī*, 332.

staff during his travels.[47] Al-Mirghitī used Abū Miqraʿ's poem as a basis for his own texts, but did not hesitate to observe that his own poem, which he composed in 1040/1630–31 was superior in content and organization, emphasizing that the reason he wrote it was that a companion of his had asked him to do so.[48] He begins it with a discussion of astrology:

Know that the people call this science astrology (*ʿilm al-tanjīm*). Its source is that a man practices it when he looks at the stars, so becoming an astrologer. For the most part this science is based on the paths of the stars, the sun, the moon, the constellations (*al-burūj*), and the planets (*al-darārī*). It is necessary for anyone who desires this science to gaze at the stars and to know them in their particulars and according to their names. This is what is called the science of astrology. The root of knowing this science that is first needed is the knowledge of the sun and its mansion (*burj*) and knowing the first day of January. The first who studied the science of astrology was our Elevated Lord Idrīs, our prophet, prayers and peace upon him. His knowledge was sound, down to [11] the time of Noah. He wrote it down on clay tablets, fired them, and buried them so that the flood would not put an end to this knowledge. They were found after this and it remained sound to the time of Jesus. The Jews entered unto him to kill him, and he said to them: "How have you sought evidence against me?" They said to him: "Through the science of astrology." He said: "O God, guide them astray in this science." And since this time, the science of astrology has been imperfect, none but those of complete reason and sound mind attaining it. The benefit of the science of astrology rests in four things: through it one ascertains the right way to go on land and sea as God the Almighty said: "Let them be guided by them" [al-Anʿam 97], knows the number of years and mathematics as God the Almighty said: "Let them know the number of years and mathematics" [Yunis 5]. He also said: "They will ask you about the new moons." [al-Baqara 189], and driving away the devils who seek to overhear, as God the Almighty said: "We have made them shooting stars to be cast at devils [al-Mulk 5]." And as an adornment of the heavens of this world . . .[49]

Al-Mirghitī continues by giving an overview of how all the benefits of astrology relate to a better understanding of God's arrangement of the heavens and a contemplation of His creation. Nevertheless, he hastens to inform his readers that anyone who believes that anything but God – be it

[47] On the identity of this enigmatic scholar on whose poem of timekeeping al-Jādirī had previously written a commentary (see his biography in al-Kattānī, *Salwat al-Anfas*), see G. S. Colin and H. P. J. Renaud, "Note sur le 'muwaqqit'."

[48] Al-Mirghitī, *al-Mumtiʿ fī sharḥ al-muqniʿ*, 7, 10. He gives the composition date of the *Muqniʿ* on page 95. The companion was an Abū ʿAbbās Aḥmad b. ʿAbd al-Sadīq al-Firaklī, a nisba that refers to Firaklah, a village in the Sūs. Al-Mirghitī also wrote a summary of *al-Mumtiʿ*, which has been published: *al-Muṭliʿ ʿalā masāʾil al-muqniʿ*.

[49] Ibid., 10–11.

a planet or a star – is a cause for anything else is deeply mistaken and an unbeliever. To drive this point home, he gives several examples of God making things occur whenever He wishes, including a vivid anecdote from an Abū Sālim Sidi Ibrāhīm b. ʿAbd al-Raḥmān al-Fīlālī regarding an occasion during a draught in his region when God caused the crops to revive without having it rain.[50] This emphasis on God's ceaseless role in creation speaks both to al-Mirghitī's adherence to the standard Ashʿarī occasionalism of his day, but also to the importance of underlining this at the beginning of a work on the perceived influence of heavenly phenomena.[51]

Following this introduction on astrology, our author moves to territory similar to that of al-Rūdānī's treatise: the Arab calendar, calendar conversions between solar and lunar, intercalary years, and then the Persian solar year.[52] There are interesting asides, both with regard to timekeeping, where al-Mirghitī observes that scholars with insight grant summer one day more than the other seasons (92 instead of 91) because that was the month during which God stopped the sun for the Prophet Joshua to aid him in his battle against the giants; but also, in al-Mirghitī's repeated inclusion of medical observations, such as the humors being linked to the seasons and diseases being more prevalent in autumn.[53] More pertinent to astrology is describing how the seasons house the solar mansions – of which there are twenty-eight, seven to a season – as well as the zodiac: Spring contains Aries, Taurus, and Gemini; Summer Cancer, Leo, and Virgo (*al-sunbula*); Fall Libra, Scorpio, and Sagittarius; and Winter is Capricorn, Aquarius, and Pisces. All of this information (and more) is then handily provided in a square (see the following page).[54]

The most interesting parts of al-Mirghitī's discussion are where he engages explicitly with the text of Abū Miqraʿ, either to correct him, or, occasionally, to defend him. The latter is the case in the case of the vexed subject of trepidation, which arises in a discussion of Abū Miqraʿ having given the dates of the vernal and autumnal equinoxes as falling respectively on the 16th of March and the 16th of September. The discrepancy between these dates and those on which the equinoxes fall in al-Mirghitī's time – March 10th

[50] Ibid., 13.
[51] Compare with al-Yūsī's views on astrology being compatible with Ashʿarī conceptions of God's causal role, as discussed in Chapter 2.
[52] Ibid., 13–25. [53] Ibid., 27–29.
[54] Image taken from the Hathi Trust copy of this text, p. 8 (https://hdl.handle.net/2027/mdp.39015081447024), originally from Max Meyerhof's collection, and a copy that is dated to 1809.

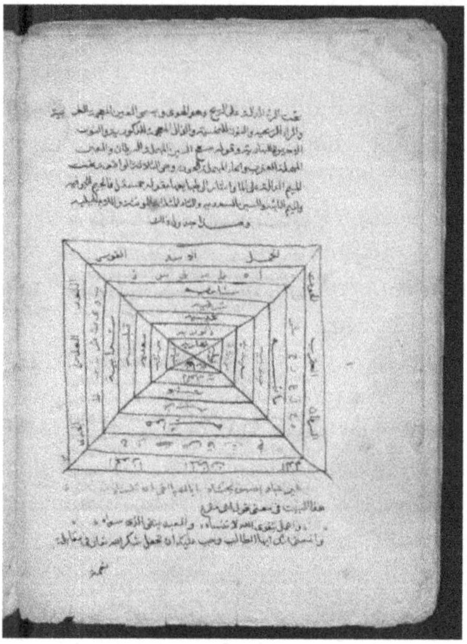

FIGURE 4.3 Al-Mumtiʿ fī sharḥ al-muqniʿ (Hathi Trust).

and September 11th – is not a mistake on the part of Abū Miqraʿ, he explains, but a result of trepidation or precession and recession:

Precession (*ḥarakat al-iqbāl*) is the movement of the sphere from east to west by itself so that every degree of the orb it moves from its place on the orb's sphere a degree ahead toward the west. Recession (*ḥarakat al-idbār*) is the opposite of precession. This occurs every 66 years, or it is said that the sphere moves a degree every 72 years. This means that the difference between the solstice and the equinox which are on the sixteenth and the solstice and the equinox on the tenth and eleventh in the speech of Abū Miqraʿ might be imagined to be a mistake. But there is neither doubt nor error here: Abū Miqraʿ said this according to the motion of accession being small at that time. Yet the motion of accession for us today is great due to the length of time from when Abū Miqraʿ based what he said. If you know this, then the reason for doubt is removed. the matter is clarified, and the reason for the difference made clear. Both the first and the second position is true, each for its time. The reason for this is that the cause of the equinox is the presence of the sun at the apex of the arc of the night and day (*qaws al-layl wa-l-nahār*), which are at the apex of Aries and Libra ....[55]

[55] Al-Mirghitī, *al-Mumtiʿ*, 32–33. The discussion continues through ibid., 36.

Considering that Abū Miqraʿ composed his poem some 300 years before al-Mirghitī's time, or roughly 5 times 66, the lower estimate for when recession occurs, this argument is persuasive. Yet it is also worth noting that al-Mirghitī continued to adhere to the theory of trepidation, a theory which had come under question in the Maghrib with the introduction of Eastern *zīj*s in the ninth/fifteenth century, and would ultimately be widely replaced by a theory of standard procession in the twelfth/eighteenth century.[56]

Quite a bit further along in his treatise, after passing over subjects such as date conversion and how to determine the day of the week in the Arab calendar when one has the non-Arab day and month, he comes to the subject of how to calculate the house of the sun (*manzilat al-shams*). To follow the narrative here, and to contextualize al-Mirghitī's own intervention, one must first consider his criticism of Abū Miqraʿ: the latter argued that one can know the house of the sun or moon by dividing the twenty-eight houses into the twelve equal mansions, with each mansion containing two and one-third house. Al-Mirghitī argues that the one who devised this method was not interested in the places of the houses in the mansions and their movements in the spheres and their distance from each other, although he had summarized it despite its lack of precision (*lā taʿṭay ghāyat al-tahqīq*).[57] The correct, precise fashion of working out the house of the sun was the following:

If you want to know the house of the sun, know how much of the season of the four seasons that you are in has passed down to your day, add to it the number of al-jīm, which is three. Then divide this into the houses of that season as was previously mentioned, and these were seven in each season. Then take for each house of this season the number *ya-jīm*, which is 13 days and begin with the first of the houses of your season. Reckon it 13, then the one that follows it and so on you reach the number of houses you have that you have calculated and which you have reached and the sun is in that one according to the amount that remains before you reach 13. An example of this: we sought the house of the sun, so we took what had passed of the season in which we were, Spring, which was 85 days. We added 3 to it, resulting in 88. We added to it what was absent from the aforegoing (*fa-ʿaṭaynā min-hā li-faragh al-muqaddam*) – thirteen – because it was the first of the houses of spring, and for the one after it the same, the same for Pisces (*baṭn al-ḥūt*) until the calculation reaches Taurus. There, the division stops and there does not remain in it more than ten days, so that we know that the sun is in that house.[58]

[56] See Samsó, "An Outline of the History of Maghribī Zijes," 93–96.
[57] Al-Mirghitī, *al-Mumtiʿ*, 55. [58] Ibid., *al-Mumtiʿ*, 56.

In his subsequent discussion al-Mirghitī notes that the sun moves through roughly one degree each day, with all houses having thirteen degrees, with the exception of *al-jabha*, which, as al-Jādirī said, has fourteen degrees, and, not as Abū Miqraʿ said, two degrees.[59] In contrast to Abū Miqraʿ, al-Jādirī appears as a trustworthy authority in al-Mirghitī's survey of astrological matters, as exhibited in a section following a description of how to calculate the lunar houses, where he is quoted at length to explain the waxing and waning of the moon.[60]

In the last section of *The Enjoyable Commentary*, al-Mirghitī takes up cosmology in a relatively standard fashion, turning to the seven planets that are encompassed within nine spheres, and the length of time that each spends in each mansion. He stresses alternatingly the authority of the philosophers, the sages, and the natural philosophers regarding this subject, and compares the spheres to subtle bodies of a round shape, like the whirl of a spindle or the skin of an onion, with the earth in the middle, the closest sphere to us being the heavens of the earth, which is the smallest sphere. The first 8 spheres are each separated by 500,000 years, with the eighth sphere of the fixed stars a thousand years removed from the outmost ninth sphere within which the other 8 rotate. The distance of the planets from the earth correlates with the length of time they spend in each mansion, with that time lengthening the further away from the earth one goes. Here too, al-Mirghitī takes issue with Abū Miqraʿ, here calling his values for Mars and Venus a scandal, and invoking the authority of his own teacher in this science, Abū ʿAbdallāh Muḥammad b. ʿImrān al-Tuwātī (d. 1031/1621–22).[61] Al-Mirghitī then provides a table with the helpful information of which planet is ascendant during which hour of the night on a given day, following it up a few pages later with a table of the mansions of the planets.[62] This is followed by a square linking the elements of the zodiac to the four elements, with each elements containing three different signs. In the Hathi Trust manuscript, the equivalence is slightly different, with the signs and mansions being grouped around the humors, instead of the corresponding elements (see the following page).[63]

In his conclusion a few pages later, al-Mirghitī stresses the importance in all astrological observations and predictions of acknowledging God as the only actor and of standing in obedience to Him. This is, in fact, he

[59] Ibid. [60] Ibid., 68. [61] Ibid., 81–87. [62] Ibid., 88, 91.

[63] Hathi manuscript, 66. The broader shared theoretical basis of medicine, astrology, and the occult in Arabic writings from the third/ninth century onwards is clearly laid out in Liana Saif, "Between Medicine and Magic."

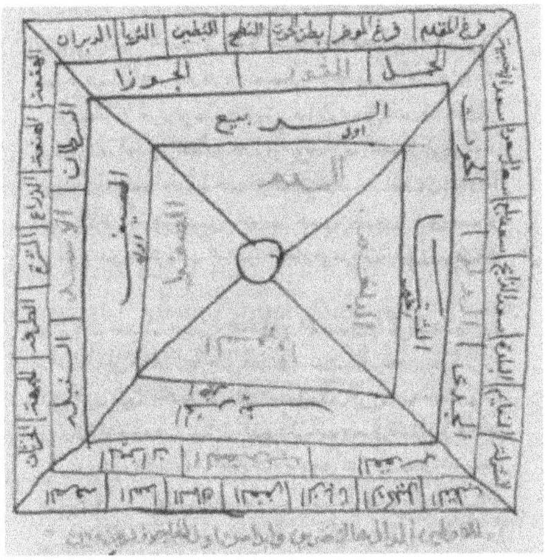

FIGURE 4.4 Al-Mumtiʿ fī sharḥ al-muqniʿ (Hathi Trust).

notes, what differentiates the practitioner of astrology from than of magic or soothsaying, and this difference is what leads to him judging the first of these to be rank unbelief and the latter polytheism. Both of these, for him, are disciplines that speak to the practitioner's belief in forces other than God or ability to access the unseen without Divine aid.[64]

In her work on the overlapping realms of medicine and the occult in Islamic thought, Liana Saif has stressed how in the seventh/thirteenth century the occult properties of the heavens and the planets ceased being thought of as the domain of the physician and were reinterpreted as the work of demonic forces.[65] Yet the occult encompassed both the demonic – jinn principally – and the divine, accessible through the names of God and of angels, and through their invocation in prayer. As seen in the discussion of his *fahrasa* in Chapter 2, al-Mirghitī alternated easily between astronomy, medicine, and lettrism in his discussion of how to remedy illnesses or protect oneself. The astrological observations at the end of *The Enjoyable Commentary* show a similar facility with shifting between astronomical measurement and prediction and a contemplation of the connection between celestial conjunctions and earthly physical compositions. As we shift our focus to medicine, we will see that this openness to the occult existed there as well.

64 Al-Mirghitī, *al-Mumtiʿ*, 93–94. Al-Mirghitī goes into greater detail on his understanding of the occult in his treatise on defending himself against sorcery – *Risāla fī ibṭāl al-siḥr* – a work I hope to discuss in a future publication.

65 Saif, "Between Medicine and Magic," 337–38.

4.4 MEDICINE BETWEEN MAN AND GOD

Our understanding of early modern medicine in the Middle East remains poor in comparison with the Medieval period. Still, in recent years it has received much more attention, with interesting developments having been in charting the reception of Paracelsus' works in the writings of Ottoman physicians, or in the influence of Indian medicine on the Persian Islamic medical tradition.[66] Far less attention has been paid to North Africa and especially Morocco, which although tied to the Ottoman empire by important economic, cultural, and intellectual links, lay beyond its political control. This is not to say that it possessed its own medical tradition, but that its geographic location facilitated it having a regional character in its scholarship. Thus, while the works of the great sixteenth-century Egyptian doctor Dāwūd al-Anṭākī (d. 1008/1599) were read and studied in Morocco – indeed al-Ṣāliḥī refers in passing to al-Anṭākī alongside a number of other authorities and ʿAbd al-Wahhāb b. Aḥmad Addarāq al-Fāsī (d. 1159/1746) wrote a commentary on al-Anṭākī's *al-Nuzha al-mubhija* – they may not have had the same influence there that they did in the Eastern Mediterranean.[67] The case of al-Anṭākī deserves remark, for while he is today the most famous early modern Muslim author of medical texts, and was widely read in the Maghrib as can be seen in current manuscript holdings,[68] no monograph has been devoted to him and even a basic understanding of whether his works engaged with the occult or not remains contested.[69] Unlike the Moroccan author discussed below, al-Anṭākī doesn't mention prayers, invocations, or the use of magic squares or talismans as medicinal remedies, although the supplement to the *Tadhkira,* written by an unnamed

[66] Bachour, *Oswaldus Crollius*; Fabrizio Speziale, "Rasāyana and Rasaśāstra in the Persian Medical Culture of South Asia." Ibn Sallūm al-Ḥalabī (d. 1081/1670)'s *Ghāyat al-ittiqān fī tadbīr badn al-insān*, which is the focus of Bachour's work, was known in Morocco as seen in a copy of it from 1767 being in the Royal Library (al-Khaṭṭābī, *Fahāris al-khizāna al-malikiyya*, vol. 2). The Royal Library also possesses a translated copy of Paracelsus' work with the title *al-Ṭibb al-kīmyāʾī al-jadīd*, although the date of the copy is not given (al-Khaṭṭābī, *Fahāris al-khizāna al-malikiyya*, vol. 5, 225).

[67] On Addarāq's commentary, which does not seem to be extant, see the editor's comments in Dāwūd al-Anṭākī, *al-Nuzha al-mubhija fī tashḥīdh al-adhhān wa taʿdīl al-amzija,* 14.

[68] The Moroccan Royal Library alone possesses twenty-two copies of *al-Tadhkira*, some of which were copied in Morocco. See Muḥammad al-ʿArabī al-Khaṭṭābī (ed.), *Fahāris al-khizāna al-malikiyya*, vol. 2, 57–70.

[69] On al-Anṭākī, see Julia Bray, "Dāwūd ibn ʿUmar al-Anṭākī." Bray's comments on al-Anṭākī's illuminationist tendencies and inclusion of magic, astrology, and lettrism in his work are suggestive and yet I have found nothing to bear out her characterizations (ibid., 49, 51–52).

author who refers to al-Anṭākī as his teacher, does takes up the question of letter magic (*sīmiyā*) and the making of secrets (*nawāmīs*).[70] The absence of prayers and talismans is also characteristic of the writings of the famed ninth/sixteenth-century Egyptian physician al-Qalyūbī (d. 1069/1659), a copy of whose *Summary of the Science of the Bodies* (*Mukhtaṣar ʿilm al-abdān*) is found in the library of the Tamgrout lodge in southern Morocco.[71]

In the Moroccan context, we know that under the last Saʿdī ruler, Ahmad al-Manṣūr, physicians were organized through a type of centralized licensing under the supervision of the scholar and physician al-Wazir al-Ghassānī (fl. 1603), who himself authored a popular treatise on *materia medica* entitled *The Garden of Flowers on the Essence of Plants and Herbs* that distinguished itself by offering an innovatory classificatory system for plants.[72] It is unclear whether this licensing system persisted following al-Manṣūr's demise. In addition, we know of works on individual sicknesses, such as ʿAbd al-Ghanī al-Zammūrī's (d. 1030/1621) treatise on gallstones (*al-Qānūn al-mufīd fī ʿalāj al-ḥasā bi-qawl sadīd*), referred to briefly above in Chapter 1, as well as a collection of recipes set to verse by Abū Qāsim al-Ghūl al-Fishtālī (d. 1059/1649), entitled *Preservation of the Temperament and Bestower of Diverse Treatments* (*Ḥāfidh al-mizāj wa lāfidh al-amshāj bi-l-ʿalāj*), entailing lines of verse that he completed in 1038/1629.[73] Al-Fishtālī was a judge and jurist living in Fez who wrote works on, among other things, the correct response to the plague (a versification of the treatise by the sixteenth-century al-Ḥaṭṭāb), and a poem on the fifth order magical square with an empty center – a subject that al-Mirghītī himself would write on – a commentary on a poem dealing with the division of water into canals. He was also an important teacher of al-Mirghitī's, and instructed him in astrology, numerology (*al-tarqīm*), and

[70] See al-Anṭākī, *al-Tadhkira*, 459.

[71] See Shihāb al-Dīn al-Qalyūbī, *Mukhtaṣar ʿilm al-abdān*. 184 ayn. Nasiriyya Lodge, Tamgrout. This work has been printed as *Tadhkīrat al-Qalyūbi fi-l-ṭibb wa-l-ḥikma*.

[72] For a reference to the licensing of doctors in Saadian Morocco, and their having a leader – during the reign of al-Manṣūr it was al-Wazīr al-Ghassānī (fl. 1012/1603) – who oversaw their affairs and tested each one of them, see Ḥajjī, *al-Ḥaraka al-fikriyya*, vol. 1, 160, where he is relying on ʿAbd al-Ghanī al-Zammūrī's (d. 1030/1621) *al-Qānūn al-mufīd fī ʿalāj al-ḥasā bi-qawl sadīd*.

[73] On al-Fishtālī's writings see al-Mirghitī, *al-ʿAwā'id al-muzriyya*, vol. 1, 48. For my reading of *Ḥāfidh al-mizāj*, I have relied on the following manuscript copies: The National Library of Morocco, q960 [I referred to the copy kept by the Jumʿa al-Majid library 579336], but I also consulted The Royal Library of Morocco 12669 [Jumʿa al-Majid library 575374], Tangier 11 [Jumʿa al-Majid library 581582], and Marrakesh 15 [Jumʿa al-Majid library 580797].

letter magic (*al-sīmiyā'*).[74] Al-Fishtālī's long poem, *Preservation of the Temperament*, contains little in the way of medical theory, but its structure is similar to the poem of al-Ṣāliḥī, and covers the afflictions of the body, beginning with the head and proceeding downwards.

In addition to these works, the sixteenth–seventeenth centuries in Morocco were witness to the composition of several poems on syphilis, perceived during the ninth/sixteenth–tenth/seventeenth centuries to be a new disease in need of explanation.[75] Of these three poems – written by ʿAbd al-Karīm b. Muʾmin b. Yaḥyā al-ʿIlj (fl. 1557–94), the already encountered polymath ʿAbd al-Raḥmān al-Fāsī, and the aforementioned ʿAbd al-Wahhāb Adarrāq – the second drew explicitly on al-Anṭākī, and can be productively read against the long poem on the medical properties of foods by ʿAbd al-Qādir Ibn Shuqrūn that he had completed in 1113/1701–02.[76] Ibn Shuqrūn had studied medicine in Fez with ʿAbd al-Wahhāb Adarrāq's father, Abū ʿAbdallāh Aḥmad Muḥammad Adarrāq, and may have been one of Moulay Ismāʿīl's own physicians.[77]

Yet none of these works contained overviews of medicine as a whole, and that was where al-Ṣāliḥī's *The Worthy Gift of Medicine* distinguished itself. Along with that of al-Fishtālī, al-Ṣāliḥī's poem was purposefully targeted toward a student audience and should be considered an introductory work. Together with the author's auto-commentary, which must have been directed to advanced students and scholars of medicine, it offers us a productive entry into Moroccan scholarship on medicine during this period.

4.4.1 *The Worthy Gift of Medicine and Its Pearls*

Abū al-ʿAbbās Sīdī Aḥmad al-Ṣāliḥī al-Darʿī (d. 1144/1731), the head of the Ṣāliḥiyya Sufi lodge in the far south of Morocco in the late seventeenth and early eighteenth centuries, achieved renown through his long poem on

[74] Ibid., vol. 1, 46 and compare with al-Yūsī, *Fahrasa* (Rabat, 2004), 76, where the category of letter magic (*al-sīmiyā'*) is used to encompass that of magical squares (*al-jadāwil*).

[75] See H. P. J. Renaud and G. S. Colin, *Documents marocains pour servir à l'histoire du "Mal Franc."*

[76] See Abdelhadi Tazi and Badr Tazi (ed. and trans.), *La medicine arabe au XVIIIe siècle à travers al "Urdjuza ash-Shakruniyya."* On Ibn Shuqrūn, see Lakhdar, *La vie littéraire au Maroc*, 161–66, where his talents as a poet are emphasized.

[77] See Tazi and Tazi, *La medicine arabe au XVIIIe Siècle*, 21–22, but compare with Ellen Amster's excellent overview of the Adarrāq family in her entry for "Adarrāq" in EI3.

medicine.[78] *The Worthy Gift of Medicine* (*al-Hadīya al-maqbūla fī l-ṭibb*) drew on a wide range of sources including the Arabic-Galenic tradition and Prophetic medicine, and not unusual for his time, al-Ṣāliḥī wrote a long commentary to fully explain it. This section will discuss this poem and its extensive commentary, *Strewn Pearls on the Worthy Gift of Medicine* (*al-Durar al-maḥmūlah ʿalā al-hadīya al-maqbūla*), while reflecting on the ways that a great deal of the prior and current historiography on medicine in the early modern Islamicate world has been focused on defining medicine as the received Galenic Tradition while categorizing prayers and spells as supplemental to medicine itself.

The Ṣāliḥiyya lodge, into which our author al-Ṣāliḥī was born in 1624, is far from any urban center of learning and close to the beginning of the Sahara proper.[79] It lies in the Draʿ valley and is not as well-known as the Nāṣiriyya lodge in Tamgrout a little further north.[80] Yet like the Nāṣiriyya lodge the Ṣāliḥī lodge must have benefited tremendously from the changes in the trade routes that had taken place a generation before al-Ṣāliḥī's birth when Aḥmad al-Manṣūr's troops marched directly south across the Sahara to conquer Timbuktu, thus opening a new and more direct trade route between West and North Africa and linking the Ṣāliḥī lodge much more directly with centers of learning in West Africa. This realignment contributed not only to the economic well-being of the lodge but also helps explain why al-Ṣāliḥī's works subsequently enjoyed broad diffusion in West Africa. Thus, while the location where our author was born and spent much of his long life – he lived until he was 107, dying in 1731 – may look marginal on a map of Morocco today, he did not experience it as such, for as we have seen in Chapters 1 and 2, during his lifetime Moroccan learning south of the Atlas was undergoing a flourishing revival. His first teacher was his father, who was the head of the lodge, and at the Ṣāliḥiyya lodge he studied logic, astronomy, mathematics, medicine, and pharmacy alongside the linguistic and religious sciences.[81] Following his father, he took over the administration of the lodge, and in 1692 the second ʿAlawite

[78] On al-Ṣāliḥī see the extensive entry in al-Darʿī, *al-Durar al-muraṣṣaʿa bi-akhbār aʿyān daraʿa*, vol. 1, 231–47. Compare with Zirikili, *al-Aʿlām*, vol. 1, 138.

[79] One of the things that is unclear from the biobibliographic literature is whether this was a fully independent lodge or whether it fell under the authority of the Nāṣiriyya lodge; al-Ziriklī identifies our author with a lodge in Aktāwah, which other sources describe as one of a series of oases between Tamgrout and Mhamid.

[80] On the beginnings of the Nasiriyya lodge and its growth, see Gutelius, "Between God and Men: The Nāṣiriyya and Economic Life in Morocco, 1640–1830," chapters 3 and 4.

[81] Al-Ṣāliḥī, *al-Hadīya al-maqbūla fī-l-ṭibb*, 8. According to one modern scholar, al-Sāliḥī's father authored alchemical recipes that were recorded by his contemporary Sīdī ʿAbdallāh al-Fīlālī (Semlali, *Histoire de l'alchemie*, 60).

ruler Moulay Ismail (rl. 1672–1727) issued an edict confirming his position as head of the Ṣāliḥiyya lodge. He wrote *The Worthy Gift of Medicine* in 1103/1691–92 when he was in his sixties,[82] and modern narratives have him dedicating it to Moulay Ismail in the hope that students would have a thousand-line poem to memorize for their initial studies of medicine, similar to the *Alfiyya* of Ibn Mālik (d. 672/1274) on Arabic grammar, a popular study text of students of Arabic.[83]

The poem itself is preceded by a long prose introduction in which al-Ṣāliḥī first praises medicine as a noble art and then lays out a basic introduction to humoral theory, explaining the connection between the elements, temperaments, and humors. He is especially interested in clarifying to the reader the basic principle – already invoked by al-Rūdānī above – that sicknesses need to be treated with remedies that encompassed the opposing humoral nature:

> An example of what we had referred to in terms of the natures of foods, medicines, and sicknesses is that if for example someone complained of there being heat inside them (*bi-ḥarārat jawfihi*), someone might say to him: your remedy is garlic, cloves, olive oil (*al-salīṭ*), stew (*al-maraq*), pigeon meat, and so on of the medicines for fever. The one who says this is completely ignorant due to his having broken the principles of the discipline and adding things of like nature. Instead, this is treated by what has been stipulated in accordance with the ruling of the opposite of the given nature, such as blessed greens including purslane, lentils, goat milk, Indian fruit (*al-thamar al-hindī*) and other medicines of coldness . . .[84]

Administration of the correct medicine is, naturally, dependent on the proper diagnosis of the patient's temperament. This depends not only on the individual humoral composition of a given person, but also of their age – childhood, youth, maturity, and old age – each of which is ruled by a different combination of heat, cold, humidity, and dryness, and well as the season in which a sickness takes place. If you want to ascertain the nature of a person's temperament make sure the person is neither hungry nor full, and that he has eaten only lightly after midafternoon, containing

[82] For this dating see al-Ṣāliḥī, *al-Durar al-maḥmūlah ʿalā al-hidaya al-maqbūla*, 7. I am grateful to Caitlyn Olson for helping me acquire a copy of this text. Al-Ṣāliḥī writes that he began his studying medicine with the books of the ancestors in Darʿ, after which he headed to Fez in 1099/1687–88, where he became sick. What happened then is unclear, but he returns via Sijilmasa to Darʿ, where all the pain he had been in ended through the grace of God, leaving him astounded; this too was one of the three positive incentives for him to benefit Muslims (ibid., 6).

[83] The Moroccan documentary on him by the Wizārat al-Awqāf makes these claims (which I have not been able to substantiate in any sources on his life from the eighteenth century): www.youtube.com/watch?v=OrPpB88gnWI

[84] Al-Ṣāliḥī, *al-Hadīya al-maqbūla fī-l-ṭibb*, 14.

nothing that will color his urine, like saffron, henna and so on. When he wakes up, let him urinate in a clean bowl and place a drop of olive oil (*salīṭ*) in it. If it spreads and grows until it covers the bowl, then his sickness is hot. If it remains it its place and does not move, then it is cold.[85] The introduction contains multiple references to the power of God and how all remedies work through the permission of the Divine, who could cure without them if He chose to do so. Al-Ṣāliḥī concludes, as if anticipating the poem's popularity, with instructions to future copyists on how they should write the names of sickness and illnesses in red ink, with the verses after them written in black.[86]

The poem proper begins stressing the importance of health for the faith, the fact that the Prophet relied on medicines, and how the author summarized what he had read regarding medicine in verse for students to memorize (p. 17). Following a brief section on how to measure out medications, it gives an overview of the different parts of the body, their ailments and the medicines for them. In this, *The Worthy Gift of Medicine* is similar in structure to both the earlier work of al-Fishtālī's, *Preservation of the Temperament*, and to one of al-Ṣāliḥī's sources, the Egyptian physician Shihāb al-Dīn al-Qalyūbī's (d. 1069/1659) *al-Qalyūbī's Memorandum on Medicine and Wisdom* (*Tadhkīrat al-Qalyūbī fi-l-ṭibb wa-l-ḥikma*).[87] While the different treatises contain varying numbers of chapters – 24, 10, and 15 – the order of the topics addressed and the nature of the topics addressed contains a great deal of overlap, although this is only partially clarified by the following table, which, for clarity's sake does not contains the breakdown of topics within each chapter (see Table 4.3).

A substantial number of verses in *The Worthy Gift* are devoted to sex and reproduction, in terms of aiding and preventing pregnancy on the one hand, and increasing sexual pleasure on the other – along with traditional warnings about the perils of too much sex (pp. 44–56). The dangers of sex preoccupied al-Ṣāliḥī, for he had mentioned it in a previous poem as well:

85 Ibid., 15. Compare with the use of oil in water to divine the future, as recounted by Leo Africanus (Jean-Léon l'Africain) in his description of Fez (*Description de l'Afrique*), vol. 1, 216–17.

86 Al-Ṣāliḥī, *al-Hadīya al-maqbūla fī-l-ṭibb*, 16. The poem's continuing relevance in Morocco is borne out by a commentary on it written by a Muḥammad ibn Ibrāhīm al-Nadhifī (d. 1285/1868–69) entitled *Mashāriq al-anwār fī riyāḍ al-azhār*, of which there is a copy in the Moroccan Royal Library, #2072 (Semlali, *Histoire de l'alchemie*, 290).

87 The title page of one of the extant copies of al-Fishtālī's *Ḥāfidh al-mizāj* (The National Library of Morocco, q960 [Jum'a al-Majid library 579336]) bears an ownership stamp of the Nasiriyya lodge on its title page, though it is unclear when exactly this copy was made or whether al-Ṣāliḥī had access to it (he did not cite it).

TABLE 4.2 Overview of the structure of al-Fishtālī's *Preservation of the Temperament*, al-Qalyūbī's *Memorandum on Medicine and Wisdom*,[88] and al-Ṣāliḥī''s *The Worthy Gift of Medicine*

Structure	al-Fishtālī	al-Qalyūbī	al-Ṣāliḥī
Chapter 1	Afflictions of the head	Knowledge of the origins of sicknesses	On measurements of ingredients
Chapter 2	Afflictions of the eye	Afflictions of the head	Afflictions of the head and its parts
Chapter 3	Afflictions of the ear[89]	Afflictions of the eye	Afflictions of the throat
Chapter 4	Afflictions of the nose	Afflictions of the ears, nose, and mouth	Afflictions of the chest
Chapter 5	Afflictions of the mouth and teeth	Afflictions of the throat, neck, chest, and breasts	Afflictions of the stomach
Chapter 6	On tumors and swellings	Afflictions of the back and stomach	Afflictions of the back, difficulty with urination, dysentery
Chapter 7	Afflictions of the chest, lungs, coughing	Afflictions of the buttocks and reproductive organs	On the stomach, digestion, quenching thirst, and purging
Chapter 8	Pains and fluttering of the heart	Afflictions of the hips and remaining parts of the body	On foods that aid and harm sex

88 I have referred to al-Qalyūbī, *Tadhkīrat al-Qalyūbī*. The copy of this work preserved in the Nasiriyya lodge at Tamgrout today bears the title *Mukhtaṣar ʿilm al-abdān* (call number 184 ayn).

89 A page seems to be missing from the National Library of Morocco, q960 [Jumʿa al-Majid library 579336] copy. I referred here, to the Marrakesh 15 [Jumʿa al-Majid library 580797] copy.

TABLE 4.2 continued

Structure	al-Fishtālī	al-Qalyūbī	al-Ṣāliḥī
Chapter 9	Afflictions of the stomach, vomit, cholera	On diseases affecting the entire body (plague, poisons)	On sex and reproductive organs
Chapter 10	Afflictions of the liver	On beautification	On wombs, vaginas, and preventing envy
Chapter 11	Pain of the spleen	Conclusion	On pregnancy and preventing it
Chapter 12	Pain of the bowels, using colic		On preserving the fetus, giving birth
Chapter 13	Pains of the liver		Afflictions of the buttocks, nerve pain, joints
Chapter 14	Pains of the womb, causes of infertility, abortion		On abscesses, tumors, sores
Chapter 15	On the testicles and penis, delayed puberty, pain during sex		On snakes and scorpions
Chapter 16	Afflictions of the back, hips, gout, joint pain		Conclusion
Chapter 17	On fevers, madness, sleep, nightmares		
Chapter 18	On poisons, bites, snakes, scorpions, rabid dogs		
Chapter 19	Afflictions of the skin, leprosy		
Chapter 20	Afflictions of the breasts		
Chapter 21	On beautification		
Chapter 22	On bloodletting		
Chapter 23	On laxatives and purgatives		
Chapter 24	Miscellaneous matters/ conclusion		

Beware of fucking (*al-nayk*), you who are struck with desire
for there is difficulty and harm in fucking.
It hastens aging, drives away the light
from one's countenance, and weakens the memory.
It extracts the marrow from the young man's leg
doubtlessly it weakens the sight.
....
The physician Hippocrates said nothing
but harm will come from fucking.
There is no shame in the word fucking
my ignorant brother, reflect!
The most eloquent of Arabs, the Prophet of God
addressed men with it, as has been related.[90]

Striking as his choice of vocabulary may be, the topic of sexual hygiene is a common one in humoral medicine, as al-Ṣāliḥī's reference to Hippocrates notes correctly. As *The Worthy Gift of Medicine* concludes, the author stresses again the danger of giving medicines too much importance and to remember that all power rests with God (p. 68). This invocation reflects the importance placed by the Ashʿārī school of theology – dominant in West and North Africa during this period – on occasionalism and on God being the only causal Agent.

The remedies recommended in *The Worthy Gift* include drinking the urine of a ram or menstrual blood to prevent pregnancy forever (p. 54), or smearing civet musk on one's penis to prevent pregnancy on a one-off basis (p. 53). These have parallels with other contemporary West and North African texts, but are distinct from works such as al-Anṭākī's, which stress the Galenic tradition more and are more reticent regarding the benefits of individual recipes.[91] The contrast between eastern and western traditions should not be emphasized too much, however, as al-Ṣāliḥī's Egyptian contemporary, al-Qalyūbī, in his discussion of remedies for male sexual problems, recommended, among other options, smearing the penis with goose fat to help achieve an erection.[92] The natural properties of *materia*

[90] This poem is given in in al-Darʿī, *al-Durar al-muraṣṣaʿa,* vol. 1, 231–47. In a tradition related by ʿAbdallāh b. ʿAbbās, the Prophet uses the word in conversation with Māʿiz b. Mālik when confirming that the latter had committed fornication (Bukhārī, *Ṣaḥīḥ*, 6824; a similar account is related in Abū Dāwūd, *Ṣaḥīḥ*, 4428).

[91] Compare with al-Anṭākī, *al-Nuzha*, 178–82. Yet note the presence of alchemical texts ascribed to al-Anṭākī in the Moroccan Royal Library (al-Khaṭṭābī, *Fahāris al-khizāna al-malikiyya* vol. 5, 42, 242).

[92] See al-Qalyūbī, *Tadhkīrat al-Qalyūbī,* 80–82.

medica covered a broad range of phenomena, including some that fall under the occult, as we have already seen in the *fahrasa* of al-Mirgīthī.

Al-Ṣāliḥī broadened the scope of his understanding of medicine further in his commentary on this poem, the *Strewn Pearls*, which he wrote shortly after *The Worthy Gift* over a period of five years: from 1694 to 1699.[93] The *Pearls,* which in the manuscript I have used runs to 653 pages, repeatedly quotes from *The Worthy Gift*, yet is organized along decidedly different principles.[94] The *Pearls'* seven chapters, along with an introduction and conclusion, depart from the structure laid out in Table 4.3 above, which used the body as an organizing principle, to take a much more expansive look at medicine:[95]

- A commentary on the initial verses and an explanation of the vocabulary of the poem
- On the kinds of illnesses and sicknesses
- On the kinds of indications and their construction
- On the relevant animals and birds and so on
- On trees, plants, and herbs
- On seeds, metals, and the like
- Types of sap

These chapter headings clarify that while the commentary draws on the poem, its own structure stretches thematically to include, among other things, long discussions of the natures and properties of both animals and metals, but also qualitatively in terms of the types of remedies it presents, with much more time being given here to the healing virtues of amulets and spells, while also warning against the at the time popular science of alchemy. There are certainly parallels between the two – the *Pearls* spends over a hundred pages on matters related to sex and reproduction[96] – but the differences highlight the two different audiences they were meant for, one for the students who populated the growing number of urban and rural centers of learning at the time, and another for scholars and physicians, revealing a deep reading of both medical and occult sources.

[93] Al-Ṣāliḥī, *al-Durar al-maḥmūla*, 652.

[94] Al-Ṣāliḥī alludes to how he went about structuring the commentary, doing so on the advice of one of his teachers, an Aḥmad Maḥammad al-ʿUmārī, known as al-Ḥabīb al-Lamṭī al-Sijilmāsī, with whom he had studied Qur'anic readings many years earlier in 1091/1680.

[95] These titles are also written down on the cover sheet of the version of the manuscript I consulted.

[96] Al-Ṣāliḥī, *al-Durar al-maḥmūla*, 530–637.

The scope of the *Pearls* as a text makes a summary of it here impossible, but it is worth making a few impressionistic points: the first is the importance al-Ṣāliḥī gives to lettrism, that is, the occult science of the nature and use of the Arabic letters, here chiefly in the context of amulets. This discussion, which stretches for over fifty pages, falls in the book's third chapter on the places of Indication (*mawaqiʿ al-isharāt*) and is broken up into three different investigations (*mabāḥith*) having to do with understanding the written word: the first deals with the science of letters, the second with enunciating letters (*al-nuṭq bi-l-ḥurūf*), and the third on bearing talismans with writing in them or drinking these.[97] In framing his discussion, al-Ṣāliḥī provides an introduction that functions also as a justification: the science of letters began with Jābir ibn Ḥayyān, the science of treasures (*al-kunūz*), and that of amulets (*ḥurūz*), citing the authority of an Abu ʿAbdallāh al-Shaykh Muḥammad al-Shāfī, were believed in by prominent scholars, including Abu ʿAbbās al-Mursī (d. 686/1287), Abu al-ʿAbbās al-Sabtī (d. 601/1204), al-Ghazālī (d. 505/1111), and al-Shādhilī (d. 656/1258). These authorities, along with al-Shiblī (d. 334/945) and al-Ḥallāj (d. 309/922) and other wise men (*ghayrahum min al-ḥukamā'*) span multiple centuries and show a bias toward the Islamic West, but are largely best known as having been prominent Sufis. Both here, and through the *Pearls*, al-Ṣāliḥī stresses that fact that the occult properties of letters and associated amulets, prayers, and invocations were placed there by God and when they refer to the names of God can they be effective. Further clarifying the demarcation between licit and illicit forms of knowledge, he claims that all prayers to jinn should be avoided and the practitioner should also stay away from alchemy, which he derides as red poverty (*al-faqr al-aḥmar*), a pun on red sulfur (*al-kibrīt al-aḥmar*), which symbolically also represents the goal of Sufi spiritual practice.[98] The repudiation of alchemy is striking, for prominent contemporaries of al-Ṣāliḥī such as al-Mirghitī whose astronomical works we discussed above, wrote favorably in his scholarly autobiography of both amulets and alchemy, and composed works on alchemy while condemning magic in the harshest terms. Despite their disagreement regarding alchemy, al-Mirghitī and al-Ṣāliḥī coincided in focusing on an understanding of God as the underlying power behind the remedies that they prescribed, be they plants, magical squares, amulets, or prayers.

[97] Al-Ṣāliḥī, *al-Durar al-maḥmūla*, 98. By comparison, al-Qalyūbī only discusses amulets in the context of warding off plague, and does so without any larger discussion of the science of letters (*Tadhkīrat al-Qalyūbī*, 95–96).

[98] Al-Ṣāliḥī, *al-Durar al-maḥmūla*, 111–12.

Other chapters in al-Ṣāliḥī's *Pearls*, such as the sixth on grains and metals, are essentially comprised of *materia medica*, lists of the healing properties of individual substances and how they should be used. The fourth, in which he spends over seventy pages discussing animals, is distinct.[99] Here, al-Ṣāliḥī offers a wide range of reflections on animals, including mankind within this category as when he offers a taxonomy of animals based on reproduction: 1) there are those animals that reproduce through parents like mankind, horses, and livestock; 2) then there are those that reproduce through themselves like worms, woodworms (*sūs al-khashab*), flies, and so on; and 3) and there are those who sometimes do one thing and sometimes another, such as scorpions, which copulate and generally reproduce, but can also emerge from yellow eggplants.[100] I have struggled with identifying the numerous sources al-Ṣāliḥī uses in the *Pearls*, but even an impressionistic list reveals the breadth of his reading, which included, unsurprisingly given the above discussion, a large number of nonmedical texts and authorities:

al-Ashrafī's *Tuḥfat al-fawā'id* (many references). I was not able to identify this work. **Theology.**
al-Nadrūmī (d. ca. 810/1408), *Qabas al-anwār wa jāmiʿ al-asrār.* **Lettrism.**
Jalāl al-Din Muḥammad b. ʿAbd al-Raḥmān b. Muḥammad al-Ḥubayshī (d. 786/1384)'s *Kitāb al-baraka*.
al-Anṭākī (d. 1008/1599), unclear which work. **Medicine.**
Suyūṭī's (d. 911/1505) *Maqāmāt al-Yāqūṭiyya* (many references). **Literature/Medicine.**
Shihāb al-Dīn al-Qalyūbī (d. 1069/1659), unclear which work. **Medicine.**
Shihāb al-Dīn ʿAbd al-Bāqī (d. 808/1405)'s commentary on al-Buṣīrī's (d. 696/1295) *Burda*. **Sufism.**
al-Qastillānī (many references), unclear which Qastillānī this is or which work is cited.
Ibn al-ʿArabī (d. 638/1240), unclear which work. **Sufism.**
Ibn al-Ḥakīm's *al-Rashīdiyya* (multiple references). I was not able to identify this work.
Jābir b. Ḥayyān (d. 195/810), unclear which work. **Alchemy.**
Abū ʿAbdallāh al-Shaykh Muḥammad al-Shāfī. **Lettrism.**
Ahmad Zarrūq (d. 899/1493), unclear which work. **Sufism.**
al-Nawawī on prayer (d. 676/1277). **Ritual.**
al-Shāfiʿī (d. 204/820) on animals, unclear which work.
Ibn Saba'in (d. 669/1269). **Sufism.**
Galen (d. 216). **Medicine.**

[99] Ibid., 182–258. [100] Ibid., 183–84.

These sources are remarkable for the absence of classical medical authorities such as Ibn Sīnā or Ibn al-Nafīs, not to mention some prominent local medical authorities such as the Moroccan al-Wazir al-Ghassānī, although he does reference others such as al-Anṭākī and al-Qalyūbī. What kind of medicine, then, are we dealing with? What does it mean that al-Ṣāliḥī's canon was so different from the one that has defined the traditional history of Graeco-Arabic medicine?

No non-Moroccan scholar has discussed al-Ṣāliḥī's texts while local scholars have not engaged with the specifics of his writings, and have characterized him with local pride as a respected scholar from the Moroccan rural south. If an earlier generation of Orientalists had read these texts, they would likely have glossed *The Worthy Gift* as a derivative summary and its commentary, the *Pearls*, as shot through with popular religion and superstition. Similar concerns have shaped the reservations that have been expressed regarding the medical writings of al-Ṣāliḥī's contemporary Aḥmad al-Raqqādī al-Kuntī (d. 1096/1684 or 1116/1704), who ran a Sufi lodge north of Timbuktu, and whose *The Book on Curing the External and Internal Illnesses to which the Body is Exposed* (*Kitab shifa' al-asqam al-ʿarida al-zahir wa al-batin min al-ajsam*), parallels al-Ṣāliḥī's in being an intriguing combination of letter magic, prayers, Galenic humoral medicine, and relevant Prophetic Traditions.[101]

I have suggested here that *The Worthy Gift* and its character as a mnemonic resource for students reflects the degree that literacy and education had developed new institutional roots in the Saʿdi period in rural areas that continued to grow in importance through the tumultuous years of the seventeenth century into the beginning of the ʿAlawite dynasty at the end of that century. *Pearls*, for its part, appears as a rich collection of material and spiritual recipes addressed to scholars who are themselves well versed in the texts and practices of medicine, Sufism, and the occult. For this audience, the fact that God had given black cumin or aluminum healing properties was no more wondrous than the fact that He had done so with the letters of his name, or, for the elect, through prayer itself.[102] We can thus use al-Ṣāliḥī's text as a mirror to reexamine our analytical categories, redefine the corpus of texts that relate to medicine, and to move toward a more nuanced history of medicine and healing in the Early Modern Islamicate world. If we can do this rigorously and the cumulative

[101] See my review of this book's edition and introduction in the *Journal of African History*, 458–59.

[102] Compare al-Ṣāliḥī, *al-Durar al-maḥmūla*, black cumin (11–12), aluminum (27–28), letters (104–05), prayers (155–70).

benefit of numerous microstudies such as this one, we should be able to move toward elucidating more fully the medical paradigm that al-Ṣāliḥī worked within, one that was held to represent an intellectual dead end to physicians working within the colonial period and the product of a social structure that was in the process of being eroded, but which during his time still held the promise of providing Moroccan society with a comprehensive vision of healing body and soul.[103]

4.5 THE NOBLE CRAFT: ALCHEMICAL WRITINGS IN SEVENTEENTH-CENTURY MOROCCO

The cautionary comments I have made in the previous two sections regarding the limitations of our understanding of astronomy, astrology, and medicine in Early Modern Morocco are equally valid for alchemy: the bulk of research into alchemy in the Arabic tradition has been focused on initial translation of Greek alchemical material and its Arabic reception as well as on authors in the Eastern Mediterranean.[104] Our lack of knowledge of early modern Muslim alchemical practices and scholarship is additionally exacerbated by the sometime contested nature of alchemy as a natural science. That is, in a qualitatively different fashion than medicine or astronomy, although closer to astrology, Muslim scholars repeatedly raised questions regarding the legitimacy of studying alchemy.[105] This statement is fair, and yet it could also give the mistaken impression that alchemy was disreputable within the social and intellectual worlds of Moroccan scholars, rulers, and the broader population. As we will see, while many of the details are still unclear, the study and practice of alchemy was

103 For an excellent analysis of the practice of medicine in Morocco during the nineteenth and early twentieth centuries, see Ellen Amster, *Medicine and the Saints*.

104 See inter alia, Kevin van Bladel, *The Arabic Hermes: From Pagan Sage to Prophet of Science*, the translated works appearing in the Corpus Alchemicum Arabicum, including most notably, Zosimos of Panopolis, *The Book of the Keys of the Work* (*Kitāb Mafātīḥ aṣ-ṣan ʿa*); Matteo Martelli, *The Four Books of Pseudo Democritus*, and the articles edited by Regula Forster in the themed issue of *Al-Qanṭara* 37 (2016) entitled "Arabic Alchemy. Texts and Contexts." This body of scholarship builds on the writings of scholars of an earlier generation such as Pierre Lory and his *Alchimie et mystique en terre d'Islam*, which is focused on the writings of Jābir b. Hayyān in the eighth and ninth centuries, and the extensive survey and study of Manfred Ullmann, *Die Natur und Geheimwissenschaften im Islam,* 145–270. There is simply no such parallel body of multigenerational scholarship for the Early Modern period or the Islamic West.

105 Compare with William R. Newman and Anthony Grafton, "Introduction: The Problematic Status of Astrology and Alchemy in Premodern Europe."

vibrant in sixteenth–seventeenth-century Morocco and was championed by prominent scholars.

While in recent years, a series of studies have appeared dealing with the study and practice of alchemy in the eastern Mediterranean in the late medieval and early modern periods – a welcome sign that the study of alchemy in the post-formative period in the Muslim world may be witnessing a resurgence – comparatively little attention has been paid to the study of alchemy in Morocco, despite its reputation during these centuries as a center of alchemical study.[106] This reputation is based on travelers' accounts, which provide a vivid, when not lurid, impression of the place of alchemy in the imagination of Early Modern century North African society. Here, the account of the enigmatic scholar and traveler al-Ḥasan al-Wazzān/Leo Africanus offers the most striking description of the practice of alchemy in early tenth/sixteenth-century Morocco, one given additional weight by his having grown up in Fez after his family was forced to leave Granada at the end of the fifteenth century.[107] In his *Description of Africa,* it is located in his lengthy description of Fez itself:

> Do not think that it has no alchemists! On the contrary, there are a great number of those who have devoted themselves to the study of this vain and senseless art. These are the filthiest and most foul-smelling men in the world due to the sulfur and other stinking substances that they use. Almost every day, in the evening, they gather in the great temple and discuss their intended inventions. They own a large number of works regarding this art, written by excellent authors. The principle one is that of Geber who lived a hundred years after Muhammad's death, and was, according to what they say, a Greek convert. This entire work, even the recipes that it offers, are set down in allegorical form. There is another author named Attoghrehi, who was a secretary of one of the sultans of Baghdad; we have spoken regarding him at greater length in the life of Arab philosophers. Another alchemical treatise was composed in poetical form, containing all of the art's details by a master named Mugairibi who was originally from Baetica. This work was commented on by a Mamluk from Damascus, a man who was extraordinarily competent in this art,

[106] For recent studies of alchemical writings in the Mashriq in the late Medieval and Early Modern periods, see Tuna Artun, "'Hearts of Gold and Silver': The Production of Alchemical Knowledge in the Early Modern Ottoman World"; Nicholas G. Harris, "In Search of ʿIzz al-Dīn Aydamir al-Ğildakī, Mamlūk Alchemist." Recent studies on alchemy in Morocco include Richard Todd, "Alchemical Poetry in Almohad Morocco: The *Shudhūr al-dhahab* of Ibn Arfaʿ Ra's," the invaluable, if eclectic, study of Semlali, *Histoire de l'alchemie* and now al-Hilālī, "Al-kimiyyāiyūn fī al-Maghrib al-Aqṣā khilāl al-ʿaṣr al-wasīṭ."

[107] On the life of Leo Africanus, much of which remains frustratingly mysterious, the best account is Natalie Zemon Davis' elegant *Trickster Travels: A Sixteenth-Century Muslim between Worlds.*

> but the commentary is harder to understand than the text. The alchemists are of two kinds: one group pursues research into the elixir, which is the substance that lends its hue to any metal or mineral. The other are devoted to experiments in multiplying a number of metals through alloys. Yet I have observed that the goal that these people are pursuing most often leads them to produce false coin. Also, most of those whom I met in Fes lack a hand.[108]

Along with a skeptical tone, the passage offers a scattershot sketch of Arabic alchemy, beginning with the third/ninth-century figure Jābir b. Ḥayyān – whose historical existence and supposed study with the sixth Shi'a Imam Jaʿfar al-Ṣādiq (d. 148/765) has been persuasively critiqued – and the corpus of texts associated with him, and then proceeding to the Seljuk scribe al-Ṭughrāʾī (d. 515/1121), whom Leo Africanus did indeed devote a biographical entry to in his *De Viris quibusdam Illustribus apud Arabes*, written in Italy in the 1520s, during the same years in which he wrote his *Description of Africa*, finished in 1526.[109] The poem by a certain al-Maghribī is likely *The Fragments of Gold* (*Shudhūr al-dhahab*) by Ibn Arfaʿ Raʾs (d. 593/1197), who was indeed born in Jaén in al-Andalus (roughly the same region in southern Iberia as the Roman province Baetica), though he lived much of his life in Fez, where he also died.[110] The Mamluk commentator, who goes unnamed here – curiously so, considering his stature – is likely the famed if enigmatic Aydimir al-Jildakī (d. ca 743/1342), who wrote at least six extant commentaries on *The Fragments of Gold*, and whose writings profoundly influenced the subsequent Arabic alchemical tradition.[111]

Leo Africanus was hardly alone in associating alchemy with debased coinage, forgery or political intrigue. Similar concerns and fantasies circulated in the Early Modern Ottoman Empire, and, with regard to the actual transmutation of metals, José Rodríguez Guerrero has detailed at length how the lure of promised alchemical knowledge enabled agents of the Wattasid ruler Muḥammad al-Burtughālī (rl. 1505–24) to retake a strategically important fortification on Morocco's northern coast from

[108] I have consulted a French translation of the Italian original. See Jean-Léon l'Africain, *Description de l'Afrique*, vol. 1, 226–27.

[109] Davis, *Trickster Travels*, 90–94. For a brief history of alchemy in the Muslim world, including a discussion of the Jābir problem, see Regula Forster, "Arabic Alchemy. Texts and Contexts," in Encyclopaedia of Islam, 3rd ed.

[110] See his biography in Semlali, *Histoire de l'alchemie*, 156–63; Todd, "Alchemical Poetry in Almohad Morocco," 119.

[111] See Semlali, *Histoire de l'alchemie*, 160, where he gives the catalog information for al-Jildakī's commentaries in Morocco's Royal Library.

the Spanish at the beginning of the sixteenth century.[112] Alchemy's sometime reputation as a questionable if not treacherous art was, of course, nothing new and it can be related to the science's doubly marginal place in Early Modern Islamic scholarship: unlike astronomy and medicine, Muslim scholars did not teach alchemy in institutional settings (despite Leo Africanus' intriguing observation that the alchemists of Fez met in the evenings in what we can assume to have been the Qarawiyyīn mosque, I have found no other such examples), and additionally the art itself cultivated a reputation for secrecy and mystery that distinguished it from other natural sciences. I will turn below to the alchemical tradition's own self-presentation, but wish to first include the critical view of alchemy of a prominent twelfth/eighteenth-century Moroccan scholar. The famed chronicler of the Saʿdīs and biographer of the ʿAlawite ruler Moulay Ismāʿīl (d. 1722), al-Ifrānī (d. 1157/1745), had the following to say in his *The Singer's Delight regarding the Affairs of the Kings of the Eleventh Century* (*Nuzhat al-ḥādī bi-akhbār mulūk al-qarn al-ḥādī*) regarding the Saʿdī ruler's restoration of the Ibn Yusūf college in Marrakesh. While long, it is worth quoting in its entirety:[113]

> There is a widespread story that the Sultan Moulay ʿAbdallāh built this college by means of alchemy, and that the virtuous Skaykh Abū al-ʿAbbās Aḥmad b. Mūsā taught it to him when, as preceded, the Sultan studied with him. This is an ignorant falsehood. What has been related of the Shaykh Sīdī Aḥmad b. Mūsā is that a man came to him and asked him to teach him the art of alchemy. The Shaykh said to him: "There are five letters that pertain to alchemy and five is the number of fingers on a hand. If you desire it, my brother, then you need to take up farming and agriculture as that is the alchemy of the people, not the alchemy of lead and copper." It is also related that the Shaykh was one of the greatest of the saints, and that he did not wish to open one of the great doors of dissent for a Muslim, one of the great causes of tribulation. For this art is one of the great doors of dissent. The Shaykh would recite these lines of poetry to those who visited him:
>
> Seek the mean in all matters –
> Do not mount a docile beast, nor a wild one.
>
> The Friends of God were united in warning against taking up and seeking after alchemy, and this for any one of the following reasons:

112 See Artun, "Hearts of Gold and Silver," 55–58, and José Rodríguez Guerrero, "Some Forgotten Fez Alchemists and the loss of the Peñon de Vélez de la Gomera in the Sixteenth Century," 298–301.

113 See al-Ifrānī, *Nuzhat al-ḥāḍī*, 108–10, and compare with the French translation given in al-Ifrānī, *Nozhet-Elhâdi: Histoire de la Dynastie Saadienne au Maroc (1511–1670)*, 94–97.

1) That, as mentioned by Ibn Sīnā, it belonged to those things that were impossible, drawing on the saying of the Almighty: {There is no alternative to God's Creation.} Just as it is not in the power of one of God's creations to change a monkey into a human or a wolf into a gazelle, so it is also not in his power to change lead into gold, or copper into silver. There were two men arguing about this. The one supporting alchemy said: "Do you deny that dyeing changes a red body into a yellow one or white into black?" The one denying it said: "I don't deny this paint, for it is not changing the essence. Rather, I deny that the art of dyeing can change a robe of white wool into cotton or red or green silk." As for paint, doubtlessly copper becomes white without this changing its essence, just like dyeing wool doesn't remove the designation of wool from it.
2) That its existence is possible but it doesn't exist in this world. This was the position of Abū al-Faraj Ibn al-Jawzī, God have mercy on him. He said: "There are three things that are generally agreed to exist and that the people of the East and the West agreed that they haven't seen: alchemy, ghouls, and griffons. Everything that has been related about them is hearsay and attribution. Stories about it are like what has been said about dumb beasts and minerals."
3) That with regard to its existence and study, it is forbidden to engage in it or to buy and sell by means of it. Abū Isḥāq al-Tūnisī was asked about it, God have mercy on him. He was asked, "Is it permitted if the result is pure?" He answered, "If silver or another metal is treated so that it becomes gold, without any doubt, but when the seller then doesn't mention to those buying it that it had been silver or whatever metal, but then I treated it so that, as you see, it became gold – then it is fraud and deceit." He said, "When he does explain what he did, no one will buy it from him for cash." They will say to him, "Others than you have treated it and it reverted to its original nature." Whoever does not clarify what he has done falls under what the Prophet said, peace be upon him: "He who deceives us is not one of us." Thus, practicing this art is forbidden. Nāṣir al-Dīn al-Mishdālī said: "I don't know anyone who says that alchemy is permitted." Ibn ʿAbd al-Barr related from the judge Abū Yūsuf that he had said: "Whoever seeks faith through speech is a heretic and whoever seeks money through alchemy will find poverty." Abū Muḥammad Ṣāliḥ used to say: "Avoid three things so that you don't take up three other things: 1) Don't drink fruit juice so that you don't take up drinking wine; 2) Don't busy yourselves with alchemy so that you don't fall into fraud and deceit; 3) Don't sit with old women so that you don't start sitting with younger ones – it is a dangerous amusement." He said, "A donkey was asked, 'Why don't you chew your cud?' He answered, 'I hate chewing what is useless.' He then recited:

I said to my companions, the light of the sun
Is close, but a distant goal to reach."

In sum, there is no foundation to the stories that have spread regarding Moulay ʿAbdallāh regarding this issue. When it was built, the pious avoided praying in the communal al-Ashrāf mosque for a while. It was said that the place of the mosque had been a Jewish cemetery, may God curse them. God knows best.

This passage is rich and speaks both to the stories surrounding the college that had initially been built by the Merinids, as well as to the prevalence of anti-Judaic tropes in Early Modern Moroccan chronicles.[114] More pertinent for us is that it gives an extensive overview of why the study of alchemy was suspect in some Moroccan scholarly circles: al-Ifrānī marshals a number of famous authorities including the famed Persian philosopher and physician Ibn Sīnā (d. 428/1037), the Baghdadi Ḥanbalī jurist Ibn al-Jawzī (d. 597/1200), the Andalūsī Mālikī jurist and traditionist Ibn ʿAbd al-Barr (d. 463/1070), and an array of lesser known North African figures to cast aspersions on the legitimacy and permissibility of alchemy.[115] The one well-known Moroccan scholar mentioned, whose reputation frames the anecdote, Aḥmad b. Mūsā (d. 971/1563), was a prominent Sufi who spread the Jazūlī order in the Sūs, and who left a political legacy in the Tazerwalt region in the Ilīgh lodge.[116] He was not known in the contemporary biographical literature for having studied alchemy, and in al-Ifrānī's narrative functions largely rhetorically to introduce the longer critique of alchemy itself.[117]

Together with Leo the African's account of alchemy in Fez, al-Ifrānī's narrative could be read as a narrative of the marginality of alchemy in Early Modern Morocco. That such an impression would be mistaken is supported by Kacem Aït Salah Semlali's recent study of alchemy in Morocco, in which he makes a strong case for alchemy having been studied and practiced continuously from the Almoravids onwards in the sixth/twelfth century, with an early milestone being Ibn Arfaʿ Ra's' (d. 593/1197–98) *Fragments of Gold*, written under the auspices of the Almohads.[118] Important here is not just that alchemical texts were written, but that respected scholars such as the mathematician Ibn al-Bannā' (d. 721/1321) wrote them, and that

[114] Compare with al-Yūsī, *The Discourses*, 87, 117, 233.

[115] I have not been about to identify Abū Isḥāq al-Tūnisī, Nāṣir al-Dīn al-Mishdālī, or Abū Muḥammad Ṣāliḥ. Compare with the above discussion above of Qāḍī ʿIyāḍ's defense of alchemy in Chapter 3.

[116] On Aḥmad b. Mūsā, see al-Tamanārtī, *Fawā'id al-jamma*, 176–91; al-Ḥudaygī, *Ṭabaqāt al-Ḥudaygī,* vol. 1, 1–10. I have not found mention of him being a teacher of the Saʿdī ruler Moulay ʿAbdallāh, although there is mention of the latter visiting him in the Tazerwalt to pay his respects (al-Sūsī, *Īlīgh qadīman wa ḥadīthan*, 22–23).

[117] Yet compare with Semlali, *Histoire de l'alchemie*, 205–07.

[118] Semlali, *Histoire de l'alchemie*, 42–43; Todd, "Alchemical Poetry in Almohad Morocco."

chroniclers repeatedly noted its practice, as did the Granadan vizier Ibn al-Khaṭīb (d. 776/1374) of the Fassi scholar Aḥmad b. Muḥammad b. Shuʿayb al-Kiryānī (d. 749/1349) – although he did not refrain from expressing his own low opinion of the art.[119] Semlali's study, which draws extensively on manuscripts in private collections that are not reflected in the catalogs that form the basis for the list of natural science manuscripts given in the appendix to this chapter, argues for the increased importance of alchemy in Moroccan scholarship from the tenth/sixteenth century onwards. The first major alchemical text of this period is Muḥammad ibn Aḥmad ibn ʿAbd al-Malik al-Ḥasanī al-Maṣmūdī's *The Comprehensive Volume on the Sufficient Preparation* (*Kitāb al-wāfī fī al-tadbīr al-kāfī*), which the author completed in 1492 and copies of which remain widespread in public and private libraries in Morocco today.[120] We know little of al-Maṣmūdī's life beyond that he studied alchemy in Egypt and settled in Tlemcen, in what is now Western Algeria, but his *nisba* suggests that he was of Amazigh and Moroccan origin.[121] In the introduction to *The Comprehensive Volume* he gives an overview of the books he studied in Egypt with a certain ʿUmar al-Qayūmī [al-Fayūmī?], which include Hermes, the tenth century Ibn Umayl, the seventh/thirteenth century Abū Qāsim Muḥammad al-ʿIrāqī, al-Jildakī, Ibn Wahshiyya, and poems of Khālid b. al-Yazīd, a grandson of Muʿāwiya, the founder of the Umayyad dynasty who was famed as the founder of the tradition of Islamic alchemy, and many others.[122] A few decades after the completion of this text, a scholar in the Draʿ valley, ʿAbdallāh ibn Muḥammad al-ʿAnnābī al-Darʿī (d. ca. 924/1518), was active as an alchemist, including providing the early generations of the Saʿdīs with gold to support their growing political ambitions. In the chronicle of Ibn ʿAskar (d. 976/1578), the Saʿdī ruler al-Ghālib Abū Muḥammad ʿAbdallāh – the same ruler whose renovation with the proceeds of alchemy of the Ibn Yūsuf college al-Ifrānī had disputed – relates:[123]

[119] Ibn al-Khaṭīb, *al-Iḥāṭa*, vol. 1, 134.

[120] The Royal Library in Rabat possesses six copies of it (al-Khaṭṭābī, *Fahāris al-khizāna al-malikiyya* vol. 5, 266–69).

[121] Semlali, *Histoire de l'alchemie*, 200–02. I have consulted the British Library manuscript digitized by the Qatar National Library (Or 13006, ff 113r–158r), which was copied in Algeria in 1177/1764.

[122] Al-Maṣmūdī, *Kitāb al-wāfī fī al-tadbīr al-kāfī*, 113v.

[123] See Ibn ʿAskar's *Dawḥat al-nāshir*, 9 (and compare with al-Ḥudaygī, *Tabaqāt al-Ḥudaygī*, vol. 1, 276–77, where the same story is preserved two centuries later). Ibn ʿAskar continues by describing how after the fall of Granada in 895/1492, al-ʿAnnābī wished to ransom captured Muslims, and that after asking for and receiving money from the Waṭṭasid ruler Abū Zakariyya in Fez, he set out to al-Andalus to do so, only to drown when the boat capsized.

My ancestor met the Shaykh ʿAbdallāh ibn Muḥammad al-ʿAnnābī al-Darʿī and complained to him about the weakness of his affairs and his lack of ability to carry out the claim his royal parents had given him. The Shaykh said, "The Prophet's descendants are the most worthy of victory! Give me whatever iron you have." My ancestor went and, not finding anything else, brought him an iron bar. He said, "Do you have any more than this?" My ancestor answered, "This is all I found." The Shaykh said, "Wait until I come back." He entered his house and returned after an hour with the bar in his hand. "Take this!" He said, and the bar was pure gold. Al-Ghālib said, "I still have my mother's anklets, which were made from that bar."

There are a series of scholars who wrote alchemical works during the Saʿdī period, including the prolific Muḥammad ibn ʿAlī al-Ḥājj al-Shuṭaybī (d. 963/1556), who, while perhaps best known to later scholars for his Qur'anic exegesis and writings on Sufism, wrote several works on alchemy as well as a commentary on the mathematician Ibn al-Bannā's *al-Mabāḥith al-aṣliyya.*[124] Others included a descendant of the putative founder of the Shādhilī Sufi order Ibn Mashīsh, ʿUmar ibn ʿIsā ibn ʿAbd al-Wahhāb al-ʿAlamī, who dedicated an alchemical treatise to Muḥammad al-Kharrūbī (d. 962/1555), himself chiefly known for his involvement in a tempestuous debate around the precise meaning of the *shahada*;[125] finally, there was also Aḥmad ibn al-Ḥasan ibn Yūsuf, known as Ibn ʿArdūn al-Zaghlī (d. 992/1584), who served as judge in Chefchaouen and who wrote two poems on alchemy.[126] It is often hard to gain much precise information about what it meant to be described as an alchemist. Thus, while we are told by the chronicler Ibn ʿAskar that Abū ʿAbdallāh Muḥammad al-Andalusī (d. 984/1577) was fascinated with the science of letters, alchemy, medicine, and geometry we are given no further details – nor is there any clarity regarding the relationship, if any, between his study of the occult sciences and the revolt of his followers in Marrakesh which led to his being arrested by the Sultan and subsequently crucified by the masses.[127] That

124 On al-Shuṭaybī, see Ibn ʿAskar, *Dawḥat al-nāshir*, 16–17 (reproduced with a few changes in al-Ḥudaygī, *Tabaqāt al-Ḥudaygī*, vol. 1, 269–70), and Semlali, *Histoire de l'alchemie*, 208–14. I have not been able to consult any of his alchemical works.

125 Semlali, *Histoire de l'alchemie*, 54. On al-Kharrūbī, see El-Rouayheb, *Islamic Intellectual History*, 222, and the references given there. Caitlyn Olson's dissertation – "Creed, Belief, and the Common Folk: Disputes in the Early Modern Maghrib (9/15–11 /17 c.)" – deals with this controversy in greater depth. More information on al-ʿAlamī can be found in the works of Muḥammad al-Mahdī al-Fāsī (d. 1109/1698), *Tuḥfat ahl al-ṣadīqiyya* and *Mumtiʿ al-asmā'*, neither one of which I have been able to consult, and both of which relate to the history of the Jazuliyya.

126 See Semlali, *Histoire de l'alchemie*, 54. These poems seem to be only extant in private libraries, such as the ones that Semlali himself consulted.

127 Ibn ʿAskar, *Dawḥat al-nāshir*, 109.

some saw a connection between alchemy and political unrest is clear in the example of the messianic rebel Ibn Abī Maḥallī, who was killed in Marrakesh in 1613 and whose views on smoking we have already discussed in Chapter 3. Both Semlali and Berque have speculated that he was an alchemist, drawing on rumors that he financed his rebellion with gold, but he does not appear to have written a work on the art itself, nor do any of the contemporary biographical sources link him with alchemy.[128] In addition, it is curious that his reputation as a supposed alchemist traveled abroad to Europe, where we find the German alchemist Michael Maier (d. 1622) writing in 1617 in his *Symbola Aureae Mensae Duodecim Nationum* in a chapter on the presence of Rosicrucians in Germany, of Ibn Abī Maḥallī as a prophet and his movement leading to the introduction of Rosicrucianism into Spain:

> At that time [1613], certain miraculous innovations were brought back from Barbary through it, in that near Morocco and Fes a certain prophet by the name of Moulay Om Hamet Ben Abdela arose from among a number of wise men. He distinguished himself through numerous occult signs, and having prepared a large enough army, overthrew and defeated the king of that region, Moulay Sidan, who was nearly unarmed and with a small number of troops. Thus, he obtained the kingdom for himself. In addition, these brothers of inconstant reputation [i.e., the Rosicrucians] were said to have come out of Barbary to Spain and were considered to be of the same art and institution as this prophet of Barbary.[129]

As intriguing as the stories regarding Ibn Abī Maḥallī as founder of Rosicrucianism are, they tell us little of the actual practice of alchemy in Morocco in the seventeenth century. Here, two works by the aforementioned al-Mirghitī are more useful: the first is a short auto-commentary on a poem on alchemy in which he decodes alchemical terminology, while the second, *On Alchemy* (*Fī ʿilm al-kīmyāʾ*) is a substantial treatise giving an overview of the art in its entirety.[130] Together, they offer an impression of Moroccan alchemy as conceived of and practiced by one of the century's most prominent Moroccan scholars, who was admittedly

[128] Berque, *Ulémas, fondateurs, insurgés du Maghreb*, 76; Semlali, *Histoire de l'alchemie*, 244–45. Berque notes that Ibn Abī Maḥallī addresses alchemy in his work *Salsabīl al-ḥaqīqa wa al-ḥaqq fī sabīl al-sharīʿa li-l-khalq* – but I have not been able to consult this work (Berque, *Ulémas*, 277).

[129] Michael Maier, *Symbola Aureae Mensae Duodecim Nationum*, 290. There is some irony in that the alchemist that Maier presents as representing the Arabs was Avicenna, who was not only Persian, but also a staunch critic of alchemy. My attention was drawn to Maier's work by Semlali, *Histoire de l'alchemie*, 242.

[130] I have consulted al-Mirghitī, *Sharḥ abyāt al-kīmyāʾ* ("Commentary on Verses on Alchemy") and al-Mirghitī, *Fī ʿilm al-kīmyāʾ*.

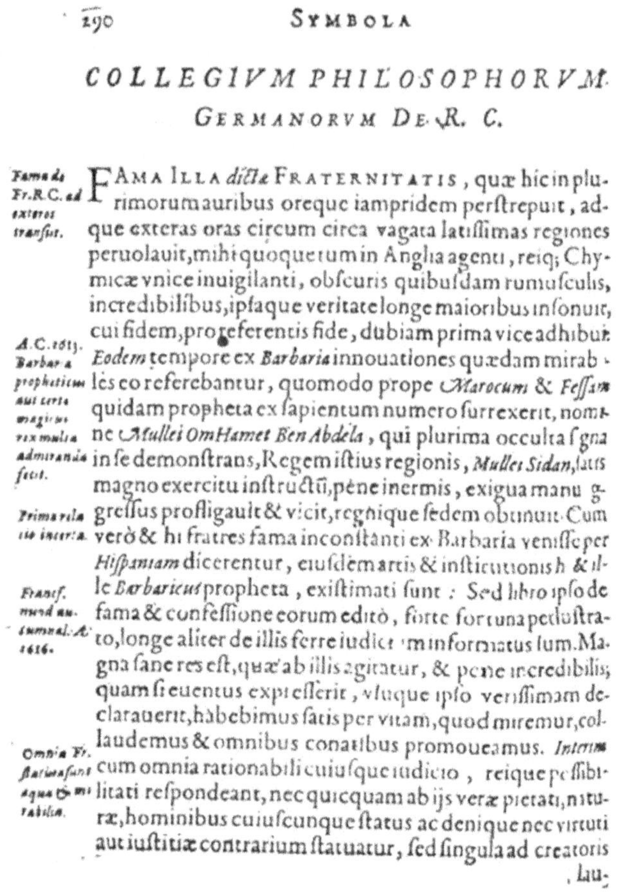

190 SYMBOLA

COLLEGIVM PHILOSOPHORVM GERMANORVM DE R. C.

Fama de Fr. R. C. ad exteros transfut. FAMA ILLA dicta FRATERNITATIS, quæ hic in plurimorum auribus oreque iampridem perstrepuit, adque exteras oras circum circa vagata latissimas regiones peruolauit, mihi quoque tum in Anglia agenti, reiq; Chymicæ vnice inuigilanti, obscuris quibusdam rumusculis, incredibilibus, ipsaque veritate longe maioribus insonuit, cui fidem, pro referentis fide, dubiam prima vice adhibui. A.C. 1613. Barbaria prophetico aut certe magico vix mulia admiranda fecit. Eodem tempore ex Barbaria innouationes quædam mirabiles eo referebantur, quomodo prope Marocum & Fessam quidam propheta ex sapientum numero surrexerit, nomine Mullei Om Hamet Ben Abdela, qui plurima occulta signa in se demonstrans, Regem istius regionis, Mullei Sidan, satis magno exercitu instructū, pene inermis, exigua manu aggressus profligauit & vicit, regnique sedem obtinuit. Prima rela ita incerta. Cum verò & hi fratres fama inconstanti ex Barbaria venisse per Hispaniam dicerentur, eiusdem artis & institutionis h & ille Barbaricus propheta, existimati sunt: Franc. mund au. tumnal. A. 1616. Sed libro ipso de fama & confessione eorum edito, forte fortuna perlustrato, longe aliter de illis ferre iudicium informatus sum. Magna sane res est, quæ ab illis agitatur, & pene incredibilis; quam si euentus expresserit, vtique ipso verissimam declarauerit, habebimus satis per vitam, quod miremur, collaudemus & omnibus conatibus promoueamus. Omnia Fr. statuta sunt æqua & mirabilia. Interim cum omnia rationabili cuiusque iudicio, reique possibilitati respondeant, nec quicquam ab ijs veræ pietati, naturæ, hominibus cuiuscunque status ac denique nec virtuti aut iustitiæ contrarium statuatur, sed singula ad creatoris lau-

FIGURE 4.5 *Symbola aureae mensae duodecim nationum* (Printed 1617).

chiefly known for his prominence in astronomy. Before turning to them, though, it is worth quoting a definition of alchemy by al-Mirghitī's most famous student al-Yūsī, from the latter's *al-Qānūn* (discussed previously in Chapter 2):

The ninth natural science is alchemy, which is the science by which one strips metallic substances of their attributes and confers upon them the attributes of other substances, as with turning lead, for example, into silver or gold. The benefit of this is clear if the process is proven. Know that this science is built on the principle of all metallic substances sharing an essential reality while differing in their transient attributes. They display some hesitation as to this being possible and of it actually taking place. The speaker spoke rightly when he said:

> Neither sufficient wealth nor sufficient alchemy
> Can be achieved. Rid your ego of this desire.
> Many have spoken of the nature of both of these
> I do not believe in the existence of either.[131]

Al-Mirghitī did not share his student's skepticism, and in *On Alchemy,* offers an introduction to the Great Art, along with an overview of the beginning of his own study of the science. The treatise, he notes in its opening, was a summary of what he had learned of the "great divine art" (*al-sina'a al-'uzma al-ilahiyya*), and especially of the Philosopher's Stone, a work that he structured according to the letters dotted in the Arabic alphabet (p. 2). This does not seem to have been an idle desire of his, as he writes eloquently both of the difficulty of studying alchemy and of seeing others fail to practice it properly:

> Know, O brother, that people differ on its nature and means (*māhiyya wa kayfiyya*) and what it is. Know also that I exerted myself for a long time in the science of the art without discovering anything, that is, of the truth. Then, God Most High opened its door for me, and taught me what kept it veiled (*mā inghalaqa 'alayhā ḥijābu-hā*). I saw it then in truth and saw what the scholars had said regarding it was true. I contemplated what was in it with the effort of my insight (*bi l-naẓr jahdī*), expending myself and my thought in the matter. Morning and evening I devoted my study to the books of the philosophers and to experiment, and its wonders became evident to me, its rivers flowed for me, and the blossoms of its flowers opened for me. I saw people working in it deviate from its path, its trace being hidden from them. Thus, they desired me to speak clearly to them of this. I was not able to do this, out of fear for myself. I left the speech with them in its entirety . . .. I did not wish to mention their like here . . . God willing, the intent is for me to clarify to you everything you need of the art. Let us begin first with the knowledge of it. They had said that the one who seeks this science must know three qualities (*khiṣāl*): its definition, what it is made of, how it is composed. If you know this truly, then you will be successful in it and will have reached your goal in this science (pp. 2–3).

While it remains unclear precisely what experiments al-Mirghitī undertook, later chapters in the book describing different types of needed glass vessels (pp. 44, 58) along with detailed descriptions of making vinegars (p. 35), sulfur (p. 74), salts (p. 76), soaps (p. 81), dyes (p. 82), as well as the use of the scales to weigh substances (p. 127) strongly suggest that alchemy

[131] Al-Yūsī, *al-Qānūn*, 163–64.

was not only a theoretical science for the author, but an imminently practical one as well.

The purpose of these experiments was the transmutation of substances and the production of the Philosopher's Stone (*al-ḥajar al-mukarram*), resting on an understanding of physical reality that overlaps in significant ways with that put forth within the Galenic medical tradition in which this temporal world of generation and corruption is composed of the four primal elements. This much is laid out in the first section of al-Mirghitī's treatise, although he begins it, as he had with his discussion of astrology, with a genealogy of the science that stretches back to Adam, whom God taught the names of everything (p. 3). Citing a book of an unnamed scholar (*baʿḍ kutub al-ḥukamāʾ*), he notes that of the ten sciences that God possessed, one was that of the art and another that of establishing the proper proportions (*al-taʿdīl*). After Adam's death, his son Seth inherited these from him, and set them down on two lead tablets, which, knowing of the flood to come, he buried in a box of lead from which they were later recovered by Noah (p. 4). Subsequently philosophers and scholars debated the nature of the lower world and its composition out of the four elements, regarding whether it was metal, vegetal, or animal in nature, each party arguing for a different type of elemental composition. This situation did not change until God revealed the Philosopher's Stone to them, which he composed out of all four elements (p. 5).

The second section takes up the question of the method of the art, beginning with an explanation of how the four elements are interrelated with flesh and blood themselves being a product of a progression of degrees from mineral to vegetal to animal. As far as the Philosopher's Stone or elixir is concerned, this is, as the sages (*al-ḥukamāʾ*) agree, a humble, subtle substance found in garbage heaps, markets, and roads, and it is the sulfur (*al-kibrīt*) that is present in animals, both those that speak and those who do not (p. 6). The humble nature of the stone is borne out by citations of poetry of both the famed Sufi Dhū al-Nūn al-Miṣrī (d. in 245/859 or 248/862) and Khālid b. Yazīd (d. 85/704), two authorities whose words form the basis for al-Mirghitī's subsequent commentary. The stone is characterized by its balanced nature, in which no element dominates, and the author spends some time reflecting on how its color also reflects its balanced elemental composition (p. 8).

The body of the treatise drifts from one subject to another, and can be roughly divided into discussion of matters related to theoretical issues and to practical questions of how to perform alchemical processes, especially with reference to particular vessels or instruments. The beginning of the treatise takes up introductory issues, with the root of the art itself, al-Mirghitī notes, lies in the creation of waters of gold and silver – a process he describes in detail – to use as base ingredients (p. 9); along with lead, these are the three earths that are the foundation of the art. The following pages describe the processes of "whitening" and "blackening" as necessary processes of the art. After covering several other processes, and along with the previous authorities quoting al-ʿIrāqī and al-Jildakī the author takes up the question of how to multiply the amount of elixir that one possesses (p. 28). While the elixir itself will be returned to later in the treatise, the contrast with the discussion of instruments is a good example of the practical aspects offered by the treatise.

The chapter on instruments begins on page 43 of the treatise, with numerous images aiding in understanding precisely how they are to be used. It is important that they be made of sturdy pure white glass – ones out of fired clay are a distinct second – (p. 44) while in size they can be from one to two and a half feet (*shibr wāḥid ilā shibrayn wa niṣf*) in height and from half a foot to a foot and a quarter in width. The author then cites many of the already named authorities, including a large number of lines of poetry by Khālid b. Yazīd (p. 45), but also the preferences for particular sizes of vessels of the above mentioned Muḥammad ibn Aḥmad ibn ʿAbd al-Malik al-Ḥasani al-Maṣmūdī (p. 46), showing an awareness of his North African antecedents alongside his knowledge of earlier Mashriqi alchemical authorities.

The manuscript page below is a good example of how al-Mirghitī uses illustrations to clarify his description of a vessel – in this case of two kinds of alembic:

> Take heed! Whoever wishes for the aforementioned vessel to be clear-sighted (*baṣīra*), let him place the spout of the vessel and the pipe above its mouth – inside it in this fashion [indicating the first drawing above]. If you wish the vessel to be blind and clear-sighted, then place over the side of the spout next to its mouth a serrated ring (*ṭawqan akhr mafruzan*) at the height of one finger. Into it empties a smaller alembic (*al-inbīq*) with a *nadam* on it to facilitate dripping (*li-l-taqṭīr*) – in this fashion [referring to the second drawing]. Know my brother that you will find no investigation into this instrument, which is the foundation of the art, nor anyone more eloquent in describing it or in mentioning it than this clear and sufficient clarification – save those who bury it in symbols, divide its description into the furthest places and refer to it with innumerous varied names (pp. 26–27).

Al-Mirghitī then gives a whole series of such names that are synonymous with alembic before describing the alembic itself as a hollowed-out ring

FIGURE 4.6 Al-Mirghitī, Fī ʿilm al-kīmyāʾ (Jumʿa al-Mājid Center).

(*tawq*), like a dome, preferably made out of glass. Its edge touches on the edge of the vessel, surrounding it a little and being one of two types, blind and clear-sighted – the latter having a spout and pipe from which the distilled liquid emerges, it is a foot and a third in length, its head consisting of a foot and its neck of a third of a foot. This neck guides the distilled liquid into a vessel while its end enters the mouth of the initial vessel, as seen in the illustration on the bottom of the left page.

The alembic being a well-known alchemical instrument since the writings of Zosimos in the third century, what is interesting in al-Mirghitī's writings is not its novelty, but that he addresses the workings of alchemy as openly as he does and that he does this in great detail (although this too is arguably a trope of some alchemical writing). His self-presentation as a trustworthy commentator on an occult tradition is similarly found in his auto-commentary on a poem on alchemy. We unfortunately do not know when the poem and its commentary were written or whether they preceded or followed the treatise on alchemy itself.

Al-Mirghitī opens the commentary by explaining that the Ḥusaynid Sharīf Sīdī ʿAbdūn ʿAbd al-Raḥmān asked him to comment on the verses he had written on alchemy, specifically with regard to the herbs and medicines mentioned therein.[132] He then gives the fourteen verses in question, which are structured as an independent poem and difficult to parse, and, befitting a commentary addressed to a descendant of the Prophet, proceeds to praise the Prophet's family extensively. He follows this with a gloss of individual phrases and terms that fall under what contemporary scholarship on premodern alchemy refers to as "Decknamen," or coded terms used to preserve the secrecy of alchemical procedures, before providing an overview of the procedure as a whole. The reader thus learns that "your washed soap" (*ṣābūnuka al-maghsūl*) means for the people of this art *(fann)* "table salt," with the word "washed" having the particular meaning of proven or reliable (*thābit*) (p. 4). Among other examples, the phrase "feed your eagle," for its part, refers to mixing the liquid in question with ammonia (*al-nushādhir*) (p. 5). After the individual glosses, he gives an overview of the poem's meaning, the first part of which reads as follows:

This is a depiction of the matter, and God willing an explanation of it. Take, with the blessings of God the Almighty, a portion of pure (*al-thābit*) white table salt (*milḥ al-ṭaʿam*), a portion of pure alkaline salt, a portion of un-dissolved (*ghayr al-masqī*) lime (*al-jīr*). Mix them all together well, then divide it into five parts. Then add a drop of sour vinegar to the first part, and take a drop from the result and place it in the second, and so on through all the parts. Preserve what results from this in a small glass vial or vessel with a narrow mouth .... Then place in it five amounts of quality burned eggshell – it causes anyone who smells it to sneeze – and add the hair of adolescent boys, that is those who are close to attaining maturity, free-born, not sick, nor of ruddy complexion and leave it until it dissolves. Render it clear through boiling (*al-ḥarqa*), and take what is clear of it, place it on the fire or in the sun, if it is summer. Divide it, and mix it with stable white eagle (i.e., ammonia). Grind it up well, place it in a glass vial, close its top and place it in a moist place (*al-nadāwa*) until it dissolves into liquid. Then, take the sulfur and grind it black and white (*balīqan*). Place it in a tajine made of new pottery and set a fire under it after having poured over it the aforementioned dissolved liquid until it covers it to the extent that you can drink it. Add more water in this fashion, while maintaining a low fire under it and keep an eye on the coals until they go out and no smoke raises and no flame brings the coal to life. Do not place it on a fire, and do not blow on it with the bellows. When you know it to be pure, remove it and its sun [i.e., gold], grind it and take the moon [i.e., silver] and sun, flatten it and scatter the aforementioned dust over it so that it is covered by it. Then close it over and bury it in the coals until morning, when you will find it to have become a quality (*maḥmūdan*) calcified salt (*baḥran*)[133] that can be ground. Then, take the same amount of mercury (*zawwāq*),

132 Al-Mirghitī, *Sharḥ abyāt al-kīmyā'* ("Commentary on Verses on Alchemy"), 1.

133 On the difficulty of correctly interpreting *baḥr*, Siggel notes that it can refer to anything that proceeds from the Stone, including magnesium (*Decknamen*, 26–27).

grind it together extremely finely and scatter the aforementioned mercury over it so that it is covered with it and bury it until morning when you will find it to have become solid.[134]

The explanation of the poem's meaning continues for another page, in which the author differentiates between producing a red or a white substance (p. 9), followed by a final section in which al-Mirghitī offers pieces of advice on details of the procedure or on how to ensure slightly different outcomes. There are a few points of interest here: he repeatedly refers to using the ovens of bakers or potters to carry out the necessary cooking processes (p. 10), and perhaps to clarify his own intention and the proper practice of the art, he stresses to his Sharifian reader the importance of the next world over this one, and that since he has explained how to make pure gold there is no need for any further operation that would involve mixing pure with impure gold as this would involve fraud, which the Prophet had spoken strongly against (p. 9). Unfortunately, the purpose of the operation described, in which the production of gold seems almost tangential to the overall process, remains unclear (to this reader at least).

Even with al-Mirghitī's stated desire to speak plainly regarding the art, which does make his work on the art reasonably approachable to the uninitiated, this brief overview of two of his alchemical writings is insufficient as a summary of his thought on the subject and to Moroccan alchemical writings in the seventeenth century in general. What is clear is that an established and influential Moroccan scholar who was renowned for his astronomical writings and who had taught thinkers of the stature of al-Rūdānī and al-Yūsī, was a public advocate and explicator of the Great Art and, as far as we can tell from the detailed descriptions in his writings of both instruments and procedures, was a practicing alchemist himself. Numerous questions remain, chiefly related to the ways on which alchemical knowledge was taught and transmitted and the degree to which it was studied and practiced. To answer these would require evidence beyond the anecdotal accounts given at the beginning of this section and access to more manuscripts contemporary to al-Mirghitī himself. Only then will we be able to set these texts and other like them in conversation with, for example, the numerous Ottoman poems on alchemy that Tuna Artun has recently discussed being in circulation in Istanbul in the same century.[135]

134 Al-Mirghitī, *Sharḥ abyāt al-kīmyā'* ("Commentary on Verses on Alchemy"), 6–8.

135 It is worth noting that the Moroccan Royal Library contains numerous manuscripts of the Ottoman alchemist ʿAlī Chalabi al-Aznaqī (d. 1018/1609) (al-Khaṭṭābī, *Fahāris al-khizānah*, vol. 5, 23 and then frequently afterwards), whom Artun discusses extensively throughout "Hearts of Gold and Silver."

4.6 CONCLUSION

The scholarship on the natural sciences in Early Modern Morocco was rich and varied, covering astronomy, astrology, medicine, and alchemy as well as other natural sciences not discussed here, such as lettrism. As seen in Chapter 2, the presence and importance of the study and teaching of these sciences is reflected in the biobiographical literature of the period, but this literature, with its emphasis on the transmitted religious and spiritual sciences arguably did not reflect the true social and intellectual presence of the natural sciences. Reviewing, even if only as briefly as I have done here, examples of these sciences and drawing attention to the ways in which their authors were connected to the central scholarly and spiritual networks of the time suggests both a more dynamic scholarly production in these fields and a more influential social presence than previous scholarship on this period had described. In this, this chapter's findings are in line with the argument laid out in Chapters 2 and 3 that the natural and mathematical sciences played an important role in the intellectual, spiritual, and social worlds of Early Modern Moroccan scholarship. What this means, and how it nuances our larger narratives of Islamic intellectual history are questions that I will take up in this book's Conclusion.

It is worth remembering that while the natural sciences were important in Early Modern Morocco, they were not uncontested. As seen in the above discussion the status of alchemy was especially debated, with al-Ṣāliḥī and al-Mirghitī, respectively attacking and defending its legitimacy, and in al-Yūsī's famed *Discourses* we find anecdotes attacking the discipline's validity although he himself defended it in his *al-Qānūn*. The broader point is that Moroccan scholars not only produced scholarship in the natural sciences, they also debated the nature, effectiveness, and validity of individual sciences, often with an eye to theological anxieties regarding fellow believers losing sight of God's omnipotence and imminence. Still, in a sense this observation rests on a false juxtaposition. It is useful to remind ourselves, yet again, that while, following our seventeenth-century Moroccan predecessors, we can productively distinguish between natural and religious sciences – I myself have done so at times – the two bodies of knowledge bled into each other constantly, regardless of the boundary work that some scholars carried out. To draw on an argument made by Andrew Cunningham with regard to natural philosophy in Early Modern Europe, when the scholars discussed in this chapter wrote on natural science, they were writing about God and about the ways in which His Habit was manifest in this temporal world. Their study of that world and their marvel at the natural processes within it was carried out with a deep awareness – emphasized repeatedly in their own writings – of God's constant creation of it and of the much more important world to come. This awareness did not lessen or hamper their appreciation of the processes of this world, or detract from their drive to discover how to manipulate these processes, as some have implied. Instead, their engagement with these sciences, as was true of their contemporaries in the eastern Mediterranean, was cultivated within social and intellectual networks that were deeply religious in nature.

Excursus

Sufism and the Spiritual Life or Balancing the Exoteric and Esoteric Sciences

With regard to Sufism, the shaykh of the community, Junayd, God be content with him, said of his time:

> The people of Sufism have all passed away –
> Sufism has become a shadow of itself.

These lines are very well known. What then of our own time? This shadow has itself become a shadow, and people's morals continue to decline. As was observed long before this: Pretense is widespread, weakness is everywhere. Today, prayers are veiled from God. The words of the people of great authority, such as the luminaries of the Qādiriyyah and the Shādhiliyyah, God be pleased with them, have reached the ears of the commoners, as has the speech of those who have mastered spiritual states from every period. The commoners' egos have been infatuated with them. The masses have been drawn to them, and have dived headlong into imitating them. When you wish to meet someone ignorant, who is exceedingly focused on his ego without knowing the external aspects of the revealed law – much less acting according to it, reaching its internal aspects, experiencing a spiritual state, or having attained a station on the path – you will find him belligerent, gossipy, and opposed to both the traditional and intellectual sciences.

Al-Yūsī, *The Discourses*[1]

There are few aspects of Islamicate civilization that are today as misunderstood by Muslims and non-Muslims as Sufism. Much as the position of the position of the natural sciences in Muslim societies was decisively recast in the nineteenth-century debates around reform in the Muslim world, the various reform movements of the eighteenth and nineteenth centuries redefined the role and nature of Sufism. For the first time in many

[1] Al-Yūsī, *The Discourses*, 315–17.

centuries, it became possible to think of Sufism as distinct from Islam, as, depending on one's ideological bent, a later deviation from an originally pure Islam or a foreign importation from Christianity. This impression was heightened in Europe and especially in the United States in the first half of the twentieth century, where for perhaps the first time it became possible to conceive of oneself as Sufi without also being a Muslim. One effect that this separation of Sufism from Muslim thought had for some historians and modernist thinkers alike was that they could assign Sufism a specific role in intellectual, social, and political developments in the premodern period. Thus, Sufism could be described as the factor holding Muslim societies back due to its esoteric, fatalistic, or obscurantist nature; it could be flagged as the social force that gave rise to anti-colonial uprisings, or that which distinguished lower from more elite social classes; it could be glossed as popular as opposed to orthodox Islam, itself usually identified with the legal tradition. Different as these characterizations are, they all give the impression that we can (and should) distinguish between Sufism and Islam itself, an impression that affects our study of the natural sciences in the premodern Muslim Middle East insofar as we could conceive of Sufism supporting or hindering their study.

It may help here to return first to definitions and to gloss Sufism generally as a tendency toward taking the example of the Prophet Muhammad as an inspiration for an ascetic and spiritual orientation to religious practice of a type that strove to subdue the desires of the lower self and to focus on the immanence of God. As such, it was one of a number of pietistic movements within Muslim societies and cannot be reduced, as it often is, simply to the category of mysticism, itself largely unhelpful. While Sufi discourses did stress the importance of personal self-discipline in order to control the body's physical desires as well as a constant focus on the divine, these aspects should be balanced with the social presence of Sufism as a set of orders structured around the authority of individual teachers with respective lodges and gatherings with which one could be affiliated to varying degrees. These orders were differentiated by distinct ritual practices, most of which related to the nature of the prayers that were recited in communal gatherings, the degree to which such gatherings encouraged ecstatic states, and the style of their leaders' instruction and interaction with their disciples. Orders were not exclusive and individual scholars could and did belong to multiple orders and studied with multiple shaykhs – spiritual authorities. The pervasive nature of Sufism should not be taken to mean that all aspects of Sufi social practice were universally accepted. And, as the quote from al-Yūsī adduced above suggests, every generation of Sufi scholars brought with it those who lamented that they were not able to match the sincerity of those before them. Similarly speaking to the diversity of practices and attitudes covered by the term, many of the debates that dealt with Sufis in the premodern period addressed specific practices associated with individual Sufi groups – veneration of saints, ecstatic sayings and states, and the possibility of saintly intercession being particularly open to debate – and not the broader pietistic-ascetic orientation or social organization of Sufism as such. In this way, our understanding of Sufism in the premodern period has been distorted by our projecting of contemporary tensions and debates onto past scholars and communities.

It is tempting here to forestall the modern idea that Sufism or certain tenets or practices falling under its umbrella functioned to obstruct scientific inquiry by invoking the European parallel of natural philosophers also being practicing Christians – Newton being perhaps the paradigmatic example of a figure central to the history of Early Modern European astronomy who also spent extensive periods of time pursuing his religious interests. There is much to recommend this comparison: recent generations of historians of European science have decidedly done away with the conflict thesis in which religious belief or organizations are held to be obstacles to the study of science. Whatever conflicts arguably did occur between practitioners of the natural sciences and representatives of religious traditions (often overlapping categories) have been contextualized to dispel notions of essential conflict and shown to have been contingent upon specific political and social circumstances. The much-discussed case of Galileo has been a powerful one in this regard. Historians of the Muslim Middle East have been slower in coming to the same conclusions, but recent publications on the thirteenth-century Kurdish scholar al-Āmidī have argued similarly that al-Āmidī's opponents who accused him of heresy were motivated politically and that the episode cannot be used as support to claim broader claims of the post-formative Islamic world having been characterized by an anti-rationalist religious orthodoxy.

Yet the comparison of the Middle East with Europe, and the awkward Islam and Christianity corollary that goes along with it differ precisely in terms of the presents from which they look back. Bluntly: because of the dominance of Europe and its former colonies in the production of the natural sciences today it is easier to differentiate the role of religion in its past between beneficial and harmful, or as a generation of social scientists put it at the beginning of the twentieth century, including Max Weber and Robert K. Merton, between Protestant and Catholic. The neatness of this binary is questionable, and historians are currently involved in a new stage of parsing the significance of the Protestant Reformation and the full extent of the effects of the Black Legend regarding the negative influence of Catholic countries on Europe's intellectual development. In the context of the Muslim Middle East, however, there has been the assumption that, something went wrong with Islam at some point and thereafter it functioned to hold back scientific progress. That is to say, evoking the example of Europe and the possibility of religion prompting or spurring on scientific engagement does little to help reassure the reader to the compatibility of Islamic piety in general, and Sufism in particular, with investigation into the natural sciences. Hence, Sufism does require commentary.

In line with the recent general interventions on the nature of Islam itself by Bauer and Ahmed, and the excellent analytical overview of Knysh, we are increasingly understanding Sufism not only to have been pervasive in the premodern Muslim world but to have been characteristic of it ever since the emergence and spread of Sufi orders in the eleventh–twelfth centuries. This is especially the case for Morocco, where as described in Chapter 1, the transmission and study of knowledge, both within and outside of urban and rural institutions, was carried out by scholars who almost without exception were affiliated with Sufi orders. What does this mean?

From the perspective of intellectual history, adherence to a Sufi order did not limit or characterize a given figure's scholarly production and since the vast majority of scholars were Sufis, it comes as no surprise that Sufis wrote on every subject. Educational institutions linked to Sufi orders similarly did not determine the limits of what could or could not be studied. To be sure, when a scholar who was a Sufi wrote on an aspect of Sufism itself, he could take up Sufism as an explicit category. This happened in an especially clear fashion with biographical dictionaries dedicated to Sufis or to the followers of a particular order. Three immediate examples for the Moroccan context that come to mind are al-Tādilī's (d. 627/1230) *Gazing upon the Men of Sufism*, Muḥammad al-Makkī ibn Mūsa ibn Nāṣir al-Dar'ī's (d. after 1166/1752) *The Set Pearls of the Accounts of the Notables of the Dar'a Valley* (*al-Durar al-muraṣṣa'a bi-akhbār a'yān Dar'a*), or Aḥmad b. Mubārak al-Lamaṭī's hagiographic biography (d. 1156/1743) *Pure Gold from the Words of Sayyidī 'Abd al-'Azīz al-Dabbāgh*. But scholars could bring the pietistic and ethical concerns they have from Sufism into their other works as well, be these scholarly autobiographies, legal opinions, works on theology, or works of *adab*. Similarly, when it came to the study of the natural sciences, belonging to a Sufi order – as did all the authors of the works in natural science discussed in this book – influenced at most their concern for believers to maintain the proper attitude toward God and His role in nature, not their engagement with the sciences in question. My point here is similar to that raised above in the excursus on causality (after Chapter 2): the spiritual precommitments Sufi authors brought to their study of nature did not impede their ability to investigate God's habitual order of Creation, and since in the absence of a strong state in the seventeenth century, Sufi orders provided the central social and intellectual ties that bound Moroccan intellectuals together, Sufism can be said to have actively supported the study of the natural sciences during this time.

CONCLUSION

The Significance of a Landscape of Sciences in Seventeenth-Century Morocco

The speaker did well when he said:
Say to the one who thinks little of his contemporaries
  And who believes that the ancients were superior.
That that ancients were once new,
  And that what is new will become ancient.

The preferable course of action for man is not just to submit, but to be content with his age. In this way, through proper comportment he will achieve success with Exalted God the Wise, the Knowing, who is the Lord of the First and the Last. He will achieve success through his thanks and gratitude, will find peace in his heart from anticipation and aspiration, and will possess an open heart for the people of his time, undertaking to establish their rights, thinking well of them, and benefiting all of them. He will perceive the virtues of the age and ignore its vices and so on. Exalted God granted the Companions a virtuous age. They used to recount the types of evil that had taken place in previous times including the worshipping of idols, the committing of vile acts and how difficult that time had been. They praised God the Exalted and thanked him. This is the attitude that the believer needs to adopt toward the goodness God the Exalted has granted him in his time in both religious and temporal matters, acknowledging how Exalted God has preserved him from present and past evils. Let him thank God for this.

Al-Yūsī, *The Discourses*

FAILINGS AND CONTRIBUTIONS

This book falls short of its mark in two major ways. The first consists of it having provided only a schematic overview of the place of the natural and exact sciences in Early Modern Morocco. This failing is directly related to the problem of sources, of how on the one hand the vast majority of our materials remain in manuscript, are laborious to access and read, and how on the other the archival resources of relevant educational institutions that could have shed more light on the natural sciences' *Sitz im Leben* are not extant. The problem of sources is a serious one and explains a good deal of the lag of Middle East intellectual history behind that of its European counterpart: a great deal of work remains to be done simply describing extant materials before broader analytical narratives can be established. But the book's second failing is more substantive and stems from a second obstacle that is more challenging: the difficulty of articulating a compelling narrative of a sustained engagement with the natural sciences that took a different path than that of the European one. This second failing may be somewhat less apparent at this moment, following the previous chapters and my discussion of Thomas Kuhn's work. But I fear that once this book's covers are closed and the reader ponders the broader sweep of Middle East history again, the evidence presented here could become incorporated into a narrative of intellectual stasis, stupor, or even decline, as even specialists in the field still tend to do with the Early Modern period.[1] These failings were, to some degree unavoidable, even as I hope that they will be remedied in the future by local case studies on the natural sciences in other parts of the Muslim world.

In the hope that my fears regarding this second failing are overblown, let us revisit the central contributions this book has attempted to make. It has set out three principal overlapping claims: 1) in seventeenth-century Morocco, the natural sciences were studied, transmitted, and practiced by

[1] For a recent example of how the intellectual history of the Middle East between the sixteenth and eighteenth centuries can still be reduced to the dismissive rubrics of esotericism and scholasticism, see Ahmed El Shamsy, *Rediscovering the Islamic Classics* – Chapter Two "*Postclassical Book Culture*." El Shamsy's depiction of Islamic intellectual history in the Early Modern period can to some extent be explained because he adopts the point of view of the nineteenth- and twentieth-century Arab intellectuals that are the focus of his often eloquent and insightful examination of the impact of printing and editing on the creation of a new Islamic canon in the nineteenth century. Nonetheless, it does the intellectual production of the Early Modern period a great disservice, and his sweeping, albeit Egyptocentric, characterization is striking considering the now substantial body of secondary literature on this period's intellectual vibrancy that has appeared in the past decade and which *Revealed Sciences* has relied upon.

scholars working within a number of religious institutions and networks; 2) this activity was decentered, took place in both urban and rural centers of learning, and while never a majority pursuit, was widely accepted; and 3) the natural sciences was an emic category that included the occult sciences and which scholars such as al-Yūsī characterized as revealed and sanctioned by God. In doing so, it has offered an initial and at times schematic overview of the place of the natural sciences in Moroccan scholarship of the long seventeenth century. Its scope was limited chronologically and geographically, even as it has argued the need to refrain from generalizations regarding the intellectual landscape of the early modern Muslim world and to focus first on detailed case studies.

The title *Revealed Sciences* speaks to the third of these claims and al-Yūsī's articulation of a framework that explains the natural sciences not as something external to Muslim practice or study, but as part of it. This description covers the range of scholarly practices described in A. I. Sabra's justly influential appropriation thesis, in which he suggested that a productive way of understanding the decline of the philosophical sciences in Islamic thought in the postclassical Muslim world was that only those aspects that informed ritual practice were actively preserved, studied, and transmitted.[2] Sabra's argument explains, as seen in the survey of biographical dictionaries in Chapter 2, how mathematics and inheritance law were often studied together, or as seen in Chapter 3, how jurists drew on medicine to address a novel legal problem. It is less satisfying as an explanation for al-Yūsī's broadly instrumentalist definition of all knowledge that benefited the Muslim community being revealed, the production of a large number of medical, astronomical, and alchemical works in eleventh-/seventeenth-century Morocco, and the widespread incorporation of the occult sciences of lettrism, astrology, and alchemy into other works such as medicine and scholarly autobiographies. To take up again the discussion of the secular introduced in the Introduction with reference to Rushain Abbasi's recent exploration of the secular in the thought of al-Ghazālī, any binary division of Islamic thought into religious and secular knowledge misses the degree to which the later was implicated in a full understanding of the former and the reverse. The natural sciences were studied throughout Morocco's educational networks, networks that were anchored in the religious institutions of the mosque, college, and lodge,

[2] A. I. Sabra, "The Appropriation and Subsequent Naturalization of Greek Science in Medieval Islam."

that drew on the economic support of pious endowments, patronage, and in the cases of the Sufi lodges, commerce and community support. The majority of scholars who studied these sciences in and beyond these institutions encountered them first through introductory works, often in the form of mnemonic poems, but there was a smaller number who went on to write, transmit, and study more deeply in the fields of medicine, astronomy, astrology, alchemy, and lettrism. The occult sciences and the natural sciences more broadly were seen by the majority of those who did not study them as worthy and legitimate objects of study – this is attested to by their presence in the biographical dictionaries and categorizations of the sciences. There were individual scholars who objected to alchemy – as did al-Ifrānī – or approvingly described other scholars as having given up medicine out of piety – as happened with al-Mirghitī – but these voices testify as much to the widespread practice of these sciences as they do to opposition to them. *Revealed Sciences* has laid out this picture of the status and nature of the natural sciences for a specific time period – the long seventeenth century that witnessed the fragmentation of the Saʿdī dynasty at its beginning and the rise of the ʿAlawite dynasty at its end – but it has not addressed how long it lasted and when it changed. Indeed, this would require another monograph. Nonetheless, we have a few hints.

Following the death of Moulay Ismāʿīl in 1139/1727, Morocco slipped into another period of political instability that ended with the rise to power of Muḥammad b. ʿAbdallāh in 1757. During his thirty-three-year rule he centralized political and military power but also carried out an ambitious project to reform Morocco's Mālikī religious establishment by declaring the equality of all four Sunni law schools and adopting many elements of the recently appeared Wahhabi reform movement.[3] In the current historiography, it is this struggle between Mālikī scholars and the centralizing efforts of the sultanate both under Muḥammad b. ʿAbdallāh (rl. 1757–90) and son and ultimate successor Sulaymān (rl. 1792–1822) that has defined the narrative of the period's intellectual history.[4] In his magisterial survey of the intellectual life of the Middle East in the seventeenth century, El-Rouayheb notes a distinct decline in the rational sciences in Morocco in the second half of the eighteenth century, and links this to the religious

[3] The best source here is Fatima Harrak, "State and Religion in Eighteenth Century Morocco: The Religious Policy of Sidi Muḫammad b. ʿAbd Allâh 1757–1790." Compare with Mohamad El Mansour, "Saints and Sultans: Religious Authority and Temporal Power in Precolonial Morocco."

[4] Along with the previous references, see Mohamed El Mansour, *Morocco in the Reign of Mawlay Sulayman*.

policies of the country's rulers.[5] In making this characterization, El-Rouayheb is thinking principally of logic and theology. The fate of Morocco's natural sciences during the eighteenth century is less clear. The survey of biographical dictionaries in Chapter 2 shows that a significant minority of scholars pursued the natural sciences well into the reign of Moulay Muḥammad. As seen in the *fahrasa* of Ibn ʿAjība, also discussed in Chapter 2, the natural sciences continued to be part of the lives of some prominent Moroccan scholars into the beginning of the nineteenth century, and extant manuscripts point to the continued writing of texts in the natural sciences during the same period (see Appendix 2). Similarly, toward the end of the eighteenth century, al-Ṭurunbāṭī, whose classification of the sciences was discussed in Chapter 2, and who drew significantly on al-Yūsī's expansive understanding of the validity of the natural sciences, enjoyed general regard among his peers and was at one point a teacher of Moulay Sulaymān himself.[6] By themselves, however, these data points remain suggestive and much more work will need to be done to elucidate the role of the natural sciences in the eighteenth and nineteenth centuries.

WHICH HISTORIES OF THE SCIENCES?

In its attempt to outline the significant energy, curiosity, and effort that Moroccan scholars devoted to the natural sciences in the seventeenth century, this book has pushed back both against the old but persistent narrative of the teleological rise of modern science and the equally old and dogged narrative of intellectual decline of the Muslim world. In doing so, it argued for the importance of richly contextualized local intellectual histories and stressed how little we still know regarding much of the intellectual history of the Early Modern Muslim Middle East and Africa. This was very much not a history of science, but a history of epistemologically varied sciences where attempts to find parallels with European history are likely to mislead as much as explain.[7] I have deliberately not focused on Moroccan-European contacts during this period and the

[5] El-Rouayheb, *Islamic Intellectual History*, 170. [6] See Chapter 2.
[7] Compare with El-Rouayheb's comment: "What this assumption elides [i.e., that intellectual activity could be measured by the degree to which Muslims interacted with European texts] is that there could have been genuine development in the realm of ideas that was neither derives from Western Europe nor paralleled Western European developments. It is this overwhelming possibility that I have attempted to flesh out in the present study" (*Islamic Intellectual History*, 357).

transmission of knowledge, which would involve the small body of translations carried out, chiefly by Moriscos in the early seventeenth century, of European medical, technical, and military texts.[8] Moroccan interest in European works in the natural sciences continued into the eighteenth century, as seen in an extant Arabic translation of John Hadley's (d. 1744) treatise on the quadrant.[9] Yet a focus on such translations risks reestablishing the centrality of Europe, with Morocco's own interest in the natural sciences being measured by its interest in what is happening further north. Such a case could be more plausibly made for the nineteenth century perhaps, but in the seventeenth century European knowledge was distinctly peripheral to Moroccan networks of scholars and institutions of scholarship and there is little anecdotal evidence that it was seen as containing much of interest.[10] Instead, I have attempted to depict the scope and content of the natural sciences within Moroccan scholarship on its own terms. After all, what is the writing of history if not a relentless interrogation of our own preconceptions of the past that can allow us insight into how our predecessors understood the world? Such an interrogation requires gaining access to worldviews that we ourselves do not share, in order to understand their claims regarding the world, and how it can best be studied and perceived, and how this knowledge can be best transmitted to others. In the writing of this history, we trouble not only the significance of the sciences, but also of categories such as religion, rationality, or culture, categories that designate areas of human experience and practice and which can be invoked to explain or to contrast given ways of doing things. The contingency of these categories and their relationship to the sciences is well reflected in the epigraph from Hodgson that opened this book: we write these histories to better understand what it means to be human, and in doing so we come to understand our world better as well.

[8] The most prominent and well-studied case of a Morisco translator is Aḥmad al-Ḥajarī, on whom see the works cited in the Introduction. For the context of such translations see Hans-Rudolph Singer, "Morisken als Übersetzer," and especially Muḥammed al-Mannūnī, "Ẓāhira taʿribiyya."

[9] See the entry in al-Khaṭṭābī, *Fahāris al-khizānah al-malakiyya*, vol. 3, 416 (in the collected volume 10381). On John Hadley's quadrant, see Charles Cotter, "The Mariner's Sextant and the Royal Society," 32–34.

[10] For the case of medicine, see the episode related by al-Yūsī on the physician al-Darrāq's meeting with a Christian physician in Tangier (*The Discourses*, 209).

APPENDIX I

List of the Sciences Given in ʿAbd al-Raḥmān al-Fāsī's *Kitāb al-Uqnūm*[1]

- ʿilm al-aqa'id wa-l-tawhid (creed)
- ʿilm al-tafsir (exegesis)
- ʿilm al-ḥadīth (Prophetic Tradition)
- ʿilm uṣūl al-fiqh (legal theory)
- ʿilm al-farā'iḍ (inheritance law)
- ʿilm al-naḥw (grammar)
- ʿilm al-taṣrīf (declension)
- ʿilm al-khaṭṭ (calligraphy)
- ʿilm al-maʿānī (rhetoric)
- ʿilm al-bayān (exposition)
- ʿilm al-badīʿ (poetical artifice)
- ʿilm al-tashrīḥ (anatomy)
- ʿilm al-ṭibb (medicine)
- ʿilm al-taṣawwuf (Sufism)
- ʿilm al-kalām (theology)
- ʿilm al-qira'āt (Qur'anic readings)
- ʿilm al-siyar (biography)
- ʿilm al-jadal (disputation)
- ʿilm al-farā'iḍ (inheritance law)
- ʿilm al-jumal (numerical value of letters)
- ʿilm al-rasm (drawing)
- ʿilm al-ḍabṭ (vocalization)

[1] The names of the sciences given at the beginning of al-Fāsī, *Uqnūm*, i–iii appear to have been added by the copyist and differ slightly from those within the text itself. I have followed the latter.

- ʿilm al-ʿarūḍ (meter)
- ʿilm al-qawāfī (rhyme)
- ʿilm al-adab (comportment)
- ʿilm al-jirāḥāt (surgery)
- ʿilm al-ʿalāj (treatment)
- ʿilm ādāb al-dīn (religious comportment)
- ʿilm al-manṭiq (logic)
- ʿilm al-tajwīd (Qur'anic recitation)
- ʿilm al-shamā'il (attributes of the Prophet)
- ʿilm al-khilāfiyyāt (divergent opinions)
- ʿilm al-fara'iḍ al-kasūra (law of fractions of inheritance)
- ʿilm al-iʿrāb (vocalization)
- ʿilm al-lugha (linguistics)
- ʿilm al-kitāba (writing)
- ʿilm al-mizān (weighing)
- ʿilm uṣūl al-qawāfī (rhyme)
- ʿilm al-musīqa (music)
- ʿilm al-baytara (veterinary medicine)
- ʿilm al-zardaqa (farriery)
- ʿilm al-adhkār al-nabawiyya (prayers for the Prophet)
- ʿilm al-mukawwanāt (ontology)
- ʿilm al-waqf wa al-ibtidā' (Pausing in Qur'anic recitation)
- ʿilm al-ḥadīth (Prophetic Tradition)
- ʿilm al-sūfisṭīqā (sophistry)
- ʿilm al-ḥisāb (mathematics)
- ʿilm uṣūl al-nahw (grammar theory)
- ʿilm naẓā'ir al-nahw (grammatical opinions)
- ʿilm asrār al-rasm (geomancy)
- ʿilm kitabat al-shiʿr (poetical composition)
- ʿilm uṣūl al-lugha (foundational linguistics)
- ʿilm qawa'id al-lugha (linguistic principles)
- ʿilm al-bayzara (falconry)
- ʿilm ṭibb al-jawāriḥ min al-ṭayr (bird medicine)
- ʿilm al-tarbiyya bi l-iṣṭilāḥ (education)
- ʿilm al-falsafa (philosophy)
- ʿilm ādāb al-qirā'a (study ethics)
- ʿilm al-tārīkh (history)
- ʿilm al-fiqh al-Mālikī (Mālikī jurisprudence)
- ʿilm al-ʿadad (arithmetic)
- ʿilm al-jabr wa l-muqābala (algebra)

- ʿilm al-taksīr (establishing surface area)
- ʿilm al-irtimāṭīqī (arithmetic)
- ʿilm al-handasa (geometry)
- ʿilm al-jaʿrāfiyā (geography)
- ʿilm al-ikhtilāj (palmistry)
- ʿilm al-firāsa (physiognomy)
- ʿilm al-siyāsa (politics)
- ʿilm al-riyāsa (rulership)
- ʿilm al-ilāhiyyāt (metaphysics)
- ʿilm qira'at al-ʿashr (Qur'anic readings)
- ʿilm al-aḥkām (legal judgments)
- ʿilm al-qaḍā (legal rulings)
- ʿilm al-futyā (legal opinions)
- ʿilm al-taʿbīr (dream interpretation)
- ʿilm al-asṭurlāb (astrolabe)
- ʿilm al-rubʿ al-mujayyab (sine quadrant)
- ʿilm al-ṣafīḥat al-zarqāliyya (Zarqālī astrolabe)
- ʿilm al-ṣafīḥat al-shakāriyya (Shakārī astrolabe)
- ʿilm al-rubʿ al-muqanṭara (astrolabic quadrant)
- ʿilm al-asṭurlāb al-junubī (southern astrolabe)
- ʿilm al-simiyā' (illusionism)
- ʿilm al-ḥikma (philosophy)
- ʿilm al-hīmyā (astrological conjuring)
- ʿilm al-tabiʿiyāt (natural potencies)
- ʿilm al-ʿādiyyāt (stallions)
- ʿilm khaṭṭ al-raml (geomancy)
- ʿilm al-katf (scapulimancy)
- ʿilm al-awqāf (pious endowments)
- ʿilm al-wathā'iq (document preparation)
- ʿilm farḍ al-nafaqāt (duty of financial support)
- ʿilm al-muʿamalāt (jurisprudence of worldly affairs)
- ʿilm al-burhān (establishing proofs)
- ʿilm al-ḥisba (market inspecting)
- ʿilm al-naẓāra (inspection of pious endowments)
- ʿilm naẓr al-mawārith (inspection of inheritance)
- ʿilm al-hay'a (astronomy)
- ʿilm al-falak (astrology)
- ʿilm al-majasṭā' (astronomy)
- ʿilm jūmṭrīqā' (astrological observation)
- ʿilm al-kalam al-ʿirfānī (inspired speech)

- ʿilm asrār al-dīn (secrets of the faith)
- ʿilm al-shahāda (profession of faith)
- ʿilm al-nawāmīs (artifices)
- ʿilm al-manāẓira (debate)
- ʿilm al-ansāb (genealogy)
- ʿilm ansāb al-shurafāʾ (genealogy of the Prophet's descendants)
- ʿilm al-masāḥa (surveying)
- ʿilm al-tawqīt (timekeeping)
- ʿilm al-irsād (celestial observation)
- ʿilm al-taʿdīl (correction of the planets' location)
- ʿilm aḥkām al-nujūm (astrology)
- ʿilm al-hay'a al-saniyya (Prophetic cosmology)
- ʿilm siyar al-nujūm (orbits of the stars)
- ʿilm al-ahilla (sighting of the crescent moon)

APPENDIX 2

Extant Manuscripts in the Natural Sciences from the Sixteenth to Eighteenth Centuries in Moroccan Libraries[1]

Muḥammad b. Muḥammad al-Ḥājj al-Baqqāl (fl. Under the Saʿdīs)	Al-Fatḥ al-rabbānī fī ṭarīqat al-imām al-Murjānī	Lettrism	Royal Library [2347][2]
Muḥammad ibn ʿAlī al-Ḥājj al-Shuṭaybī (d. 963/1556)	Sharḥ miftāḥ al-jalīl fī tadbīr al-ḥajar	Alchemy	Ben Souda Library [351][3]
	Kitāb dhāt al-faṣalayn wa khalʿi ʿan al-naʿlayn fī baḥri al-lujayn	Alchemy	Qarawiyyīn Library [1516/4][4]

[1] This list is constrained by my ability (and that of the authors of the catalogs I have drawn on) to identify the authors of individual manuscripts – I have listed only those whose authors I could identify as having lived between the sixteenth and eighteenth centuries – as well as, naturally, by whether manuscripts are extant or not. I have not listed manuscripts by authors from outside of Morocco. Considering the number of anonymous manuscripts that plausibly date from the period in question, this is a conservative approach. Consider, for example the anonymous *Tuḥfat al-aḥbāb fī māhīyat al-nabāt wa-l-aʿshāb*, the author of which likely wrote in southern Morocco in the seventeenth century: H. P. J. Renaud and G. S. Colin (eds. and trans.), *Tuḥfat al-aḥbāb: Glossaire de la matière médicale Marocaine*. There are sure to be numerous other manuscripts that I have missed as well, and in that sense this list should be seen as an impressionistic overview of extant manuscripts, not an authoritative list.

[2] See al-Khaṭṭābī, *Fahāris al-khizānah*, vol. 5, 320–21. The library has six copies of this text.

[3] Mentioned in Semlali, *Histoire de l'alchemie*, 213.

[4] Ibid.

continued

Aḥmad b. Ḥumayd al-Muṭarrafī al-Marrakushī (d. 1001/1592)	Jamʿi al-muhimmāt al-muḥtāj ilayhā fī ʿilm al-mīqāt	Astronomy	S (Salé) [1040, 1041][5]/Royal Library [6662][6]
	Sharḥ rawḍat al-azhār	Astronomy	Royal Library [4151][7]
	Al-Muqarrab fī waṣf al-mujayyab	Astronomy	Royal Library [1638, 1009][8]
ʿAbd al-Ghānī al-Zammūrī (fl. 1578–1602)	Al-Qānūn al-mufīd fī ʿalāj al-ḥaṣāh bi-qawl sadīd	Medicine	ʿAllāl al-Fāsī Library [318 ayn]
ʿAbd al-Karīm Mu'min b. Yaḥyā al-Janawī al-ʿIlj fl. under Moulay ʿAbdallāh (rl. 1557–74)	Sharḥ al-urjūza fī al-amrāḍ al-bībshiyya	Medicine	Printed[9]
ʿAbd al-ʿAziz al-Maghrāwī (d. 1014/1605–06)	Urjūza fī tarḥīl al-shams	Astronomy	Great Mosque of Wazzan
	Nihāyat tarḥīl al-shams bi-l-manāzil	Astronomy	Royal Library [6441][10]
al-Wazīr al-Ghassānī (d. 1019/1610)	Rīḥānat al-juyyūb fī khawwāṣ al-ʿaqāqīr wa l-ʿushshūb	Medicine	S (Salé) [1227][11]
	Sharḥ al-mufradāt al-ṭibbiyya	Medicine	Gannun Library

[5] See Muḥammad Ḥajjī (ed.), *Fihris al-khizānah al-ʿilmiyya*, 482–83.

[6] Al-Khaṭṭābī, *Fahāris al-khizānah*, vol. 3, 181. The library has four copies of this treatise.

[7] Ibid., 311–12. [8] Ibid., 372–73.

[9] This is one of the four works edited and translated in H. P. J. Renaud and G. S. Colin, *Documents marocains pour server à l'histoire du "Mal Franc."*

[10] Al-Khaṭṭābī, *Fahāris al-khizānah*, vol. 3, 386–87. There is some debate on the death date of the author.

[11] See Muḥammad Ḥajjī, *Fihris al-khizānah al-ʿilmiyya*, 565.

continued

	Tafsīr baʿḍ al-a'shāb wa-l-ʿaqāqīr muqtaṭaf min rīḥānat al-juyyūb fī khawwāṣ al-ʿaqāqīr wa-l-ʿushshūb	Medicine	National Library[12]
	Ḥadīqat al-azhār fī sharḥ māhīyat al-ʿushb wal-l-ʿaqqār	Medicine	Royal Library [2294, 7336, 1063]/Printed[13]
	Al-Rawḍ al-maknūn fī sharḥ rajaz Ibn ʿAzrūn	Medicine	Royal Library [569][14]
ʿAbd al-Raḥmān b. ʿAmr b. Aḥmad al-Buʿaqīlī, known as Ibn al-Muftī (d. 1006/1597 or 1020/1611)	Sabk al-ʿibāra bi-alfāẓ al-yasāra	Astronomy	S (Salé) [1099][15]
	Sharḥ rawḍat al-azhār	Astronomy	Royal Library [1691][16]
	Qaṭf al-anwār min rawḍat al-azhār	Astronomy	Great Mosque of Wazzan/ Royal Library [489][17]
	Taḥṣīl al-maṭlab min al-rubʿ al-mujayyab	Astronomy	Royal Library [10366][18]

12 See al-Khaṭṭābī, *Fahāris al-khizānah al-malakiyya* (Rabat: 1982), vol. 2, 94. This work has been edited and printed. See below.

13 Abū Qāsim al-Ghassānī, *Ḥadīqat al-azhār fī māhīyat al-ʿushab wa-l-ʿaqqār.* Muḥammad al-ʿArabī al-Khaṭṭābī (ed.) (Beirut: Dār al-Gharb al-Islāmī, 1990).

14 Muḥammad al-ʿArabī al-Khaṭṭābī, *Fahāris al-khizānah al-malakiyya* (Rabat: 1982), vol. 2, 118.

15 Muḥammad Ḥajjī, *Fihris al-khizānah al-ʿilmiyya*, 506.

16 Al-Khaṭṭābī, *Fahāris al-khizānah*, vol. 3, 311.

17 Ibid., 410–15. The library has nine copies of this text.

18 See ibid., 148–49. This treatise sets al-Māridīnī's – a *muwaqqit* of the al-Azhar mosque who died in 912/1506 – treatise on the al-rubʿ al-mujayyab to verse.

continued

	Mukhtaṣar manẓūm li-risālat Badr al-dīn al-Māridīnī fī-l-rub' al-mujayyab	Astronomy	Royal Library [10109][19]
Aḥmad b. Yūsuf al-Qaṣrī al-Fāsī (d. 1021/1612)	Kayfiyyat al-jadwal al-ma'īnī	Astronomy	Royal Library [5483][20]
Abū al-'Abbās Aḥmad b. Muḥammad, known as al-Ma'yūb (d. 1022/1613)	Shuhūr al-'arab	Astronomy	Royal Library [2803][21]
'Abd al-Raḥmān b. Muḥammad al-Fāsī (d. 1036/)	Tafsīr asmā' al-'ushshāb wa-l-'aqāqīr al-shā'i'a	Medicine	Royal Library[22] [K 267]
'Abd al-Raḥmān b. 'Īsā b. Sayyid al-'Amri al-Ḥanafī (d. 1038/1628)	Urjūza fī manāzil al-kawākib	Astronomy	S (Rabat)
Muḥammad b. Abī al-Qāsim b. al-Qāḍī, known as Ibn al-'Āfiyya (d. 1040/1631)	Tafjīr al-anhār khilāl rawḍat al-azhār	Astronomy	S (Salé) [1023][23]/ Wazzan/Royal Library [9312, 259][24]
	Jum'a al-fawā'id wa ḥaṣr al-qawā'id	Astronomy	Royal Library [315][25]
	Tuḥfat al-mawālī fī sharḥ silk al-lālī fī l-mukhammas al-khālī	Astronomy	Royal Library [5484, 9214][26]

[19] Ibid., 339–40. [20] Al-Khaṭṭābī, *Fahāris al-khizānah*, vol. 5, 303.
[21] Ibid., vol. 3, 322–23. [22] Al-Khaṭṭābī, *Fahāris al-khizānah*, vol. 2, 80.
[23] Muḥammad Ḥajjī, *Fihris al-khizānah al-'ilmiyya*, 475.
[24] Al-Khaṭṭābī, *Fahāris al-khizānah*, vol. 3, 165–66. [25] Ibid., 184–85.
[26] Ibid., vol. 5, 284–85.

continued

Aḥmad al-Maqqarī (d. 1041/1632)	Namaṭ al-akmal fī dhikr al-zamān al-mustaqbal	Astronomy	Royal Library [4505][27]
al-Hashtūkī (d. 1046/1636–37)	Risāla fī ʿilm al-falak	Astronomy	Ḥamziyya Zāwiya
Abū al-Qāsim b. Aḥmad b. Muḥammad b. ʿĪsā al-Ghūl al-Fishtālī (d. 1059/1649)	Kayfiyyat qism al-mayāh ʿalā al-sāqiyya aw al-qādūs	Engineering	S (Salé) [1160][28]
	Urjūza fī l-ṭibb	Medicine	Royal Library [12030z; 12069z][29]
	Urjūza fī l-awfāq	Lettrism	Royal Library [6675][30]
	Ghayāt al-iksīr fī ʿamal al-tawfīq wa-l-taksīr	Alchemy	Royal Library [6675][31]
Abū al-Ḥasan ʿAlī b. Ibrāhīm al-Andalusī (d. 1065/1655)	Urjūzat al-fawākih al-ṣayfiyya wa-l-kharīfiyya	Medicine	Included in al-Mirghitī's fahrasa[32]
	Urjūzat al-intiṣār li-baqlat al-anṣār	Medicine	Included in al-Mirghitī's fahrasa[33]
ʿAbd al-Qādir al-Shiblī (fl. 1070/1659)	Urjūza fī manāzil al-fuṣūl	Astronomy	Royal Library [1052][34]

[27] Ibid., 314.

[28] See Muḥammad Ḥajjī, *Fihris al-khizānah al-ʿilmiyya*, 543; Jumʿa al-Mājid Center, #577780.

[29] Al-Khaṭṭābī, *Fahāris al-khizānah*, vol. 2, 209–10. [30] Ibid., vol. 5, 276.

[31] Ibid., vol. 3, 465–66; Jumʿa al-Mājid Center, #461251.

[32] See al-Mirghitī, *al-ʿAwāʾid al-mūzriyya*, vol. 2, 658–76. According to the editor, this poem was dedicated to Moulay Zaydān's son Walīd in 1045/1635–36. It was edited by ʿAbdallāh Bennaser al-ʿAlawī and included in his *al-Shiʿr al-saʿdī: tafāʿul al-wāqiʿ wa l-fikr wa l-ibdāʿ*.

[33] See al-Mirghitī, *al-ʿAwāʾid al-mūzriyya*, vol. 2, 676–77.

[34] See al-Khaṭṭābī, *Fahāris al-khizānah*, vol. 3, 120–21.

continued

Aḥmad b. ʿAbdāllah b. Yaʿqūb al-Samlālī al-Rasmūkī (d. 1073/1663)	Masāʾil fī l-ṭibb	Medicine	Royal Library[35]
Muḥammad b. Amad al-Sabbāgh (d. 1076/1665)	Sharḥ naẓm fī ṣināʿat al-awfāq	Lettrism	Great Mosque of Wazzan
ʿAbdallāh b. ʿAbd al-Qādir Abī Shaykh al-Lakhmī (d. 1079/1668)	Bustān al-abrār fī ḥall alfāẓ rawḍat al-azhār	Astronomy	S (Salé) [990][36]/ Royal Library [4151, 7077][37]
Muḥammad b. Saʿīd b. Muḥammad al-Mirghitī (d. 1089/1678)	Tuḥfat al-muḥtāj fī ḥukm akl al-nās al-dajjāj	Medicine	Q (Fez)
	Qaṣīdat lāmiyya fī al-awfāq	Lettrism	Q (Fez)
	Qaṣīda fī aḥkām ʿaql al-dajjāj	Medicine	Tamgrout/ ʿUthmānīya Collection in Inzakān in the Sūs/Dha'ib Collection[38]
	Taqyīd fī umūr kullīya yuntafiʿ bi-hā fī ibṭāl la-aʿmāl al-siḥrīya	Magic (on the forbidden nature of)	Public Library of Tetuan;[39] Maḥjūbīya Collection[40]

35 Ibid., vol. 2, 156. 36 Muḥammad Ḥajjī, *Fihris al-khizānah al-ʿilmiyya*, 462.

37 Al-Khaṭṭābī, *Fahāris al-khizānah*, vol. 3, 139–41. The author was a *muwaqqit* in Madinat al-Qasr. There is some debate about his death date.

38 For the last two copies, see Najīb, *Fihris ma lam yufahras*, 33.

39 The manuscript is catalogued under the number 659 (224–28). My thanks to Professor Jaʿfar ben el Haj Soulami for sending me a copy of this text.

40 See Najīb, *Fihris ma lam yufahras*, 38–39.

continued

	al-Muqniʿ fī ʿilm Abī Muqriʿ	Astronomy	Imam ʿAlī Library in Taroudant[41]/ Royal Library [12248z][42]
	al-Mumtiʿ fī sharḥ al-muqniʿ	Astronomy	Maḥjūbīya Collection/ Dha'ib Collection[43]/ Royal Library [10352, 12472z][44]/ Hathi Trust[45]/ Printed
	al-Muṭlaʿ ʿalā masālik al-muqniʿ	Astronomy	Maḥjūbīya Collection/al-Tīdsīya Library[46]/Royal Library[47]/S (Salé) [1174][48]
	Sharḥ urjūza fī ṭarīq al-kīmyā'	Alchemy	Royal Library [10413][49]
	Sharḥ al-mukhtaṣar ʿalā al-Muqniʿ fī ʿilm Abī Muqriʿ	Astronomy	Royal Library [10353][50]

41 See ibid., 57.

42 Al-Khaṭṭābī, *Fahāris al-khizānah*, vol. 3, 373–76. The library has six copies of this poem.

43 See Najīb, *Fihris ma lam yufahras*, 104. The text has now also been edited and printed. See below.

44 Al-Khaṭṭābī, *Fahāris al-khizānah*, vol. 3, 349–51.

45 The Hathi Trust's copy of this manuscript is from the University of Michigan (originally from Max Meyerhof's collection). https://hdl.handle.net/2027/mdp.39015081447024

46 See Najīb, *Fihris ma lam yufahras*, 101.

47 Al-Khaṭṭābī, *Fahāris al-khizānah*, vol. 3, 343–48. The library has ten copies of the text.

48 See Muḥammad Ḥajjī, *Fihris al-khizānah al-ʿilmiyya*, 539–40.

49 See al-Khaṭṭābī, *Fahāris al-khizānah*, vol. 5, 155–57. The library has four copies of the text. This is the same commentary that I referred to in Chapter 4, based on the manuscript from the King ʿAbd al-ʿAziz Library.

50 See ibid., vol. 3, 312–13.

continued

	Fī ʿilm al-kīmyā'	Alchemy	National Library [accessed through the copy kept by the Jumaʿ al-Majid Center][51]
ʿAbdallāh b. Ḥamza b. Abī Sālim (d. 1090/1679)	Sharḥ al-yawāqīt fī ʿilm al-mawāqīt	Astronomy	Ḥamziyya Zawiya
Muḥammad b. Sulaymān al-Rūdānī al-Sūsī (d. 1094/1683)	Qalā'id al-laʾālī	Astronomy	Sulemaniye [2756]
	Maqāṣid al-ʿawālī bi-qalā'id al-la'ālī	Astronomy	Ḥamziyya Zawiya/ Royal Library [4151][52]/Hathi Trust[53]
ʿAli b. Muhammad al-Dādīsī (d. 1094/1683)	Urjūzat bidāyat al-tullāb fi ʿilm waqt al-yawm bi l-ḥisāb	Astronomy	Tamgrout
	Itḥāf dhawī al-albāb fī sharḥ bidāyat al-ṭullāb	Astronomy	Royal Library [9510][54]
	Fatḥ al-muqīt fī sharḥ al-yawāqīt	Astronomy	Royal Library [5789, 315][55]/ Hathi Trust[56]

[51] The Jumʿa al-Mājid Center copy gives the Moroccan National Library number "Marrakesh 154."

[52] See al-Khaṭṭābī, *Fahāris al-khizānah*, vol. 3, 376. This is only a chapter of the work.

[53] The Hathi Trust's copy of this manuscript is from the University of Michigan (originally from Max Meyerhof's collection). https://hdl.handle.net/2027/mdp.39015081447016

[54] See al-Khaṭṭābī, *Fahāris al-khizānah al-Ḥasaniyya* (Rabat, 1983), vol. 3, 116–17.

[55] See ibid., vol. 3, 401–02.

[56] The Hathi Trust's copy of this manuscript is from the University of Michigan (originally from Max Meyerhof's collection). https://hdl.handle.net/2027/mdp.39015081446802

continued

	Ikmāl fatḥ al-muqīt fī sharḥ al-yawāqīt	Astronomy	Royal Library [1045][57]
	Naẓm al-yawāqīt fī-l-mubtaghī min ṣinʿat al-mawāqīt	Astronomy	Ḥamziyya Zāwiya/ Royal Library [10866][58]
ʿAbd al-Raḥmān b. ʿAbd al-Qādir Abī Zayd al-Fāsī (d. 1096/1685)	al-Marqūm fī al-ʿamal bi l-nujūm	Astronomy	S (Salé) [1172][59]
	Masālik al-akhyār fī ʿilm waqt al-layl wa-l-nahār	Astronomy	S (Salé) [1173][60]/ Royal Library [6441][61]
	Urjūza fī al-ʿamal bi-l-rubʿ al-mujayyab	Astronomy	Great Mosque of Wazzan/Royal Library [7416][62]
	Tuḥfat al-ṭullāb fī ʿamal bī-l-asṭrulāb	Astronomy	Great Mosque of Wazzan/Royal Library [7106][63]
	al-Ghurra fī l-kalām ʿalā bayt al-ibra	Astronomy	Great Mosque of Wazzan/Royal Library [6662][64]
	Manẓūma fi iʿmār al-ʿaqqār	Medicine	Great Mosque of Wazzan [1548][65]
	Risāla fī ʿilm al-falak	Astronomy	Ḥamziyya Zāwiya
	Naẓm fī ʿilm al-falak	Astronomy	Ḥamziyya Zāwiya

57 See al-Khaṭṭābī, *Fahāris al-khizānah*, vol. 3, 135–37. The library has three copies of the text.

58 Ibid., 427. 59 See Muḥammad Ḥajjī, *Fihris al-khizānah al-ʿilmiyya*, 539. 60 Ibid.

61 See al-Khaṭṭābī, *Fahāris al-khizānah*, vol. 3, 341–42. 62 Ibid., 124.

63 Ibid., 121–23. The library has three copies of this text.

64 Ibid., 397–99. The library has four copies of this text.

65 See al-Ṭanjī and al-Ghāzī, *al-Fihris al-waṣfī li-makhṭūṭāt khizānat al-masjid al-aʿẓam bi-Wazzān*, vol. 2, 753.

continued

	Mishkat al-anwār fī awqāt al-layl wa l-nahār	Astronomy	Ḥamziyya Zāwiya
	Ta'līf fī l-asṭrulāb	Astronomy	Ḥamziyya Zāwiya
	Manẓūma fī l-asṭrulāb	Astronomy	Ḥamziyya Zāwiya
	Urjūza fī l-tawqīt	Astronomy	Royal Library [6678][66]
	Rajaz fī bayt al-ibra	Astronomy	Royal Library [3964, 1638][67]
	al-Ightibāṭ fī sharḥ nuzhat al-istinbāṭ li-l-marjānī	Lettrism	Royal Library [5819, 10910][68]
	al-Mathaf al-manwāl	Lettrism	Royal Library [10396][69]
	al-Maṭlab fī l-rub' al-mujayyab	Astronomy	Royal Library [6665][70]
	Urjūza fī l-ḥabb al-ifranjī	Medicine	Printed[71]
Abū 'Abdallāh Sīdī Muḥammad b. Sa'īd al-Sūsī (d. 1098/1687)	Manẓūma fī ayām al-'ām al-'arabī	Astronomy	Ḥamziyya Zāwiya
Muḥammad b. Aḥmad al-Bu'aqīlī al-Sūsī (d. 11th century)	Majmū' al-manāfi' fī 'ilm al-ṭibb al-nāfi'	Medicine	Great Mosque of Wazzan/al-Tīdsīya Library[72]/Royal Library [10088, 1941][73]/ National Library

66 See al-Khaṭṭābī, *Fahāris al-khizānah*, vol. 3, 123–24. 67 Ibid., 198–99.

68 See ibid., vol. 5, 279. This is a commentary on a text by the Tunisian scholar 'Abdallāh b. Muḥammad al-Murjānī.

69 Ibid., 306. 70 See ibid., vol. 3, 342–43.

71 This is one of the four works edited and translated in H. P. J. Renaud and G. S. Colin, *Documents marocains pour server à l'histoire du "Mal Franc."*

72 See Najīb, *Fihris ma lam yufahras*, 95.

73 See al-Khaṭṭābī, *Fahāris al-khizānah*, vol. 2, 138–39.

continued

Al-ʿArabī b. ʿAbd al-Salām al-Fāsī (d. fl. 17th century)	Shifā' al-ʿalīl fī bayan qiblat ṣāḥib al-tanzīl	Medicine	Royal Library [6588][74]
ʿUbaydallāh Aḥmad b. Ibrāhīm b. ʿAbd al-Mu'min (fl. 1106/1694–95)	Kitāb nathr al-durar ʿalā naẓm al-hadiya fī-l-ṭibb	Medicine	National Library
Muḥammad b. Habbūs al-Ṣanhājī (d. 1109/1697–98)	al-Manāfi' al-bayyina wa ma yuṣliḥ fī l-arbaʿa al-azmina	Medicine	Tamgrout
Muḥammad b. ʿAbd al-ʿAzīz al-Andalusī al-Asfī (fl. 1113/1701)	Irshād al-sāʾil ilā maʿrifat jihat al-qibla bi-l-dalā'il	Astronomy	Royal Library [1110][75]
	Sharḥ rajaz fī man-āzil al-qamr	Astronomy	Royal Library [1110][76]
	Risāla fī l-ʿamal bi-l-rubʿ	Astronomy	Royal Library [1110][77]
ʿAbd al-Salām b. Aḥmad Ibn Zākūr (fl. 1113/1701)	Kifāyat al-labīb fī l-tawqīt bi-ʿamal al-nisba min al-jayyūb	Astronomy	Royal Library [1213][78]
Muḥammad al-Ghumarī al-Ashʿarī al-Azharī (d. 1128/1716)	Maknūn ḍamā'in arbāb al-jafriyyāt wa l-ḥufriyyāt	Lettrism	S (Salé) [1262][79]

[74] See ibid., vol. 3, 321–22. This is a reply to the Algerian scholar al-Tājūrī, and in the introduction the author refers to "our teacher, ʿAbd al-Raḥmān al-Fāsī."
[75] Ibid., 130–31. [76] Ibid., 300–01. [77] Ibid., 244.
[78] Ibid., 326–27. This library has three copies of this treatise.
[79] See Muḥammad Ḥajjī, *Fihris al-khizānah al-ʿilmiyya*, 571.

continued

Aḥmad b. Muḥammad b. Muḥammad b. Ya'qūb al-Wallālī (d. 1128/1716)	Nuzhat al-anẓār fī rawḍat al-azhār	Astronomy	Royal Library [6006][80]
Aḥmad b. Sulaymān al-Rasmūkī (d. 1133/1721)	Kifāyat dhuwī al-albāb	Astronomy	S (Salé) [1152][81]
	Sharḥ nazm fī kay-fiyyat istikhrāj dukhūl yanāīr	Astronomy	Q (Fez)
	Sharḥ manẓūma fī-l-burūj wa-l-manāzil	Astronomy	Great Mosque of Wazzan
	al-Fawā'id al-marḍīya fī istikh-rāj al-manāzil wa-l-burūj al-mutawahhima wa-l-ṣūrīya	Astronomy	Maḥjūbīya Collection[82]
	Mukhtaṣar fī-l-ṭibb	Medicine	Al-Tīdsīya Library[83]
	'Umdat al-rughghāb fī ḥall alfāẓ ma'ūnat al-ṭullāb	Astronomy	Royal Library [5987, 3979][84]
Abū 'Abdallāh Muḥammad b. 'Abd al-Raḥmān b. 'Abd al-Qādir al-Fāsī (d. 1134/1721)	Risāla fī l-taqwīm	Astronomy	Royal Library [9908, 1534][85]

80 See al-Khaṭṭābī, *Fahāris al-khizānah*, vol. 3, 381.
81 See Muḥammad Ḥajjī, *Fihris al-khizānah al-'ilmiyya*, 529.
82 See Najīb, *Fihris ma lam yufahras*, 85–86. 83 Ibid., 97.
84 See al-Khaṭṭābī, *Fahāris al-khizānah*, vol. 3, 394–95. 85 Ibid., 209–10.

continued

Ṣālīḥ b. Muḥammad al-Mu'ṭī al-Sharqī (d. 1139/ 1726)	Tuḥfat al-ṭullāb fī kashf mā khufiya min al-astrulāb	Astronomy	Royal Library [7421][86]
'Abd al-Qādir b. al-'Arabī Ibn Shaqrūn al-Maknāsī (d. 1140/1728)	al-Nafḥa al-wardiyya fī-l-'ushba al-hindiyya	Medicine	Royal Library [3636][87]
	Manẓūma fī-l-aghdhiyya wa-l-ashriba wa-l-adwiyya	Medicine	Royal Library [7587][88]
	al-Urjūza al-shaqrūniyya	Medicine	Printed[89]
Muḥammad b. Ya'qūb b. Muḥammad al-Sūsī al-Tamlī al-Sajāwī (fl. 1141/1729)	Sharḥ urjūzat al-Mirghithī		Royal Library [10389][90]
Muḥammad al-'Arabī b. 'Abd al-Raḥmān al-Shafshāwnī (fl. 1143/ 1731)	Risāla fī l-'amal bi-l-asṭrulāb	Astronomy	Royal Library [5367, 1818][91]
Aḥmad b. Ṣāliḥ Abū al-'Abbās al-Dar'ī al-Ṣāliḥī Laknāwī (d. 1144/1731)	Hadīyat al-maqāl fī al-ṭibb li l-nisā' wa l-rijāl	Medicine	Great Mosque of Wazzan

86 Ibid., 177.

87 Ibid. vol. 2, 163–64. On this text see H. P. J. Renaud and G. S. Colin, *Documents marocains pour server à l'histoire du "Mal Franc,"* 27–29.

88 al-Khaṭṭābī, *Fahāris al-khizānah*, vol. 2, 207–09. The Royal Library has a total of twelve copies of this text.

89 This text has been edited and translated into French by Badr Tazi and Abdelhadi Tazi, *La medicine arabe au XVIIIe Siècle à travers al "Urdjuza Ash-Shakruniyya."*

90 See al-Khaṭṭābī, *Fahāris al-khizānah*, vol. 5, 156–57.

91 See ibid., vol. 3, 236–37.

continued

	al-Hadīya al-maqbūla fī ḥulal al-ṭibb al-mashmūla	Medicine	Imam 'Ali Library in Taroudant/al-Tīdsī Collection[92]/ Royal Library [5256][93]/S (Salé) [1239/1240]/ Printed[94]
	al-Durar al-maḥmūla 'alā al-hadīya al-maqbūla fī 'ilal al-ṭibb al-mashmūla	Medicine	Tamgrout/al-Tīdsī Collection[95]/ National Library/Royal Library [357; Z12030; 9048][96]
'Abd al-Wahhāb b. Aḥmad Adarrāq (d. 1159/1746)	Urjūza fī l-ḥabb al-ifranjī	Medicine	Printed[97]
Abū 'Abdallāh Muḥammad b. Muḥammad al-Dalīmī al-Warzāzī al-Dar'ī (fl. 1160/1747)	Sharḥ al-muqni' fī 'ilm Abī Muqri'	Astronomy	Royal Library [5889, 12472z][98]

92 See Najīb, *Fihris ma lam yufahras*, 117.

93 Al-Khaṭṭābī, *Fahāris al-khizānah* vol. 2, 212–13. The Royal Library has a total of eight copies of this text.

94 This poem has been edited by the author's descendant Moulay al-Mahdī b. 'Alī al-Ṣāliḥī, *al-Hadīya al-maqbūla fī l-ṭibb*.

95 See Najīb, *Fihris ma lam yufahras*, 54.

96 Al-Khaṭṭābī, *Fahāris al-khizānah*, vol. 2, 100–02.

97 This is one of the four works edited and translated in H. P. J. Renaud and G. S. Colin, *Documents marocains pour server à l'histoire du "Mal Franc."*

98 See al-Khaṭṭābī, *Fahāris al-khizānah*, vol. 3, 318–20. This is a commentary on al-Mirghitī's poem.

continued

Muḥammad b. ʿAlī al-Hawzālī (d. 1162/1749)	Mufīd al-rughbān sharḥ mahmāz al-ghuflān ʿalā furūʿ al-waqt wa l-adhān	Astronomy	Maḥjūbīya Collection/al-ʿUthmānīya Collection[99]
Muḥammad b. ʿAbd al-Salām b. Ḥamdūn Bannānī al-Fāsī (d. 1163/1750)	Sharḥ al-manẓūma al-maʿmūla fī muhimmāt ʿilm al-asṭrulāb	Astronomy	S (Rabat) [1118][100]/ Ḥamziyya Zāwiya/Royal Library [492][101]
Abū Madyān Muḥammad b. Muḥammad al-Manbihī (d. 1164/1751)	Urjūza fī l-ṣinaʿa	Alchemy	Royal Library [4195][102]
Muhammad b. ʿAli al-Ḥumaydī (d. 1179/1765)	Risāla fī ālat al-raṣd dhāt al-kursī	Astronomy	S (Salé) [1055][103]/ Royal Library [826, 265][104]
ʿAbdallāh b. Aḥmad b. Muḥammad al-Kaḍāḍī al-Tinzāḍī (fl. 1181/1767)	Nubdha min al-fawā'id al-ṭibiyya/ʿUnwān al-shifā maʿ ṣidq al-ṭibb wa l-wafā	Medicine	Royal Library [1160, 5518][105]
ʿAbd al-Raḥmān b. Muḥammad al-ʿArabī al-Mufarraj al-Andalusī al-Ansārī (fl. 1188/1774)	Risāla fī-l-tawqīt (wa jadāwil mul-ḥaqa bi-hā)	Astronomy	Royal Library [2523 (two copies)][106]

99 See Najīb, *Fihris ma lam yufahras*, 103.

100 Muḥammad Ḥajjī, *Fihris al-khizānah al-ʿilmiyya*, 515.

101 Al-Khaṭṭābī, *Fahāris al-khizānah*, vol. 3, 295–97. The library has a total of three copies of this text, which is a commentary on the poem on the astrolabe by ʿAbd al-Raḥmān al-Fāsī.

102 Al-Khaṭṭābī, *Fahāris al-khizānah al-Ḥasaniyya* (Rabat, 1986), vol. 5, 13.

103 Muḥammad Ḥajjī, *Fihris al-khizānah al-ʿilmiyya*, 488.

104 See al-Khaṭṭābī, *Fahāris al-khizānah*, vol. 3, 226–27.

105 See ibid., vol. 2, 160, 167–68. This work is an abridgement of al-Darʿī's *al-Durar al-maḥmūla*.

106 See ibid., vol. 3, 219–20.

continued

ʿAbd al-Salām b. ʿAlī al-Rijrājī (fl. 1190/1776)	Khulāṣat al-durar fī ālat al-ḥajar		Royal Library [1025, 1925][107]
	Sharḥ baytayn min qaṣīda		Royal Library [1025][108]
	Qaṣīda fī al-ṣinaʻa	Alchemy	Royal Library [1025][109]
Muḥammad b. al-Ṭīb b. ʿAbd al-Qādir Skīrj al-Andalusī (d. 1194/1780)	Muʿīnat al-ṭullāb ʿalā al-tawaṣṣul li-l-asṭrulāb	Astronomy	Royal Library [127][110]
Muḥammad b. al-Ḥasan Bannānī al-Fāsī (d. 1194/1780)	Risāla ḥawl al-naẓar ilā al-nujūm wa l-istidlāl bi-hā	Astronomy	S (Salé) [1049][111]
Muḥammad b. Ḥabīballāh b. al-Fāḍil al-Yaʿqūbī al-Tiyūsī (fl. 1195/1781)	Itḥāf al-Muṣannaf al-Saʻīd li-sharḥ qaṣīdat Muḥammad b. Saʿīd (al-Mirghithi)		Royal Library [7205][112]
ʿAbdallāh b. ʿAbd al-ʿAzīz al-Marrākushī (Ibn ʿAzzūz) (d. 1204/1789)	Dhahāb al-kusūf wa nafy al-ẓalma fī ʿilm al-ṭibb wa l-ṭabāʼiʿ wa l-ḥikma	Medicine	Royal Library [5799, 11887z, 314, 5774][113]/ National Library/S (Salé) [1224][114]
	Ithmād al-baṣāʼir fī maʻrifat ḥikmat al-maẓāhir		Royal Library [6891][115]

107 See ibid., vol. 5, 67–68. Compare with ibid., 204, where a fragment of an unidentified work by this author is listed.
108 Ibid., 151–52.
109 See ibid., vol. 5, 261. The library has two copies of this poem in this volume.
110 See ibid., vol. 3, 363–64.
111 See Muḥammad Ḥajjī, *Fihris al-khizānah al-ʿilmiyya*, 486.
112 See al-Khaṭṭābī, *Fahāris al-khizānah*, vol. 5, 273–74.
113 See ibid., vol. 2, 106–08.
114 See Muḥammad Ḥajjī, *Fihris al-khizānah al-ʿilmiyya*, 564.
115 See al-Khaṭṭābī, *Fahāris al-khizānah*, vol. 5, 274–75. The library has three copies of this text.

continued

	al-Amr al-wāfī wa l-tartīb al-kāfī fī l-sirr al-khāfī		Royal Library [1678, 758][116]
	al-Anwār fī sirr al-ikhtiṣār		Royal Library [31][117]
	Lubāb al-ḥikma fī l-ʿilm al-ḥurūf wa ʿilm al-asmāʾ al-ilahiyya	Lettrism	Royal Library [1081, 1514][118]
	al-Fuṣūl fī ḥall al-maʿqūd wa ʿaqd al-maḥlūl		Royal Library [4576][119]
al-Ṭayyib b. ʿAbdallāh Sāsī al-Samlālī al-Asfī (d. 1204/1789)	Risāla fī tafsīr jadāwil tawqītiyya	Astronomy	Royal Library [12288z][120]
	Riyāḍ al-azhār fī maʿrifat awqāt al-layl wa-l-nahār	Astronomy	Royal Library [1739][121]
Abū al-Rabīʿ Sulaymān b. Aḥmad al-Fishtālī al-Fāsī al-muwaqqit (d. 1208/1794)	Bughyat dhawī al-raghbāt fī sharḥ ʿawīṣ risāla al-Māridīnī fī-l-rubʿ al-jībī min al-mīqāt	Astronomy	Royal Library [10177, 10056, 11984z][122]
	Risāla fī-l-ṣafīḥa al-jāmiʿa	Astronomy	Royal Library [1009][123]
Aḥmad b. ʿAlī al-Balghīthī (d. 12008/1794)	Sharḥ manhāj al-ṭālib fī taʿdīl al-kawākib	Astronomy	S (Salé) [1120][124]
Muḥammad b. al-Tayyib b. al-Makkī al-Ajbārī al-Ḥasanī (fl. 1214/1799)	Risāla fī khaṣaṣ al-awqāt min zījay Ibn al-Shāṭir al-Dimashqī wa-l-Samarqandī	Astronomy	Royal Library [8873][125]

[116] Ibid., 276, and compare with vol. 3, 433. [117] See ibid., vol. 5, 277.
[118] Ibid., 304–05. [119] Ibid., 322–23. [120] See ibid., vol. 3, 208–09.
[121] Ibid., 279–83. The library has four copies of this work. [122] Ibid., 141–43.
[123] Ibid., 235, 379–80. [124] See Muḥammad Ḥajjī, *Fibris al-khizānah al-ʿilmiyya*, 515.
[125] See al-Khaṭṭābī, *Fahāris al-khizānah*, vol. 3, 223–24.

continued

Muḥammad b. Muḥammad al-Jinnān fl. during the reign of Sultan Sulaiman (rl. 1791–1822)	Arkān al-tawqīt al-khamsa	Astronomy	Royal Library [10388][126]
Muḥammad b. ʿAbd al-Hādī al-Ṭatwānī al-Maṣrī [fl. 12th century hijri]	Tadhkirat al-mushtāq fī l-ʿilm al-taksīr wa l-awfāq	Lettrism	Royal Library [5483, 488][127]

[126] Ibid., 131–32. [127] Ibid., vol. 5, 286.

Bibliography

ARABIC PRIMARY SOURCES

Anṭākī, Dāwūd al-. *al-Nuzha al-mubhija fī tashḥīdh al-adhhān wa ta'dīl al-amzija*. (Beirut: Mu'assisat al-balāgh, 1999).

al-Tadhkira. Ahmad Shams al-Din (ed.) (Beirut: Dar al-Kutub al-'Ilmiyya, 1971).

Aqḥiṣārī, Aḥmad al-. *An Ottoman Manifesto Against Smoking*. Yahya Michot (ed. and trans.) (Markfield: Interface Publications, 2010).

'Ayyāshī, 'Abdallāh b. Muḥammad al-. *al-Ḥukm bi-l-'adl wa-l-inṣāf al-rāfi' li-l-khilāf*. 'Abd al-'Aẓīm Ṣaghīrī (ed.) (Rabat: Wizārat al-Awqāf, 2015).

Riḥlat al-'Ayyāshī 1661–1663. Sa'īd al-Faḍlī and Sulaymān al-Qarshī (eds.) (Abu Dhabi: Dār al-suwayd l'l-nashr wa'l-tawzī'a, 2006), 2 vols.

Iqtifā' al-athar ba'd dhahāb ahl al-athar: fahras Abī Sālim al-'Ayyāshī. Nafīsa al-Dhahabī (ed.) (Rabat: Manshūrāt kulliyat al-adab wa-l-'ulūm al-insāniyya, 1996).

Azharī, Ṣāliḥ 'Abd al-Samī' al-Ābī al-. *Jawāhir al-iklīl 'alā mukhtaṣar al-imām Khalīl*. (Beirut: al-Maktaba al-'Aṣriyya, 2000), 2 vols.

Burzulī, Abū al-Qāsim al-. *Fatāwā*. Muḥammad al-Ḥabīb al-Hīlah (ed). (Beirut: Dār al-Gharb al-Islāmī, 2002), 7 vols.

Dar'ī, Muḥammad al-Makkī b. Musā b. Nāṣir al-. *al-Durar al-muraṣṣa'a bi-akhbār a'yān Dar'a*. Maḥammad al-Ḥabīb Nūḥī (ed.) (Casablanca: al-Najāḥ al-jadīda, 2014), 2 vols.

Fāsī, 'Abd al-Raḥmān al-. *Kitāb al-uqnūm fī mabādi' al-'ulūm*. Manuscript 6585 from Morocco's National Library, 272 pages.

Ghazālī, Abū Ḥamīd al-. *Faith in Divine Unity and Trust in Divine Providence*. David Burrell (trans.) (Louisville, KY: Fons Vitae, 2001).

The Incoherence of the Philosophers. Michael E. Marmura (ed. and trans.) (Provo: Brigham Young University Press, 1997).

The Remembrance of Death and the Afterlife. Tim Winter (trans.) (Louisville, KY: Fons Vitae, 1989).

Majmū'a rasā'il al-imām al-Ghazālī (Beirut: Dār al-kutub al-'ilmiyya, 1988).

Ḥajarī, Aḥmad b. Qāsim al-. *Kitāb nāṣir al-dīn ʿalā ʾl-qawm al-kāfirīn* (The Supporter of Religion against the Infidels). P. S. van Koningsfeld, Q. al-Samarrai, and G. A. Wiegers (trans. and eds.), 2nd ed. (Madrid: Consejo Superior de Investigaciones Científicas, 2015).

Ḥawwāt, Sulaymān al-. *Thamrat anīs fī taʿrīf bi-nafsī li-abī al-rabīʿ Sulaymān al-Ḥawwāt al-Shafshāwnī (1160/1747–1231/1816).* ʿAbd al-Ḥaqq al-Ḥaymar (ed.) (Chefchaouen: Wizārat al-thaqāfa markaz al-dirāsāt wa al-buḥūth al-andalusiyya, 1996).

Ḥuḍaygī, Muḥammad b. Aḥmad al-. *Ṭabaqāt al-Ḥuḍaygī.* Aḥmad Bū Mazgū (ed.) (Casablanca: Maṭbaʿat al-Najāḥ al-Jadīda, 2006), 2 vols.

Ibn al-Akfānī, Muḥammad b. Ibrāhīm. *Irshād al-qāṣid ilā asnā al-maqāsid.* ʿAbd al-Munʿim Muḥammad ʿUmar (ed.) (Cairo: Dār al-Fikr al-ʿArabī, 1990).

Ibn ʿAskar, Muḥammad. *Dawḥat al-nāshir li-maḥāsin man kāna bi-l-maghrib min mashāyikh al-qarn al-ʿāshir.* Muḥammad Ḥajjī (ed.) (Rabat: Dār al-Maghrib li-l-talīf wa l-tarjama wa l-nashr, 1977).

Ibn Ghāzī, Muḥammad b. Aḥmad. *Fihris Ibn Ghāzī.* Muḥammad al-Zāhī (ed.) (Casablanca: Dār al-Maghrib, 1979).

Ibn al-Ḥājj, Muḥammad. *al-Madkhal al-sharʿī al-sharīf ʿalā-l-madhāhib* (Cairo: Dār al-Fikr, 1981), 4 vols.

Ibn Juzayy, Muḥammad b. Aḥmad. *al-Qawānīn al-aḥkām al-sharʿiyyah* (Cairo: ʿĀlam al-fikr, 1980).

Ibn al-Khaṭīb, Lisān al-Dīn. *al-Iḥāṭa* (Beirut: Dār al-Kutub al-ʿIlmiyya, 2003), 4 vols.

Ibn Nāṣir al-Aghlānī al-Darʿī, Ḥusayn b. Muḥammad. *Fihris Ibn Nāsir.* Aḥmad al-Saʿīdī (ed.) (Beirut: Dār al-Kutub al-ʿIlmiyya, 2005).

Ibn al-Qāḍī, Aḥmad. *Durrat al-ḥijāl fī asmāʾ al-rijāl.* Muḥammad al-Aḥmadā Abū al-Nūr (ed.) (Cairo: Dār al-turāth, 1971), 3 vols.

Ibn Taymiyya, Aḥmad. *Le haschich et l'extase.* Yahya Michot (trans.) (Beirut: Les editions Albouraq, 2001).

Ifrānī, Muḥammad al-. *Safwa man intashara min akhbār ṣulaḥāʾ al-qarn al-ḥādī ʿashar* (Casablanca: Markaz al-Turāth al-Thaqāfī al-Maghrabī, 2004).

Nuzhat al-ḥādī bi-akhbār mulūk al-qarn al-ḥādī. ʿAbd al-Laṭīf al-Shādhilī (ed.) (Casablanca: Maṭbaʿa al-Najāḥ al-Jadīda, 1998).

Nozhet-Elhâdi: Histoire de la Dynastie Saadienne au Maroc (1511–1670). O. Houdas (ed. and trans.) (Paris: Libraire de la Societé Asiatique, 1889).

Fāsī, ʿAllāl al-. *Aḥādīth fī l-falsafa wa-l-tārīkh wa-l-ijtimāʿ* (Rabat: Mu'assisat ʿAllāl al-Fāsī, 2014).

al-Taṣṣawuf al-Islāmī fī-l-Maghrib (Rabat: Mu'assasat ʿAllāl al-Fāsī, 1998).

Fihrī, ʿAbdallāh b. Muḥammad b. ʿAbd al-Rahman al-Fāsī al-. *al-Iʿlām bi-man ghabara min ahl al-qarn al-ḥādī ʿashar.* Fāṭimah Nāfiʿ (ed.) (Casablanca: Dār Ibn Ḥazm, 2008).

Kattānī, Muḥammad b. Jaʿfar b. Idrīs al-. *Salwat al-anfās wa muḥādathat al-akyās bi-man uqbira min al-ʿulamāʾ wa'l-ṣulaḥāʾ bi-fās* (Casablanca: Dār al-thaqāfa, 2004), 4 vols.

Min rasāʾil al-Imām Muḥammad b. ʿAbd al-Kabīr al-Kattānī fī-l-adāb wa-l-sulūk (Amman: Dār al-Rāzī, 1998).

Ḥukm al-tadkhīn ʿinda al-aʾimma al-arbaʿ wa ghayri-him (Damascus: Maktabat al-Ghazālī, 1990).

Laqānī, Ibrāhīm al-. *Kitāb naṣīḥat al-ikhwān* (Bahrain: Jāmi'at Baḥrain, 1993).
Maghrāwī, Aḥmad b. Abī Jum'a al-. *Jāmi' jawāmi' al-ikhtiṣār wa al-tibyān*. Aḥmad Jalūlī al-Badawī and Rābiḥ Būnār (eds.) (Algiers: Al-Ḥaraka al-Waṭaniyya li l-Nashr wa l-Tawzi', ND).
Manjūr, Aḥmad al-. *Sharḥ al-manhaj al-muntakhab ilā qawā'id al-madhhab* (Beirut: 'Ālam al-Kutub, 2003), 2 vols.
Fihris Aḥmad al-Manjūr. Muḥammad al-Ḥājjī (ed.) (Rabat: Dār al-Maghrib, 1976).
Mayyāra al-Fāsī, Muḥammad b. Aḥmad. *Fahrasa*. Badr al-'Umrānī al-Ṭanjī (ed.) (Casablanca: Dār Ibn Ḥazm, 2009).
Māzarī, Muḥammad al-. *Sharḥ al-Talqīn*. Muḥammad al-Mukhtār al-Salāmī (ed.) (Beirut: Dār al-Gharb al-Islāmī, 1997).
Mirghitī, Abū 'Abdallāh Muḥammad b. Sa'īd al-. *al-'Awā'id al-muzriyya bi-l-mawā'id*. Muḥammad al-'Arabī Ashrīfī (ed.) (Rabat: Wizārat al-Awqāf, 2007), 3 vols.
al-Muṭli' 'alā masā'il al-muqni'. Ṣāliḥ b. 'Abdallāh al-Ilghī al-Ṣāliḥī (ed.) (Casablanca: Maṭba' al-Najāḥ al-Jadīdah, 1999).
al-Mumti' fī sharḥ al-muqni'. al-Mundhir 'Abd al-Raḥmān (ed.) (Casablanca: Dār al-Rashshād al-ḥadītha, ND).
Nāṣirī, Aḥmad al-. *Kitāb al-istiqsā li-akhbār duwal al-maghrib al-aqsā* (Rabat: Manshūrāt wizārāt al-thaqāfa wa-l-ittiṣāl, 2001), 9 vols.
Qādirī, Muḥammad b. al-Ṭayyib al-. *Nashr al-mathānī li-l-ahl al-qarn al-ḥādī 'ashr wa-l-thānī*. Muḥammad Ḥājjī and Aḥmad al-Tawfīq (eds.) (Rabat: Dār al-Maghrib li-l-ta'līf wa-l-tarjama wa-l-nashr, 1977), 4 vols.
Qalyūbī, Aḥmad al-. *Tadhkīrat al-Qalyūbī fi l-ṭibb wa-l-ḥikma*. Aḥmad Farīd al-Mazīdī and Munā Shablī (eds.) (Beirut: Dār al-kutub al-'ilmiyya, 2001).
Qarāfī, Aḥmad al-. *Kitāb al-furūq* (Cairo: Dār al-Salām, 2001), 4 vols.
Rasmūkī, al-. *Wafayāt al-Rasmūkī*. al-Mukhṭār al-Sūsī (ed.) (Rabat: Maṭba' al-Sāḥil, 1988).
Sāḥilī al-Andalusī, Abū 'Abdallāh al-. *Bughyat al-sālik fī ashraf al-masālik*. 'Abd al-Raḥīm al-'Alamī (ed.) (Rabat: Wizārat al-awqāf, 2003), 2 vols.
Salāmī, Muḥammad al-Mukhtār al-. *al-Ṭibb fī daw' al-īmān* (Beirut: Dār al-Gharb al-Islāmī, 2001).
Ṣāliḥī al-Idrīsī al-Laktāwī al-Dar'ī, Abū al-'Abbās Sīdī Aḥmad b. Ṣāliḥ al-. *al-Hadīya al-maqbūla fī-l-ṭibb*. al-Mahdī b. 'Alī al-Ṣāliḥī (ed.) (Warzazat: Mu'assasat al-furāt li l-ta'līm al-khāṣṣ, 2011).
Sijilmāsī, 'Abd al-Wāḥid al-. *al-Ilmām bi-ba'ḍ man laqaytuhu min 'ulamā' al-islām*. Nafīsa al-Dhahabī (ed.) (Rabat: Imprimérie Rabat Net, 2008).
Tamanārtī, Abū Zayd 'Abd al-Raḥmān al-. *Fawā'id al-jamma fī isnād 'ulūm al-umma*. Muḥammad al-Mannūnī (ed.) (Beirut: Dār al-Kutub al-'Ilmiyya, 2007).
Ṭurunbāṭī, Muḥammad b. Mas'ūd al-. *Bulūgh aqṣā al-murād fī sharaf al-'ilm*. 'Abdallāh Ramadānī (ed.) (Rabat: Markaz al-dirasāt wa l-abḥāth wa iḥyā' al-turāth, 2008).
'Umayrī, Abū Qāsim al-. "*Fahrasat al-'Umayrī*." Aḥmad Dajjūgh (ed.) PhD Dissertation, University of Muhammad Fifth, Rabat, N.D.
Urmawī, Sirāj al-Dīn al-. *Risāla fī farq bayna naw'ay al-'ilm al-ilahī wa-l-kalām*. Muḥammad Akram Abū Ghawsh (ed.) (Amman: Dār al-Nūr, 2013).

Wansharīsī, Aḥmad b. Yaḥyā al-. *al-Mi'yār*. (Rabat: Dār al-Gharb al-Islāmī, 1981).

Wazzānī, Abū 'Īsā al-. *al-Mi'yār al-jadīd al-jāmi' al-mu'rib 'an fatāwā al-muta'akhkhirīn min 'ulamā' al-maghrib*. (Rabat: Dār al-Ḥadīth al-Ḥasaniyya, 1996), 13 vols.

Yaḥṣūbī, 'Iyāḍ b. Mūsā, al-. *Tartīb al-Madārik*. (Tripoli: Maktabat al-Ḥayāt, 1968).

Yūsī, Ḥasan al-. *The Discourses: Volume One*. Justin Stearns (trans. and ed.) (New York: New York University Press, 2020).

Fahrasat al-Yūsī. Zakarīyā al-Khuthayrī (ed.) PhD Dissertation, University of Muhammad Fifth, Rabat, 2004.

al-Qānūn fī aḥkām al-'ilm wa aḥkām al-'ālim wa aḥkām al-muta'allim. Ḥamīd Ḥamānī (ed.) (Rabat: Maṭba'a Shālat al-Rabāṭ, 1998).

al-Muḥāḍarāt. Muḥammad Ḥajjī and Aḥmad al-Sharqāwī (Rabat: Dār al-Gharb al-Islāmī, 1982).

ARABIC SOURCES IN MANUSCRIPT

Fāsī, 'Abd al-Qādir al-. *Jawāb su'āl ḥawl al-tabagh "ṭāba."* 'Allāl al-Fāsī Library, Rabat. Call number: 746 ayn, 5–7.

Fāsī, 'Abd al-Raḥmān b. Muḥammad al-. *Jawāb 'an ḥukm ṭāba*. 'Allāl al-Fāsī Library, Rabat. Call number: 746 ayn, 218–23.

Fāsī, Ibn Yusūf al-. *Sahm al-iṣāba*. Allāl al-Fāsī Library, Rabat. Call number: 755 ayn, 209–217.

Fishtālī, Abū Qāsim al-Ghūl al-. *Ḥāfidh al-mizāj wa lāfidh al-amshāj bi-l-'alāj*. The National Library of Morocco, q960 [Jum'a al-Mājid library 579336].

Ghassānī, Abū Qāsim b. Muḥammad b. Abī al-Na'īm al-. *Jawāb 'ulamā' Fās 'an al-'ushba al-khabītha*. King 'Abd al-'Aziz Foundation, MS 326.M6, 74 v–91 r.

Ibn Nāṣir b. 'Amr b. 'Uthmān al-Dar'ī, Maḥammad. *Kitāb al-ajwiba al-Nāṣiriyya*. King 'Abd al-'Aziz Foundation, Number MS203_M3.

Maṣmūdī, Muḥammad ibn Aḥmad al-. *Kitāb al-wāfī fī al-tadbīr al-kāfī*. British Library manuscript digitized by the Qatar National Library (Or 13006, ff 113 r–158 r).

Mirghitī, Abū 'Abdallāh Muḥammad b. Sa'īd al-. *Sharḥ abyāt al-kimyā'* ("Commentary on Verses on Alchemy"). King 'Abd al-'Aziz Foundation, MS 552/13, 8–19 [paginated 1–13].

Fī 'ilm al-kimyā'. 134 Pages. Jum'a al-Mājid Center, originally from the Moroccan National Library with the number Marrakush 154.

Qalyūbī, Shihāb al-Dīn al-. *Mukhtaṣar 'ilm al-abdān*. 184 ayn. Nāṣiriyya Lodge, Tamgrout.

Rūdānī, Muḥammad al-. *Maqāsid al-'awālī*. Hamziyya Lodge Library, number 1787.

Maqāsid al-'awālī. Hathi Trust. https://hdl.handle.net/2027/mdp.39015081447016

Ṣāliḥī al-Idrīsī al-Laktāwī al-Dar'ī, Abū al-'Abbās Sīdī Aḥmad b. Ṣāliḥ al-. *al-Durar al-maḥmūla 'alā al-hadiya al-maqbūla*. National Library of Morocco, number 1653, 653 pages.

Tinbuktī, Aḥmad Bābā al-. *al-Lama' fī l-ishārah bi-ḥukm ṭābah*. King 'Abd al-'Azīz Foundation, Number MS 527_M11, Casablanca, 1–36.

Yūsī, Ḥasan al-. *As'ila wa ajwiba*. Jum'a al-Majid Center, 551322, p. 7 (Originally from the Tichit Library in Mauritania).

Zammūrī, ʿAbd al-Ghanī al-. *al-Qānūn al-mufīd fī ʿalāj al-ḥaṣā bi-qawl sadīd*. ʿAllāl al-Fāsī Library, call number 318 C.

SECONDARY SOURCES

Abbasi, Rushain. "Did Premodern Muslims Distinguish Between the Religious and the Secular? The *Dīn-Dunyā* Binary in Medieval Islamic Thought," *Journal of Islamic Studies* 31 (2020), 1–42.

Abdel Haleem, M. A. S. *The Qur'an* (Oxford: Oxford University Press, 2008).

Abdelajouad, Majid. "Issues in the History of Mathematics Teaching in Arab Countries," *Paedogogica Historica* 42 (2006), 629–64.

Abi-Mershed, Osama. "The Transmission of Knowledge and the Education of the 'Ulama in Late Sixteenth-Century Maghrib: A Study of the Biographical Dictionary of Muhammad ibn Maryam." In Mary Ann Fay (ed.) *Auto/Biography and the Construction of Identity and Community in the Middle East* (New York: Palgrave, 2000), 19–36.

Abū al-Ṣawāb, Ibrāhīm. "Dirāsa biblīūghrāfiyya li-baʿḍ al-makhṭūṭāt al-mawjūdah bi-l-maktabāt al-ʿāmma wa-l-khāṣṣa bi-Tārūdānt." In *Tārūdānt: ḥāḍirat Sūs* (Agadir: Manshurat kulliyat al-adab wa-l-ʿulum al-insaniyya, 1997), 137–53.

Abun-Nasr, Jamil M. *A History of the Maghrib in the Islamic Period* (Cambridge: Cambridge University Press, 1987).

Adas, Michael. *Machines as the Measure of Men: Science, Technology, and Ideologies of Western Dominance* (Ithaca, NY: Cornell University Press, 1989).

Adil, Sabahat. "Memorializing al-Maqqarī: The Life, Work, and Worlds of a Muslim Scholar." PhD Dissertation, University of Chicago, 2015.

Ahmed, Asad. "The *Sullam al-ʿulūm* of Muḥibballāh al-Bihārī: A Milestone in Arabo-Islamic Logic." In Khaled El-Rouayheb and Sabine Schmidtke (eds.) *The Oxford Handbook of Islamic Philosophy* (Oxford: Oxford University Press, 2017).

Ahmed, Asad Q. and Margaret Larkin. "The *Ḥāshiya* and Islamic Intellectual History," *Oriens* 41 (2013), 213–16.

Ahmed, Shahab. *What is Islam? The Importance of Being Islamic* (Princeton, NJ: Princeton University Press, 2016).

Akbari, Suzanne Conklin. *Idols in the East: European Representations of Islam and the Orient, 1100–1450* (Ithaca, NY: Cornell University Press, 2009).

Akkach, Samer. *'Abd al-Ghani al-Nabulusi: Islam and the Enlightenment* (Oxford: Oneworld, 2007).

'Alawī, 'Abdallāh Bennaser al-. *al-Shiʿr al-saʿdī: tafāʿul al-wāqiʿ wa l-fikr wa l-ibdāʿ* (Fez: Manshūrāt jāmiʿa Sīdī Muḥammad b. ʿAbdallāh, 2006).

Allen, Robert. *The British Industrial Revolution in Global Perspective* (Cambridge: Cambridge University Press, 2009).

Amster, Ellen. *Medicine and the Saints: Science, Islam, and the Colonial Encounter in Morocco, 1877–1956* (Austin, TX: University of Texas Press, 2013).
"Adarraq." In Kate Fleet, Gudrun Krämer, Denis Matringe, John Nawas, and Everett Rowson (eds.) *Encyclopaedia of Islam*, 3rd ed. (2007). http://dx.doi.org/10.1163/1573-3912_ei3_COM_23831
Antonius, George. *The Arab Awakening* (New York: J. B. Lippincott, 1939).
Anzali, Ata. *"Mysticism" in Iran: The Safavid Roots of a Modern Concept* (Columbia, SC: The University of South Carolina Press, 2017).
Arjona Castro, Antonio. "Las epidemias de peste bubónica en Andalucía en el siglo XIV," *Boletín de la Real Academia de Córdoba de Ciencias, Bellas Artes y Nobles Artes* 56 (1985), 49–58.
Artun, Tuna. "'Hearts of Gold and Silver': The Production of Alchemical Knowledge in the Early Modern Ottoman World." PhD Dissertation, Princeton University, 2013.
Azmeh, Aziz al-. *Ibn Khaldūn: An Essay in Reinterpretation* (London: Routledge, 1982).
Bachour, Natalia. *Oswaldus Crollius und Daniel Sennert im frühneuzeitlichen Istanbul: Studien zur Rezeption des Paracelsismus im Werk des osmanischen Arztes Ṣāliḥ b. Nasrullah Ibn Sallūm al-Ḥalabī* (Freiburg: Centaurus Verlag & Media, 2012).
Baer, Gabriel. "The Waqf as a Prop for the Social System (Sixteenth–Twentieth Centuries)," *Islamic Law and Society* 4 (1997), 264–97.
Bakar, Osman. *Classification of Knowledge in Islam: A Study in Islamic Philosophies of Science* (Cambridge: The Islamic Texts Society, 1998).
Bashir, Shahzad. "On Islamic Time: Rethinking Chronology in the Historiography of Muslim Societies," *History and Theory* 53 (2014), 519–44.
Sufi Bodies: Religion and Society in Medieval Islam (New York: Columbia University Press, 2013).
Batran, Aziz. *Tobacco Smoking under Islamic Law: Controversy over Its Introduction* (Beltsville, MD: Amana Publication, 2003).
"Ḥarb Fatāwī al-Tadkhīn Bayna al-ʿUlamā' wa-l-Muslimīn min Shamāl wa Gharb Ifrīqiyya fī-l-Aqdayn al-Awwal wa-l-Thānī min Ẓuhūr al-Tabagh." In *Fās wa Ifrīqīya* (Fez: Maʿhad al-dirāsāt al-ifrīqīya, 1993), 183–223.
Bauer, Thomas. *Die Kultur der Ambiguität: Eine andere Geschichte des Islams* (Berlin: Insel Verlag, 2011).
Bayly, C. A. *The Birth of the Modern World 1780–1914* (Oxford: Blackwell Publishing, 2004).
Bazzāz, Muḥammad al-Amīn al-. *Tārīkh al-awbi'a wa-l-majāʿat bi-l-maghrib fī l-qarnayn al-thāmin wa- l- tāsiʿ ʿashara* (Rabat: Manshūrāt Kulliyāt al-Adab, 1992).
Bazzaz, Sahar. *Forgotten Saints: History, Power and Politics in the Making of Modern Morocco* (Cambridge: Harvard Center for Middle Eastern Studies, 2010).
Beck, Herman. *L'image d'Idrīs II, ses descendants de Fās et la politique sharīfienne des sultans marīnides (656–869/1258–1465)* (Leiden: Brill, 1989).
Benchekroun, Chafik. "Écriture et réécriture de l'histoire des Idrissides." In Iyas Hassan (ed.) *La littérature aux marges du 'adab: Regards croisé sur la*

prose arabe classique (Beirut: Publications de l'Institut Français de Damas, 2017), 298–335.

"Les Idrissides: L'histoire contre son histoire," *Al-Masaq* 23 (2011), 171–88.

Benjelloun-Laroui, Latifa. *Les bibliothèques au Maroc* (Paris: Maisonneuve et Larose, 1990).

Bennison, Amira. *The Almoravid and Almohad Empires* (Edinburgh: Edinburgh University Press, 2016).

Berger, Lutz. "Ein Herz wie ein trockener Schwamm: Laqanis und Nabulusis Schriften über den Tabakrauch," *Der Islam* 78 (2001), 249–93.

Berque, Jacques. *Ulémas, fondateurs, insurgés du Maghreb: XVIIe siècle* (Paris: Sindbad, 1982).

L'intérieur du Maghreb: XVe–XIXe siècle (Paris: Gallimard, 1978).

Al-Yousi: Problèmes de la Culture Marocaine au XVIIème Siècle (Paris: Mouton & Co, 1958).

Bevilacqua, Alexander. *The Republic of Arabic Letters: Islam and the European Enlightenment* (Cambridge: Belknap Press, 2018).

Biesterfeldt, Hans Hinrich. "Arabisch-islamische Enzyklopädien: Formen und Funktionen." In Christel Meier (ed.) *Die Enzyklopädie vom Hochmittelalter bis zur frühen Neuzeit* (Munich: Wilhelm Fink, 2002), 43–83.

"Medieval Arabic Encyclopedias of Science and Philosophy." In Steven Harvey (ed.) *The Medieval Hebrew Encyclopedias of Science and Philosophy* (Dordrecht: Kluwer, 2000), 77–98.

Binebine, Aḥmed Chouqui al-. "Ẓāhirat waqf al-kutub fī tārīkh al-khizāna al-maghribiyya." In *Dirāsāt fī ʿilm al-makhṭūṭāt wa l-baḥth al-bibliyūghāfī* (Marrakesh: Maṭbaʿa wa l-warāqa al-waṭaniyya, 2004), 2nd ed., 110–34.

"Fahrasat al-makhṭūṭ al-ʿarabī: al-tajriba al-maghribiyya." In *Dirāsāt fī ʿilm al-makhṭūṭāt wa l-baḥth al-bībliyūghrāfī* (Marrakesh: Maṭbaʿa wa l-warāqa al-waṭaniyya, 2004), 2nd ed., 42–78.

Tā'rīkh khazā'in al-kutub bi l-maghrib (Marrakesh: Maṭbaʿa wa l-warāqa al-waṭṭaniyya, 2003).

Boyarin, Daniel. *Borderlines: The Partition of Judaeo-Christianity* (Philadelphia, PA: The University of Pennsylvania Press, 2004).

Bray, Julia. "Dāwūd ibn ʿUmar al-Anṭākī." In Devin Stewart and Joseph Lowry (eds.) *Essays in Arabic Literary Biography 1350–1850* (Wiesbaden: Otto Harrassowitz, 2009), 48–56.

Brentjes, Sonja. *Teaching and Learning the Sciences in Islamicate Societies (800–1700)* (Turnhout: Brepols, 2018).

"On Four Sciences and Their Audiences in Ayyubid and Mamluk Societies." In Syrinx von Hees (ed.) *Inḥiṭāṭ – The Decline Paradigm: Its Influence and Persistence in the Writing of Arab Cultural History* (Würzburg: Ergon Verlag, 2017), 139–72.

Review of Salim T. S. al-Hassani (ed.) *1001 Inventions: The Enduring Legacy of Muslim Civilization* (Washington, DC: National Geographic Society, 2012), 3rd ed. in *Aestimatio* 10 (2013), 119–53.

Travellers from Europe in the Ottoman and Safavid Empires, 16th–17th centuries: Seeking, Transforming, Discarding Knowledge (Burlington: Variorum, 2010).

"The Mathematical Sciences in the Safavid Empire: Questions and Perspectives." In D. Hermann and F. Speziale (eds.) *Muslim Cultures in the Indo-Iranian World during the Early-Modern and Modern Periods* (Berlin: Klaus Schwarz Verlag, 2010), 325–402.

"Pride and Prejudice: The Invention of a 'Historiography of Science' in the Ottoman and Safavid Empires by European Travellers and Writers in the Sixteenth and Seventeenth Centuries." Reprinted in *Travellers from Europe in the Ottoman and Safavid Empires, 16th–17th Centuries: Seeking, Transforming, Discarding Knowledge*, [2005], 229–54.

"Early Modern Western European Travellers in the Middle East and Their Reports about the Sciences" [2004]. Reprinted in *Travellers from Europe in the Ottoman and Safavid* Empires, *16th–17th Centuries: Seeking, Transforming, Discarding Knowledge* [2004], 379–420.

"Teaching the Mathematical Sciences in Islamic Societies Eighth–Seventeenth Centuries." In Karp and Schubring (eds.) *Handbook on the History of Mathematics Education* (New York: Springer, 2004), 85–105.

Brentjes, Sonja, Taner Edis and Lutz Richter-Bernburg (eds.) *1001 Distortions: How (Not) to Narrative History of Science, Medicine, and Technology in Non-Western Cultures* (Würzberg: Ergon Verlag, 2016).

Brockopp, Jonathan (ed.) *Islamic Ethics of Life: Abortion, War, and Euthanasia* (Columbia, SC: The University of South Carolina Press, 2003).

Brockopp, Jonathan and Thomas Eich (eds.) *Muslim Medical Ethics* (Columbia, SC: The University of South Carolina Press, 2008).

Buehler, Arthur F. *Recognizing Sufism: Contemplation in the Islamic Tradition* (New York: I.B. Tauris, 2016).

Bulliet, Richard. *The Patricians of Nishapur: A Study in Medieval Islamic Social History* (Cambridge: Harvard University Press, 1972).

Burke III, Edmund. *The Ethnographic State: France and the Invention of Moroccan Islam* (Oakland, CA: The University of California Press, 2014).

Prelude to Protectorate in Morocco: Precolonial Protest and Resistance, 1860–1912 (Chicago, IL: The University of Chicago Press, 1976).

"Pan-Islam and Moroccan Resistance to French Colonial Penetration, 1900–1912," *The Journal of African History* 13 (1972), 97–118.

Buskens, Léon and Baudouin Dupret, "The Invention of Islamic Law: A History of Western Studies of Islamic Normativity and Their Spread in the Orient." In François Pouillon and Jean-Claude Vatin (eds.) *After Orientalism: Critical Perspectives on Western Agency and Eastern Re-Appropriations* (Leiden: Brill, 2015), 31–47.

Butterfield, Hebert. *The Origins of Modern Science: 1300–1800* (New York: Macmillan Company, 1959).

Calderwood, Eric. *Colonial al-Andalus: Spain and the Making of Modern Moroccan Culture* (Cambridge: Harvard University Press, 2018).

"The Beginning (or end) of Moroccan History: Historiography, Translation, and Modernity in Ahmad b. Khalid al-Nasiri and Clemente Cerdeira," *International Journal of Middle East Studies* 44 (2012), 399–420.

Calero Secall, María Isabel. "La peste en Málaga, según el malagueño al-Nubāhī." In *Homenaje al Profesor Jacinto Bosch Vilá* (Granada: Universidad de Granada, 1991), vol. 1, 57–71.

Calvo, Emilia. "Two Treatises on Mīqāt from the Maghrib (14th and 15th Centuries A.D.)," *Suhayl* 4 (2004), 159–206.

Cañizares-Esguerra, Jorge. *Nature, Empire, and Nation: Explorations of the History of Science in the Iberian World.* (Stanford, CA: Stanford University Press, 2006).

Chakrabarty, Dipesh. "AHR Roundtable: The Muddle of Modernity," *American Historical Review* 116 (2011), 663–75.

Chamberlain, Michael. *Knowledge and Social Practice in Medieval Damascus, 1190–1350* (New York: Cambridge University Press, 1994).

Charette, François. *Mathematical Instrumentation in Fourteenth-Century Egypt and Syria: The Illustrated Treatise of Najm al-Dīn al-Miṣrī* (Leiden: Brill, 2003).

Clark, John F. M. "Intellectual History and the History of Science." In Richard Whatmore and Brian Young (eds.) *A Companion to Intellectual History* (Oxford: Wiley Blackwell, 2016), 155–69.

Coindreau, Roger. *Les Corsaires de Salé* (Rabat: La Croisée des Chemins, 2006 [1948]).

Colin, G. S. and H. P. J. Renaud. "Note sur le 'muwaqqit' marocain Abû Muqri' – ou mieux Abû Miqra' – al-Baṭṭîwî (XIIIe siècle J.-C.)," *Hésperis* 25 (1938), 94–96.

Congourdeau, Marie-Hélène and Mohamed Melhaoui. "La perception de la peste en pays chrétien byzantine et musulman," *Revue des Études Byzantines* 59 (2001), 95–124.

Conrad, Lawrence. "A Ninth-Century Muslim Scholar's Discussion of Contagion." In Lawrence Conrad and Dominik Wujastyk (eds.) *Contagion: Perspectives from Pre-Modern Societies* (Aldershot: Ashgate, 2000), 163–77.

"The Pilgrim from Pest: Goldziher's Study Tour to the Near East (1873–74)." In Ian Netton (ed.) *Golden Roads: Migration, Pilgrimage and Travel in Medieval and Modern Islam* (London: Routledge, 1995), 110–59.

"The Dervish's Disciple: On the Personality and Intellectual Milieu of the Young Ignaz Goldziher," *The Journal of the Royal Asiatic Society of Great Britain and Ireland* 2 (1990), 225–66.

Cook, Harold. *Matters of Exchange: Commerce, Medicine, and Science in the Dutch Golden Age* (New Haven, CT: Yale University Press, 2007).

Cooperson, Michael. "The Abbasid 'Golden Age': An Excavation," *Al-'Uṣūr al-Wusṭā* 25 (2017), 41–65.

Cornell, Vincent. *The Realm of the Saint* (Austin, TX: University of Texas Press, 1998).

Cory, Stephen. *Reviving the Islamic Caliphate in Early Modern Morocco.* (Burlington: Ashgate, 2013).

"Sharīfian Rule in Morocco (Tenth–Twelfth/Sixteenth–Eighteenth Centuries)." In Maribel Fierro (ed.) *The New Cambridge History of Islam Part 4: North and West Africa* (Cambridge: Cambridge University Press, 2010), 451–79.

Cotter, Charles. "The Mariner's Sextant and the Royal Society," *Notes and Records of the Royal Society of London* 33 (1978), 23–36.

Crone, Patricia. *Pre-Industrial Societies: Anatomy of the Pre-Modern World.* (London: Oneworld, 2015 [1989]).

Cunningham, Andrew. "Getting the Game Right: Some Plain Words of the Identity and Invention of Science," *Studies in the History and Philosophy of Science* 19 (1988), 365–89.

Dale, Stephen. *The Muslim Empires of the Ottomans, Safavids, and Mughals.* (Cambridge: Cambridge University Press, 2010).

Dallal, Ahmed. *Islam, Science, and the Challenge of History* (New Haven, CT: Yale University Press, 2010).

"The Origins and Early Development of Islamic Reform." In Robert Hefner (ed.) *The New Cambridge History of Islam Vol. 6: Muslims and Modernity: Culture and Society since 1800* (Cambridge: Cambridge University Press, 2010), 107–47.

"Appropriating the Past: Twentieth Century Reconstruction of Pre-Modern Islamic Thought," *Islamic Law and Society* 7 (2000), 325–58.

Davis, Kathleen. *Periodization and Sovereignty: How Ideas of Feudalism and Secularization Govern the Politics of Time* (Philadelphia, PA: University of Pennsylvania Press, 2008).

De Blois, F. C., Van Dalen, B., Humphreys, R. S., et al. "Taʾrīkh." In P. Bearman, Th. Bianquis, C. E. Bosworth, E. van Donzel, and W. P. Heinrichs (eds.) *Encyclopaedia of Islam*, 2nd ed. (1954). http://dx.doi.org/10.1163/1573-3912_islam_COM_1184

Delphin, G. *Fas, son université et l'enseignement supérieur musluman* (Oran: Paul Perrier, 1889).

Deverdun, G. "Ibn al-Ḳāḍī." In P. Bearman, Th. Bianquis, C. E. Bosworth, E. van Donzel, and W. P. Heinrichs (eds.) *Encyclopaedia of Islam*, 2nd ed. (1954). http://dx.doi.org/10.1163/1573-3912_islam_SIM_3227

Dols, Michael. "The Leper in Medieval Islamic Society," *Speculum* 58 (1983), 891–916.

The Black Death in the Middle East (Princeton, NJ: Princeton University Press, 1977).

El-Ariss, Tarek. *Trials of Arab Modernity: Literary Affects and the New Political* (New York: Fordham University Press, 2013).

El Hamel, Chouki. *Black Morocco: A History of Slavery, Race, and Islam* (Cambridge: Cambridge University Press, 2012).

El Mansour, Mohamed. "Saints and Sultans: Religious Authority and Temporal Power in Precolonial Morocco." In Kisaichi Masatoshi (ed.) *Popular Movements and Democratization in the Islamic World* (London: Routledge, 2001), 1–32.

"Salafis and Modernists in the Moroccan Nationalist Movement." In John Ruedy (ed.) *Islamism and Secularism in North Africa* (New York: St. Martin's Press, 1994), 53–71.

Morocco in the Reign of Mawlay Sulayman (Cambridgeshire: MENAS Press, 1990).

El-Rouayheb, Khaled. *Islamic Intellectual History in the Seventeenth Century* (Cambridge: Cambridge University Press, 2015).

"The Rise of 'Deep Reading' in Early Modern Ottoman Scholarly Culture." In Sheldon Pollock, Benjamin A. Elman, and Ku-Ming Kevin Chang (eds.) *World Philology* (Cambridge: Harvard University Press, 2015), 201–24.

Relational Syllogisms and the History of Arabic Logic, 900–1900 (Leiden: Brill, 2010).

"Opening the Gate of Verification: The Forgotten Arab-Islamic Florescence of the Seventeenth Century," *International Journal of Middle East Studies* 38 (2006), 263–81.

"Was there a Revival of Logical Studies in Eighteenth-Century Egypt?," *Die Welt des Islams* 45 (2005), 1–19.

"Sunni Muslim Scholars on the Status of Logic, 1500–1800," *Islamic Law and Society* 11 (2004), 213–32.

Elshakry, Marwa. *Reading Darwin in Arabic, 1860–1950* (Chicago, IL: The University of Chicago Press, 2014).

El Shamsy, Ahmed. "The Social Construction of Orthodoxy." In Tim Winter (ed.) *The Cambridge Companion to Classical Islamic Theology* (Cambridge: Cambridge University Press, 2008), 97–117.

Endress, Gerhard. "The Cycle of Knowledge: Intellectual Traditions and Encyclopaedias of the Rational Sciences in Arabic Islamic Hellenism." In Gerhard Endress (ed.) *Organizing Knowledge: Encyclopaedic Activities in the Pre-Eighteenth Century Islamic World* (Leiden: Brill, 2006), 103–33.

Fadel, Mohammad. "The Social Logic of *Taqlīd* and the Rise of the *Mukhtaṣar*," *Islamic Law and Society* 3 (1996), 193–233.

"Adjudication in the Mālikī Madhhab: A Study of Legal Process in Medieval Islamic Law." PhD Dissertation, University of Chicago, 1996.

Fancy, Nahyan. *Science and Religion in Mamluk Egypt: Ibn al-Nafīs, Pulmonary Transit and Bodily Resurrection* (New York: Routledge, 2013).

Feldhay, Riva and F. Jamil Ragep (eds.) *Before Copernicus: The Cultures and Contexts of Scientific Learning in the Fifteenth Century* (Montreal: McGill-Queen's University Press, 2017).

Fernández Félix, Ana. *Cuestiones legales del Islam temprano: La ʿutbiyya y el proceso de formación de la sociedad islámica andalusí* (Madrid: CSIC, 2003).

Fierro, Maribel. "Proto-Malikis, Malikis, and Reformed Malikis in al-Andalus." In Peri Bearman, Rudolph Peters and Frank E. Vogel (eds.) *The Islamic School of Law: Evolution, Devolution, and Progress* (Cambridge: Islamic Legal Studies Program, 2005), 57–76.

"The Legal Policies of the Almohad Caliphs and Ibn Rushd's *Bidāyat al-Mujtahid*," *Journal of Islamic Studies* 10 (1999), 226–48.

Fleck, Ludwik. *Genesis and Development of a Scientific Fact* (Chicago, IL: The University of Chicago Press, 1979 [1936]).

Forcada, Miquel. *Ética e ideología de la Ciencia: El Médico-Filósos en al-Andalus (siglos X-XII)* (Almería: Fundación Ibn Tufayl de Estudios Árabes, 2011).

Forster, Regula. (ed.) "Arabic Alchemy. Texts and Contexts." Themed Issue, *Al-Qanṭara* 37 (2016). In Kate Fleet, Gudrun Krämer, Denis Matringe,

John Nawas, and Everett Rowson (eds.) *Encyclopaedia of Islam*, 3rd ed. (2007). http://dx.doi.org/10.1163/1573-3912_ei3_COM_23831

Fowden, Garth. *Before and After Muḥammad: The First Millennium Refocused* (Princeton, NJ: Princeton University Press, 2014).

Empire to Commonwealth: Consequences of Monotheism in Late Antiquity (Princeton, NJ: Princeton University Press, 1993).

Gaiser, Adam. "Slaves and Silver across the Strait of Gibraltar: Politics and Trade between Umayyad Iberia and Khārijite North Africa," *Medieval Encounters* 19 (2013), 41–70.

García-Arenal, Mercedes. *Ahmad al-Mansur: The Beginnings of Modern Morocco* (Oxford: Oneworld Publications, 2009).

Messianism and Puritanical Reform: Mahdis of the Muslim West (Leiden: Brill, 2006).

"*Mahdī, Murābit, Sharīf*: l'avènement de la dynastie Sa'dienne," *Studia Islamica* 71 (1990), 77–114.

"Sainteté et pouvoir dynastique au Maroc," *Annales* 45 (1990), 1019–42.

"The Revolution of Fās in 869/1465 and the Death of Sultan 'Abd al-Ḥaqq al-Marīnī," *Bulletin of the School of Oriental and African Studies* 1 (1978), 43–66.

García-Arenal, Mercedes and Eduardo Manzano Moreno, "Idrīssisme et villes idrīssides," *Studia Islamica* 82 (1995), 5–34.

García-Arenal, Mercedes and Gerard Wiegers Martin Beagles (trans.) *A Man of Three Worlds: Samuel Pallache, a Moroccan Jew in Catholic and Protestant Europe* (Baltimore, MD: Johns Hopkins University Press, 2007).

García Gómez, Emilio. *Foco de antigua luz sobre la Alhambra* (Madrid: Publicaciones del Instituto Egipcio de Estudios Islámicos en Madrid, 1988).

Gardiner, Noah. "Forbidden Knowledge? Notes on the Production, Transmission, and Reception of the Major Works of Aḥmad al-Būnī," *Journal of Arabic and Islamic Studies* 12 (2012), 81–143.

Geertz, Clifford. *Islam Observed* (Chicago, IL: The University of Chicago Press, 1971).

Gesink, Indira. *Islamic Reform and Conservatism: Al-Azhar and the Evolution of Modern Sunni Islam* (London: I.B. Tauris, 2010).

"'Chaos on the Earth': Subjective Truths versus Communal Unity in Islamic Law and the Rise of Militant Islam," *American Historical Review* 108 (2003), 710–33.

Gimaret, D. "al-Milal wa 'l-niḥal." In P. Bearman, Th. Bianquis, C. E. Bosworth, E. van Donzel, and W. P. Heinrichs (eds.) *Encyclopaedia of Islam*, 2nd ed. (1954). http://dx.doi.org/10.1163/1573-3912_islam_COM_1184

Goldziher, Ignaz. *Stellung der alten islamischen Orthodoxies zu den antiken Wissenschaften* (Berlin: Verlag der Königl. Akademie der Wissenschaften, 1916), 3–46.

Muhammedanischer Studien (Halle: Niemeyer,1889–90), 2 vols.

Golinski, Jan. *Making Natural Knowledge: Constructivism and the History of Science* (Chicago, IL: The University of Chicago Press, 2005 [1998]).

Grant, Edward. *God and Reason in the Middle Ages* (Cambridge: Cambridge University Press, 2001).

Foundations of Modern Science in the Middle Ages (Cambridge: Cambridge University Press, 1996).
Green, Nile. *Sufism: A Global History.* (Oxford: Wiley-Blackwell, 2012).
Grehan, James. "Smoking and 'Early Modern' Sociability: The Great Tobacco Debate in the Ottoman Middle East (Seventeenth to Eighteenth Centuries)," *American Historical Review* 111 (2006), 1352–77.
Griffel, Frank. "What Do We Mean by 'Salafī'? Connecting Muḥammad 'Abduh with Egypt's Nūr Party in Islam's Contemporary Intellectual History," *Die Welt des Islams* 55 (2015), 186–220.
"Ibn Tūmart's Rational Proof for God's Existence and Unity, and His Connection to the Niẓāmiyya *Madrasa* in Baghdad." In Patrice Cressier, Maribel Fierro, and Luis Molina (eds.) *Los Almohades: Problemas y Perspectivas* (Madrid: CSIC, 2005) 2, 753–813.
Grunebaum, G. E. "The Problem: Unity in Diversity." In Gustave E. von Grunebaum (ed.) *Unity and Variety in Muslim Civilization* (Chicago, IL: The University of Chicago Press, 1955), 17–37.
Günther, Sebastian and Dorothee Pielow (eds.) *Die Geheimnisse der oberen und der unteren Welt: Magie im Islam zwischen Glaube und Wissenschaft* (Leiden: Brill, 2018).
Gutas, Dimitri. "Ibn al-Nafīs's Scientific Method," *Nazariyat* 4 (2017), 139–45.
Gutelius, David. "Between God and Men: The Nāṣiriyya and Economic Life in Morocco, 1640–1830." PhD Dissertation, Johns Hopkins University, 2001.
Hacking, Ian. "Introductory Essay." In *The Structure of Scientific Revolutions* (50th anniversary ed.) (Chicago, IL: The University of Chicago Press, 2012), vii–xxxvii.
Haider, Najam. "Contesting Intoxication: Early Juristic Debates over the Lawfulness of Alcoholic Beverages," *Islamic Law and Society* 20 (2013), 48–89.
"The Community Divided: A Textual Analysis of the Murders of Idrīs b. 'Abd Allāh (d. 175/791)," *Journal of the American Oriental Society* 128 (2008), 459–75.
Ḥajjī, Muḥammad. *al-Zāwiyah al-dilā'iyyah wa dawru-hā al-dīnī wa l-ʿilmī wa l-siyāsī* (Rabat: Kulliyāt al-Adab, 1988), 2nd ed.
(ed.) *Fihris al-khizānah al-ʿilmiyya al-ṣubīḥiyya bi-Silā* (Kuwait: Manshūrāt maʿhad al-makhṭūṭāt al-ʿarabiyya, 1985).
(ed.) *Mawsūʿat aʿlām al-maghrib* (Beirut: Dār al-gharb al-islāmī, 1980), 10 vols.
al-Ḥaraka al-fikriyya bi-l-maghrib fī l-ʿahd al-saʿadiyyīn (Rabat: Dār al-Maghrib li l-ta'lif wa l-tarjamah wa l-nashr, 1977), 2 vols.
Halevi, Leor. *Muhammad's Grave: Death Rites and the Making of Islamic Society* (New York: Columbia University Press, 2007).
Hallaq, Wael. "Was the Gate of Ijtihad Closed?," *The International Journal of Middle East Studies* 16 (1984), 3–41.
Hämeen-Anttila, Jaakko, "Ibn Waḥshiyya." In Kate Fleet, Gudrun Krämer, Denis Matringe, John Nawas, and Everett Rowson (eds.) *Encyclopaedia of Islam*, 3rd ed. (2007). http://dx.doi.org/10.1163/1573-3912_ei3_COM_32287
Hamès, Constant. (ed.) *Coran et talismans: Textes et pratiques magiques en milieu musulman* (Paris: Karthala, 2007).

Entre recette magique d'al-Bûnî et prière islamique d'al-Ghazālī: Textes talismaniques d'Afrique occidentale," *Fétiches II: puissance des objets, charme des mots* 12 (1993), 187–233.

Hanegraaff, Wouter. "Beyond the Yates Paradigm: The Study of Western Esotericism Between Counterculture and New Complexity," *Aries* 1 (2001), 5–37.

Hanna, Nelly. *Ottoman Egypt and the Emergence of the Modern World 1500–1800* (Cairo: AUC Press, 2014).

Hanssen, Jans and Max Weiss (eds.) *Arabic Thought beyond the Liberal Age: Towards an Intellectual History of the Nahda* (Cambridge: Cambridge University Press, 2016).

Ḥarakāt, Ibrāhīm. *Madkhal ta'rīkh al-'ulūm bi-l-maghrib al-muslim ḥattā al-qarn 9/15h* (Casablance: Dār al-rashād al-ḥadītha, 2000), 3 vols.

Ḥarbī, Mubārak Jazā' al-. "Namādhij min juhūd fuqahā' al-mālikiyya al-maghāriba fī tadwīn al-nawāzil al-fiqhiyya," *Majallat al-Sharī'a wa-l-Dirāsāt al-Islāmiyya* 21 (2006), 281–364.

Harrak, Fatima. "State and Religion in Eighteenth Century Morocco: The Religious Policy of Sidi Muḥammad b. 'Abd Allâh 1757–1790." SOAS Dissertation, 1989.

Harris, Nicholas G. "In Search of 'Izz al-Dīn Aydamir al-Ǧildakī, Mamlūk Alchemist," *Arabica* 64 (2017), 531–56.

Harris, Sam. *The End of Faith: Religion, Terror, and the Future of Reason* (New York: W. W. Norton, 2005).

Harrison, Peter. *The Territories of Science and Religion* (Chicago, IL: The University of Chicago Press, 2015).

The Bible, Protestantism, and the Rise of Modern Science (Cambridge: Cambridge University Press, 2001).

Hasse, Dag Nikolaus. *Success and Suppression: Arabic Sciences and Philosophy in the Renaissance* (Cambridge: Harvard University Press, 2016).

Hendrickson, Jocelyn. "A Guide to Arabic Manuscript Libraries in Morocco, with Notes on Tunisia, Algeria, Egypt, and Spain," *MELA Notes* 81 (2008), 15–88.

Hendrickson, Jocelyn and Sabahat Adil, "A Guide to Arabic Manuscript Libraries in Morocco: Further Developments," *MELA Notes* 86 (2013), 1–19.

Henriet, Parick. "Miguel Asín Palacios (1871–1944), le christianisme et l'islam: theses, idéologie, reception." In Jean-Luc Fournet, Jean-Michel Mouton, and Jacques Paviot (eds.) *Civilisations en Transition (III): Sociétés Multiconfessionnelles à Travers l'Histoire du Proche Orient* (Byblos: Publications du Centre International des Sciences de l'Homme, 2017), 181–212.

Hershenzon, Daniel. *The Captive Sea: Slavery, Communication, and Commerce in Early Modern Spain and the Mediterranean* (Philadelphia, PA: University of Pennsylvania Press, 2018).

"Traveling Libraries: The Arabic Manuscripts of Muley Zidan and the Escorial Library," *Journal of Early Modern History* 18 (2014), 535–58.

Hess, Andrew. *The Forgotten Frontier: A History of the Sixteenth-Century Ibero-African Frontier* (Chicago, IL: The University of Chicago Press, 1978).

Hīlah, Muḥammad al-Ḥabīb al-. "Classification of Andalusian and Maghribī books of Nawāzil from the Middle of the Fifth to the End of the Ninth Century AH."

In John Cooper (ed.) *The Significance of Islamic Manuscripts* (London: Islamic Heritage Foundation, 1992), 71–78.

Hilālī, Muḥammad Yāsir al-. "Al-kimiyyāiyūn fī al-Maghrib al-Aqṣā khilāl al-ʿaṣr al-wasīṭ al-muta'akhir bayna dawāfiʿ al-ishtighāl bi as-sanʿati wa mawāqif al-fuqahā' wa-aṣūfiyya," *Hespéris-Tamuda* 54 (2019), 265–313.

Hirschler, Konrad *A Monument to Medieval Syrian Book Culture – The Library of Ibn ʿAbd al-Hādī* (Edinburgh: Edinburgh University Press, 2020).

The Written Word in the Medieval Arabic Lands (Edinburgh: Edinburgh University Press, 2011).

Hodgson, Marshall. *The Venture of Islam* (Chicago, IL: The University of Chicago Press, 1974), 3 vols.

Hofer, Nathan. *The Popularisation of Sufism in Ayyubid and Mamluk Egypt* (Edinburgh: Edinburgh University Press, 2015).

Honerkamp, Kenneth. "al-Ḥasan ibn Masʿūd al-Yūsī." In Malcolm Lowry and Devin J. Stewart (eds.) *Essays in Arabic Literary Biography, 1350–1850* (Wiesbaden: Harrassowitz Verlag, 2009), 410–19.

Hottox, Ralph, S. *Coffee and Coffeehouses: The Origins of a Social Beverage in the Medieval Near East* (Seattle, WA: University of Washington Press, 1985), 57–60.

Hourani, Albert. *Arabic Thought in the Liberal Age 1798–1939*. (Cambridge: Cambridge University Press, 1962).

Hoyland, Robert. *Seeing Islam as Others Saw It: A Survey of Christian, Jewish and Zoroastrian Writings on Early Islam* (Princeton, NJ: Darwin Press, 1998).

Huff, Toby. *The Rise of Early Modern Science: Islam, China, and the West*. (Cambridge: Cambridge University Press, 2003).

Humphries, Stephen. *Islamic History: A Framework for Inquiry*. Revised ed. (Princeton, NJ: Princeton University Press, 1991).

Hunwick, John. "A New Source for the Biography of Ahmad Baba al-Tinbukti," *Bulletin of the School of Oriental Studies* 27 (1964), 568–93.

Hussin, Iza. *The Politics of Islamic Law: Local Elites, Colonial Authority, and the Making of the Muslim State* (Chicago, IL: The University of Chicago Press, 2016).

Ibrahim, Ahmed Fekry. "Rethinking the Taqlīd Hegemony: An Institutional, Longue-Durée Approach," *Journal of the American Oriental Society* 136 (2016), 801–16.

Iḥnāna, Yūsuf. *Taṭawwur al-madhhab al-Ashʿarī fī-l-gharb al-islāmī*. 2nd ed. (Rabat, Wizārat al-Awqāf wa-l-Shuʿūn al-Islāmiyya, 2017).

Jackson, Sherman. *Islamic Law and the State: The Constitutional Jurisprudence of Shihāb al-Dīn al-Qarāfī* (Leiden: Brill, 1996).

"*Taqlīd*, Legal Scaffolding and the Scope of Legal Injunctions in Post-Formative Theory *Muṭlaq* and *ʿĀmm* in the Jurisprudence of Shihāb al-Dīn al-Qarāfī," *Islamic Law and Society* 3 (1996), 165–92.

Jaques, R. Kevin. *Authority, Conflict, and the Transmission of Diversity in Medieval Islamic Law* (Leiden: Brill, 2006).

Jacques-Meunie, D. *Le Maroc Saharien des origins a 1670* (Paris: Librairie Klinksieck, 1982), 2 vols.

Jīdī, ʿUmar al-. *Mabāḥith fī al-madhhab al-mālikī* (Rabat: al-Hilāl al-ʿArabiyyah, 1993).

Josephson-Storm, Jason Ā. *The Myth of Disenchantment: Magic, Modernity, and the Birth of the Modern Sciences* (Chicago, IL: The University of Chicago Press, 2017).

Kafadar, Cemal. "How Dark Is the History of the Night, How Black the Story of Coffee." In Arzu Öztürkmen and Evelyn Birge Vitz (eds.) *Medieval and Early Modern Performance in the Eastern Mediterranean* (Turnhout: Brepols, 2014), 243–69.

Kaḥḥāla, ʿUmar Riḍā. *Muʿjam al-mu'allifīn* (Beirut: Mu'assasat al-Risāla, 1993). 4 vols.

Karamustafa, Ahmet T. *Sufism: The Formative Period* (Berkeley, CA: University of California Press, 2007).

Katz, Marion Holmes. "The Problem of Abortion in Classical Sunni fiqh." In Jonathan E. Brockopp (ed.) *Islamic Ethics of Life: Abortion, War, and Euthanasia* (Columbia, SC: South Carolina Press, 2003), 25–50.

Keddie, Nikki R. *An Islamic Response to Imperialism: Political and Religious Writings of Sayyid Jamāl ad-Dīn al-Afghānī* (Berkeley, CA: University of California Press, 1983 [1968]).

Kennedy, E. S. "Astronomical Events from a Persian Astrological Manuscript," *Centaurus* 24 (1980), 162–77.

Kennedy, E. S. and M. H. *Geographical Coordinates of Localities from Islam.* (Frankfurt: Institut für Geschichte der Arabisch-Islamischen Wissenschaften, 1987).

Kesrouany, Maya. *Prophetic Translation: The Promise of European Literature in the Egyptian Imaginary* (Edinburgh: Edinburgh University Press, 2019).

Khaneboubi, Ahmed. *Les Institutions gouvernementales sous les Mérinids (1258–1465)* (Paris: Éditions l'Harmattan, 2008).

Les premiers Sultans Mérinids 1269–1331: Histoire politique et sociale (Paris: Éditions l'Harmattan, 1987).

Khaṭṭābī, Muḥammad al-ʿArabī al-. (ed.) *Fahāris al-khizāna al-malikiyya* (Rabat: Maṭbaʿa al-Najāḥ al-Jadīda, 1982), 2.

(ed.) *Fahāris al-khizānah al-malakiyya* (Rabat: Maṭbaʿa al-Najāḥ al-Jadīda, 1982), 3.

(ed.) *Fahāris al-khizāna al-malikiyya* (Rabat: Maṭbaʿa al-Najāḥ al-Jadīda, 1982), 5.

King, David. *In Synchrony with the Heavens: Studies in Astronomical Timekeeping and Instrumentation in Medieval Islamic Civilization* (Leiden: Brill, 2004).

"On the History of Astronomy in the Medieval Maghrib." In *Etudes philosophiques et sociologiques dédiés à Jamal ed-Dine Alaoui* Fez: Université Sidi Mohamed Ben Abdallah, Publications de la Faculté des Lettres et des Sciences Humaines Dhar El Mahraz (Numéro Spécial 14, 1999), 27–61.

"On the Role of the Muezzin and Muwaqqit in Medieval Islamic Societies." In *Synchrony with the Heavens* (Brill: Leiden, 2004), 2 vols., vol. 1, 623–78 (originally published in 1996).

"The Origin of the Astrolabe According to the Medieval Islamic Sources." In *Islamic Astronomical Instruments* (London: Variorum, 1987), vol. III, 44–83.

King, David and Julio Samsó. "Zīdj." In P. Bearman, Th. Bianquis, C. E. Bosworth, E. van Donzel, and W. P. Heinrichs (eds.) *Encyclopaedia of Islam*, 2nd ed. (1954). http://dx.doi.org/10.1163/1573-3912_islam_COM_1184

Knysh, Alexander. *Sufism: A New History of Islamic Mysticism* (Princeton, NJ: Princeton University Press, 2017).

Kugle, Scott. (ed.) *Sufi Meditation and Contemplation: Timeless Wisdom from Mughal India* (New Lebanon, NY: Omega Publications, 2012).

Sufis and Saints Bodies (Chapel Hill, NC: University of North Carolina Press, 2011).

Rebel Between Spirit and Law: Ahmad Zarruq, Sainthood, and Authority in Islam (Bloomington, IN: Indiana University Press, 2006).

Kuhn, Thomas. *The Structure of Scientific Revolutions* (50th anniversary ed.) (Chicago, IL: The University of Chicago Press, 2012).

"The Trouble with the Historical Philosophy of Science." In James Conant and John Haugeland (eds.) *The Road Since Structure* (Chicago, IL: The University of Chicago Press, 2000), 105–20.

"Commensurability, Comparability, Communicability." In James Conant and John Haugeland (eds.) *The Road Since Structure: Philosophical Essays, 1970–1993 with an Autobiographical Interview* (Chicago, IL: The University of Chicago Press, 2000 [originally published in 1983]), 33–57.

Kurz, Marlene. *Ways to Heaven, Gates to Hell: Fażlīzāde ʿAlī's Struggle with the Diversity of Ottoman Islam* (Berlin: EB Verlag, 2011).

l'Africain, Jean-Léon. *Description de l'Afrique* (Paris: Maisonneuve, 1980), 2 vols.

Laḥmar, Ḥamīd. *Al-fihris al-waṣfī li-l makhṭūṭāt khizānāt al-zāwiya al-nāṣirīya* (Rabat: Manshūrāt wizārat al-awqāf, 2013), 6 vols.

Al-fihris al-waṣfī li-l makhṭūṭāt khizānāt al-zāwiya al-ḥamzīya al-ʿayyāshīya (Rabat: Manshūrāt wizārat al-awqāf, 2009), 4 vols.

Lakhdar, Mohammed. *La vie littéraire au Maroc sous la dynastie ʿalawide* (Rabat: Éditions Techniques Nord-Africaines, 1971).

Laroui, Abdallah. *L'histoire du Maghreb: Un Essai de synthèse*. 2ème ed. (Casablanca: Centre Culturel Arabe, 2001) [1982].

Lauzière, Henri. "What We Mean Versus What They Meant by 'Salafi': A Reply to Frank Griffel," *Die Welt des Islams* 56 (2016), 89–96.

The Making of Salafism: Islamic Reform in the Twentieth Century (New York: Columbia University Press, 2016).

Lévi-Provençal, E. *Les historiens des Chorfas* (Casablanca: Afrique Orient, 1991 [1922]).

Liang, Yuen-Gen, Abigail Balbale, Andrew Devereux, and Camilo Gómez-Rivas (eds.). *Spanning the Strait: Studies in Unity in the Western Mediterranean*. (Leiden: Brill, 2013).

Lockman, Zachary. *Contending Visions of the Middle East: The History and Politics of Orientalism*. 2nd ed. (Cambridge: Cambridge University Press, 2010).

Lohlker, Rüdiger. *Islamisches Völkerrecht: Studien am Beispiel Granada* (Bremen: Kleio Humanitas, 2006).

Lory, Pierre. *La science des lettres en Islam* (Paris: Dervy, 2004).

Alchimie et mystique en terre d'Islam (Paris: Gallimard, 2003 [1989]).

Love, Paul. "The Sufris of Sijilmasa: Toward a History of the Midrarids," *The Journal of North African Studies* 15 (2010), 173–88.

Madelung, Wilfred. "Notes on Non-Ismā'īlī Shi'ism in the Maghrib," *Studia Islamica* 44 (1976), 87–97.

Madgharī, Abd al-Kabīr al-'Alawī al-. *al-Faqīh Abū 'Alī al-Yūsī* (Rabat: Wizārat al-awqāf wa l-shu'ūn al-islāmiyya, 1989).

Maier, Michael. *Symbola Aureae Mensae Duodecim Nationum* (Frankfurt: Printed by Anthony Humanij, 1617).

Makdisi, George. *The Rise of Colleges: Institutions of Learning in Islam and the West* (Edinburgh: Edinburgh University Press, 1981).

Makhlūf, Muḥammad. *Shajrat al-nūr al-zakiyya fī ṭabaqāt al-mālikiyya* (Beirut: Dār al-kutub al-'ilmiyya, 2003), 2 vols.

Shajrat al-nūr al-zakiyya (Beirut: Dar al-Kitab al-'Arabi, 1930–31).

Mannūnī, Muḥammad al-. *Ḥaḍarat wādī Dara' min khilāl al-nuṣūṣ wa-l-athār* (Rabat: Manshurat Wizarat Al-Awkaf Wa Al-Shu'un Al-Islamiyah, 1973).

"Faṣla taṣif al-dirāsah bi-l-qarawiyyīn ayyām al-manṣūr al-sa'dī." In *Qabs min 'aṭā' al-makhṭūṭ al-maghribī* (Beirut: Dār al-Gharb al-Islāmī, 1999), 4 vols., vol. 3, 1125–55 [originally published in *al-Baḥth al-'ilmī*, vol. 7 (1966)].

"Ẓāhira ta'ribiyya fī l-maghrib ayyām al-sa'diyyin," *Revista del Instituto Egipcio de Estudios Islámicos* 9 (1963), 329–58.

Marín, Manuela. *Testigos coloniales: españoles en Marruecos [1860–1956]* (Barcelona: Ediciones Bellaterra, 2015).

"Los estudios árabes y el colonialism español en Marruecos (siglos XIX–XX)." In A. Agud, A. Cantera, A. Falero, et al. (eds.) *Séptimo centenario de los estudios orientales de Salamanca* (Salamanca: Universidad, 2012), 251–64.

Marchand, Suzanne L. *German Orientalism in the Age of Empire: Religion, Race, and Scholarship* (New York: Cambridge University Press, 2009).

Marcus, Abraham. *The Middle East on the Eve of Modernity* (New York: Columbia University Press, 1989).

Martelli, Matteo. *The Four Books of Pseudo Democritus* (London: Routledge, 2019).

Masud, Muhammad Khalid, Brinkley Messick and David S. Powers (eds.) *Islamic Legal Interpretation: Muftis and their Fatwas* (Cambridge: Harvard University Press, 1996).

Maslūtī, Muṣṭafā b. 'Umar al-. *Muḥammad b. Sulaymān al-Rūdānī* (Rabat: Markaz al-dirāsāt wa-l-abḥāth wa-iḥyā' al-turāth, 2010).

Masuzawa, Tomoko. *The Invention of World Religions: Or, How European Universalism Was Preserved in the Language of Pluralism* (Chicago, IL: The University of Chicago Press, 2005).

Matar, Nabil. "Confronting Decline in Early Modern Arabic Thought," *Journal of Early Modern History* 9 (2005), 51–78.

Matthee, Rudi. *Persia in Crisis: Safavid Decline and the Fall of Isfahan* (London: I. B. Tauris, 2012).

The Pursuit of Pleasure: Drugs and Stimulants in Iranian History, 1500–1900 (Princeton, NJ: Princeton University Press, 2005).

Mayeur-Jaouen, Catherine. "'À la poursuite de la réforme': Renouveaux et débats historiographiques de l'histoire religeuse et intellectuelle de l'islam, XVe–XXIe siècle," *Annales* 73 2 (2019), 317–58.

Maziane, Leïla. *Salé et ses corsaires (1666–1727): Un port de course marocain au XVIIe siècle* (Caen: Presses Univeritaies de Caen, 2007).

McVaugh, Michael. "The 'Humidum Radicale' in Thirteenth-Century Medicine," *Traditio* 30 (1974), 259–83.

Mediano, Fernando. *Familias de Fex (ss. XV–XVII)* (Madrid: CSIC, 1995).

Melchert, Christopher. *The Formation of the Sunni Schools of Law, 9th–10th Centuries C.E.* (Leiden: Brill, 1997).

Melvin-Koushki, Matthew. "*Taḥqīq* vs. *Taqlīd* in the Renaissances of Western Early Modernity," *Philological Encounters* 3 (2018), 1–44.

"In Defense of Geomancy: Sharaf al-Dīn Yazdī Rebuts Ibn Khaldūn's Critique of the Occult Sciences," *Arabica* 64 (2017), 346–403.

"(De)colonizing Early Modern Occult Philosophy," *Magic, Ritual, and Witchcraft* 12 (2017), 98–112.

"Afterword: Conjuncting Astrology and Lettrism, Islam and Judaism," *Magic, Ritual, and Witchcraft* 12 (2017), 89–97.

Merton, Robert K. *Science, Technology and Society in Seventeenth Century England* (New York: Howard Fertig, 2001 [1938]).

Michon, Jean-Louis. *The Autobiography of the Moroccan Sufi Ibn ʿAjība* (translated from the French by David Streight) (Louisville, KY: Fons Vitae, 1999).

Michot, Yahya. *L'opium et le café* (Beirut: Les editions Albouraq, 2009).

Miège, Jean-Louis. *Le Maroc et l'Europe (1830–1894)* (Paris: Presses Universitaires de France, 1961), 3 vols.

Mladenović, Bojana. *Kuhn's Legacy: Epistemology, Metaphilosophy, and Pragmatism* (New York: Columbia University Press, 2017).

Moin, A. Afzar. *The Millennial Sovereign: Sacred Kinship & Sainthood in Islam* (New York: Columbia University Press, 2012).

Montgomery, James E. "Islamic Crosspollinations." In Anna Akasoy, James E. Montgomery, and Peter E. Pormann (eds.) *Islamic Crosspollinations: Interactions in the Medieval Middle East* (Exeter: Gibb Memorial Trust, 2007), 148–94.

Moosa, Ebrahim. "Shaykh Aḥmad Shākir and the Adoption of a Scientifically-Based Lunar Calendar," *Islamic Law and Society* 5 (1998), 57–89.

Morrison, Robert. *The Light of the World: Astronomy in al-Andalus* (Berkeley, CA: University of California Press, 2016).

"A Scholarly Intermediary between the Ottoman Empire and Renaissance Europe," *Isis* 105 (2014), 32–57.

Islam and Science: The Intellectual Career of Niẓām al-Dīn al-Nīsābūrī (New York: Routledge, 2007).

Munson Jr., Henry. *Religion and Power in Morocco* (New Haven, CT: Yale University Press, 1993).

Musallam, Basim. *Sex and Society in Islam: Birth Control before the Nineteenth Century* (Cambridge: Cambridge University Press, 1983).

Najīb, Aḥmad ʿAbd al-Karīm. *Fihris ma lam yufahras min al-makhtutat al-ʿarabiyya fi al-khizanat al-khaṣṣa bi-l-mamlaka al-maghribiyya* (Casablanca: Markaz Najībawaih, 2008).

Navarro Brotóns, Victor and William Eamon (eds.) *Más allá de la Leyenda Negra: España y la Revolución Científica* (Valencia: Instituto de Historia de la Ciencia y Documentación López Piñero, 2007).

Newman, William R. and Anthony Grafton. "Introduction: The Problematic Status of Astrology and Alchemy in Premodern Europe." In William Newman and Anthony Grafton (eds.) *Secrets of Nature: Astrology and Alchemy in Early Modern Europe* (Cambridge: The MIT Press, 2001), 1–38.

Nirenberg, David. *Anti-Judaism: The Western Tradition* (New York: W. W. Norton and Company, 2013).

Ohlheiser, Abby. "A Short History of Richard Dawkins vs. The Internet," in *The Atlantic*, August 8, 2013. www.theatlantic.com/national/archive/2013/08/short-history-richard-dawkins-vs-internet/312350/

Olson, Caitlyn. "Beyond the Avicennian Turn: The Creeds of Muḥammad b. Yusūf al-Sanūsī (d. 895/1490)," *Studia Islamica* 115 (2020), 101–40.

"Creed, Belief, and the Common Folk: Disputes in the Early Modern Maghrib (9 /15–11/17 c.)." PhD Dissertation, Harvard University, 2020.

Osterhammel, Jürgen. *The Transformation of the World: A Global History of the Nineteenth Century* (Princeton, NJ: Princeton University Press, 2014).

Parker, Geoffrey. *Global Crisis: War, Climate Change, & Catastrophe in the Seventeenth Century* (New Haven, CT: Yale University Press, 2013).

Pellat, C. "Iyās b. Muʿāwiya." In P. Bearman, Th. Bianquis, C. E. Bosworth, E. van Donzel, and W. P. Heinrichs (eds.) *Encyclopaedia of Islam*, 2nd ed. (1954). http://dx.doi.org/10.1163/1573-3912_islam_SIM_3717

Pellat, Charles. "L'astrolabe sphérique d'ar-Rūdānī," *Bulletin d'études orientales* 28 (1975), 83–165.

"Muhammad al-Rudani: al-Naqiʿa ʿala al-ala al-nafiʿah," *Bulletin d'études orientales* 26 (1973), 7–82.

Perho, Imerli. *The Prophet's Medicine – A Creation of the Muslim Traditionalist Scholars* (Helsinki: University of Helsinki, 1995).

Pettigrew, Erin. "The History of Islam in Mauritania." In *The Oxford Research Encyclopedia of African History.* https://oxfordre.com/africanhistory/view/10.1093/acrefore/9780190277734.001.0001/acrefore-9780190277734-e-628

Pingree, David. "Hellenophilia versus the History of Science." In Isabelle Pingree and John M. Steele (eds.) *Pathways into the Studies of Ancient Sciences* (Philadelphia, PA: The American Philosophical Society Press, 2014), 3–12.

Pormann, Peter and Emilie Savage-Smith. *Medieval Islamic Medicine* (Georgetown, WA: Georgetown University Press, 2007).

Pormann, Peter. "The Physician and the Other: Images of the Charlatan in Medieval Islam," *Bulletin of the History of Medicine* 79 (2005), 189–227.

Porter, Geoffrey. "At the Pillar's Base: Islam, Morocco, and Education in the Qarawiyin Mosque, 1912–2000." PhD Dissertation, NYU, 2002.

Powers, David. *Law, Society, and Culture in the Maghrib, 1300–1500* (Cambridge: Cambridge University Press, 2002).

"The Maliki Family Endowment: Legal Norms and Social Practices," *International Journal of Middle East Studies* 25 (1993), 379–406.

"Orientalism, Colonialism, and Legal History: The Attack on Muslim Family Endowments in Algeria and India," *Comparative Studies in Society and History* 31 (1989), 535–71.

Principe, Lawrence. *The Secrets of Alchemy* (Chicago, IL: The University of Chicago Press, 2001).

Ptolemy, Claudius. *Geography.* Edward Luther Stevenson (trans.) (New York: Dover Publications, 1991 [1932]), 160.

Qaddūrī, ʿAbd al-Majīd al-. *Ibn Abī Mahallī al-faqīh al-thā'ir wa riḥlatu-hu al-iṣlīt al-khirrīt* (Rabat: Manshurat 'Ukāẓ, 1991).

Rabinow, Paul. *Reflections on Fieldwork in Morocco* (Berkeley, CA: University of California Press, 2007 [1977]).

Racine, Matthew T. *A Most Opulent Iliad: Expansion, Confrontation & Cooperation on the Southern Moroccan Frontier 1505–1542* (San Diego, CA: Lake George Press, 2012).

Ragep, Jamil. "al-Battānī, Cosmology, and the Early History of Trepidation in Islam." In Josep Casulleras and Julio Samsó (eds.) *From Baghdad to Barcelona: Studies in the Exact Sciences in Honour of Prof. Juan Vernet* (Barcelona: Anuari de Filologia, 1996), vol. 1, 267–98.

Ragep, Sally R. *Jaghmīnī's Mulakhkhaṣ: An Islamic Introduction to Ptolemaic Astronomy* (Cham: Springer, 2016).

Raj, Kapil. *Relocating Modern Science: Circulation and Construction of Knowledge in South Asia and Europe, 1650–1900* (New York: Palgrave Macmillan, 2007).

Reichmuth, Stefan. "The Arabic Concept of *Dīn* and Islamic Religious Sciences in the 18th Century: The Case of Murtaḍā al-Zabīdī (d. 1791)," *Oriens* 44 (2016), 94–115.

Reid, Megan. *Law and Piety in Medieval Islam* (New York: Cambridge University Press, 2013).

Reinhart, A. Kevin. "Islamic Law as Islamic Ethics," *The Journal of Religious Ethics* 11 (1983), 186–203.

Renaud, H. P. J. and G. S. Colin, *Documents marocains pour servir à l'histoire du "Mal Franc"* (Paris: Librairie Larose, 1935).

(eds. and trans.) *Tuḥfat al-aḥbāb: Glossaire de la matière médicale Marocaine* (Paris: Librairie Orientaliste Paul Geuthner, 1934).

Richter-Bernburg, Lutz. "Medicine in Islam: Contested Autonomy." In Annette Weissenrieder and Gregor Etzelmüller (eds.) *Islam, Religion and Illness* (Eugene: Wipf and Stock: 2016), 44–57.

Rius, Monica. *La Alquibla en al-Andalus y al-Magrib al-Aqṣà* (Barcelona: Institut "Millás Vallicrosa" d'História de la Ciencia Arab, 2000).

Robinson, Chase. "Conclusion: From Formative Islam to Classical Islam." In Chase Robinson (ed.) *The New Cambridge History of Islam* (Cambridge: Cambridge University Press, 2011), 683–95.

Rodríguez Guerrero, José. "Some Forgotten Fez Alchemists and the Loss of the Peñon de Vélez de la Gomera in the Sixteenth Century." In Miguel López-Pérez (ed.) *Chymia: Science and Nature in Medieval and Early Modern Europe* (Cambridge: Cambridge University Press, 2010), 291–309.

Rofail Farag, F. "The Muslims' Medical Achievements," *Arabica* 25 3 (1978), 292–309.

Rouighi, Ramzi. *Inventing the Berbers: History and Ideology in the Maghrib* (Philadelphia, PA: University of Pennsylvania Press, 2019).

The Making of a Mediterranean Emirate: Ifrīqiyā and Its Andalusis, 1200–1400 (Philadelphia, PA: University of Pennsylvania Press, 2011).

"The Andalusi origin of the Berbers?," *Journal of Medieval Iberian Studies* 2 (2010), 93–108.

Rosenthal, Franz. *The Herb: Hashish versus Medieval Muslim Society* (Leiden: Brill, 1971).

Saad, Elias. *Social History of Timbuktu: The Role of Muslim Scholars and Notables 1400–1900* (Cambridge: Cambridge University Press, 1983).

Sabra, A. I. "Situating Arabic Science: Locality versus Essence," *Isis* 87 (1996), 654–70.

"Science and Philosophy in Medieval Islamic Theology: The Evidence of the Fourteenth Century," *Zeitschrift für Geschichte der arabisch-islamischen Wissenschaften* 9 (1994), 1–42.

"The Appropriation and Subsequent Naturalization of Greek Science in Medieval Islam: A Preliminary Statement," *History of Science* 25 (1987), 223–43.

Said, Edward. *Orientalism*. (New York: Vintage, 1978).

Saif, Liana. "What is Islamic Esotericism?," *Correspondences* 7 1 (2019), 1–61.

"Between Medicine and Magic: Spiritual Aetiology and Therapeutics in Medieval Islam." In Siam Bhayro and Catherine Rider (eds.) *Demons and Illness from Antiquity to the Early Modern Period* (Leiden: Brill, 2017), 313–38.

Saliba, George. *Islamic Science and the Making of the European Renaissance* (Cambridge: The MIT Press, 2007).

"Seeking the Origins of Modern Science?" *Bulletin of the Royal Institute for InterFaith Studies* 1 (1999), 139–52.

"The Role of the Astrologer in Medieval Islamic Society," *Bulletin d'Etudes Orientales* 44 (1992), 45–67.

Samsó, Julio. "An outline of the history of Maghribī zijes from the end of the thirteenth century." In *Astronomy and Astrology in al-Andalus and the Maghrib* (Variorum: Aldershot, 2007) (originally in *Journal for the History of Astronomy* 29 (1998), 93–102).

Sanseverino, Ruggero Vimercati. *Fès et sainteté, de la foundation à l'avènement du Protectorat (808–1912)* (Rabat: Centre Jacque-Berque, 2014).

Savage-Smith, Emilie. "Medicine." In Roshdi Rashed (ed.) *Encyclopedia of the History of Arabic Science* (London: Routledge, 1996), 3 vols., vol. 3, 903–62.

Scharbrodt, Oliver. "The Salafiyya and Sufism: Muḥammad ʿAbduh and his *Risālat al-Wāridāt* (Treatise on Mystical Interpretations)," *Bulletin of the School of Oriental and African Studies* 70 (2007), 89–115.

Schmidl, Petra. *Volkstümliche Astronomie im islamischen Mittelalter. Zur Bestimmung der Gebetszeiten und der Qibla bei al-Aṣbaḥī, Ibn Raḥīq und al-Fārisī* (Leiden: Brill, 2007).

Schwartz, Randy. "Issues in the Origin and Development of *Hisab al-Khata'ayn* (Calculation by Double False Position)." Conference Paper Presented at Eighth North African Meeting on the History of Arab Mathematics (Huitième Colloque Maghrebienne sur l'Histoire des Mathématiques Arabes) [COMHISMA 8] Radès, Tunisia, December 18–20, 2004.

Schwartz, Stuart B. *All Can Be Saved: Religious Tolerance and Salvation in the Iberian Atlantic World* (New Haven, CT: Yale University Press, 2008).

Segalla, Spencer D. *The Moroccan Soul: French Education, Colonial Ethnology, and Muslim Resistance, 1912–1956* (Lincoln, NE: University of Nebraska Press, 2009).

Semlali, Kacem Aït Salah. *Histoire de l'alchemie et des alchemists au Maroc* (Rabat: Self-published, 2015).

Serrano Ruano, Delfina. "Later Ashʿarism in the Islamic West." In Sabine Schmidtke (ed.) *The Oxford Handbook of Islamic Theology* (Oxford: Oxford University Press, 2016), 515–33.

"Por qué llamaron los almohades antropomorfistas a los almorávides?" In Patrice Cressier, Maribel Fierro, and Luis Molina (eds.) *Los almohades: Problemas y perspectives* (Madrid: CSIC, 2005), vol. 2, 815–52.

Shaham, Ron. *The Expert Witness in Islamic Courts: Medicine and Crafts in the Service of Law* (Chicago. IL: The University of Chicago Press, 2010).

Shamsy, Ahmed. *Rediscovering the Islamic Classics: How Editors and Print Culture Transformed an Intellectual Tradition* (Princeton, NJ: Princeton University Press, 2020).

Shank, J. B. "Special Issue: After the Scientific Revolution: Thinking Globally about the Histories of the Modern Sciences," *Journal of Early Modern History* 21 (2017), 377–93.

The Newton Wars and the Beginning of the French Enlightenment (Chicago, IL: The University of Chicago Press, 2008).

Shapin, Steven. *The Scientific Life: A Moral History of a Late Modern Vocation.* (Chicago, IL: The University of Chicago Press, 2009).

The Scientific Revolution (Chicago, IL: The University of Chicago Press, 1996).

A Social History of Truth (Chicago, IL: The University of Chicago Press, 1994).

Shatzmiller, Maya. *The Berbers and the Islamic State: The Marīnid Experience in Pre-Protectorate Morocco* (Princeton, NJ: Marcus Wiener, 2000).

Siggel, Alfred. *Decknamen in der Arabischen Alchemistischen Literatur* (Berlin: Akademie Verlag, 1951).

Simenel, Roman. "Le livre comme trésor: Aura, predation et secret des manuscrits savants du Sud marocain," *Terrain* 59 (2012), 48–65.

Singer, Hans-Rudolph. "Morisken als Übersetzer." In Dietrich Briesemeister (ed.) *Sprache, Literature, Kultur: romanistische Beiträge* (Bern: Peter Lang, 1974), 37–49.

Smith, J. Z. "Religion, Religions, Religious…." In Mark C. Taylor (ed.) *Critical Terms for Religious Studies* (Chicago, IL: The University of Chicago Press, 1998).

Spadola, Emilio. *The Calls of Islam: Sufis, Islamists, and Mass Mediation in Urban Morocco* (Bloomington, IN: Indiana University Press, 2013).

Speziale, Fabrizio. "Rasāyana and Rasaśāstra in the Persian Medical Culture of South Asia," *History of Science in South Asia* 7 (2019), 1–41.

Stearns, Justin. Review of Ahmad b. ʿUmar al-Raqqadi al-Kunti, *Kitab shifa' al-aqsam al-ʿarida al-zahir wa al-batin min al-ajsam*. Floréal Sanaugustin (ed.) 3 vols. *Journal of African History* 54 (2013), 458–59.

Infectious Ideas: Contagion in Premodern Islamic and Christian Thought in the Western Mediterranean (Baltimore, MD: Johns Hopkins University Press, 2011).

"Writing the History of the Natural Sciences in the Pre-Modern Muslim World: Historiography, Religion, and the Importance of the Early Modern Period," *History Compass* 9 (2011).

"The Legal Status of Science in the Muslim World in the Early Modern Period: An Initial Consideration of *fatwa*s from Three Maghribi Sources." In Asad Q. Ahmed, Benham Sadeghi, and Michael Bonner (eds.) *The Islamic Scholarly Tradition: Studies in Islamic History, Law, and Thought in Honor of Professor Michael Allan Cook on His Seventieth Birthday* (Leiden: Brill, 2011), 265–90.

"Contagion in Theology and Law: Ethical Considerations in the Writings of Two 14th-Century Scholars of Nasṛ id Granada," *Islamic Law and Society* 14 (2007), 109–29.

"Review of Khaled El-Rouayheb, Islamic Intellectual History in the Seventeenth Century: Scholarly Currents in the Ottoman Empire and the Maghreb," *Journal of the American Oriental Society* 137 (2017), 437–40.

Strohmaier, Gotthard. "Galenism Caught between Faith in God and 'Prophetic Medicine'." In Annette Weissenrieder and Gregor Etzelmüller (eds.) *Islam, Religion and Illness* (Eugene: Wipf and Stock: 2016), 32–43.

Stroumsa, Guy. *A New Science: The Discovery of Religion in the Age of Reason.* (Cambridge: Harvard University Press, 2010).

Sūsī, Muḥammad al-Mukhtār al-. *Īlīgh qadīman wa ḥadīthan* (Rabat: al-Maṭbaʿah al-Mālikiyya, 2005 [1966]), 4th ed.

Ṭanjī, Badr al-ʿAmrānī al-. and Muḥammad Saīd al-Ghāzī. *al-Fihris al-waṣfī li-makhṭūṭāt khizānat al-masjid al-aʿẓam bi-Wazzān*. ʿAbd al-Laṭīf al-Jīlānī (ed.) (Rabat: Wizārat al-Awqāf, 2008), 3 vols.

Tannous, Jack. *The Making of the Medieval Middle East: Religion, Society, and Simple Believers* (Princeton, NJ: Princeton University Press, 2018).

Targhī, ʿAbdallāh al-. *Fahāris ʿulamā' al-maghrib mundhu al-nasha' ilā nihāyat al-qarn al-thānī ʿashr li-l-hijra* (Tetuan: Manshūrāt kulliyat al-ādāb wa l-ʿūlum al-insāniyya, 1999).

Tazi, Abdelhadi and Badr Tazi (ed. and trans.) *La medicine arabe au XVIIIe siècle à travers al "Urdjuza ash-Shakruniyya"* (Cairo: L'Organisation Egyptienne Générale du Livre, 1984).

Terem, Etty. *Old Texts, New Practices: Islamic Reform in Modern Morocco* (Stanford, CA: Stanford University Press, 2014).

Tezcan, Baki. *The Second Ottoman Empire* (Cambridge: Cambridge University Press, 2010).

Todd, Richard. "Alchemical Poetry in Almohad Morocco: The *Shudhūr al-dhahab* of Ibn Arfaʿ Ra's," *Oriens* 44 (2016), 116–44.

Tolan, John. *Saracens: Islam in the Medieval European Imagination* (New York: Columbia University Press, 2002).

Toomer, G. J. (ed. and trans.) *Ptolemy's Almagest* (Princeton, NJ: Princeton University Press, 1998).

Touati, Houari. *Entre Dieu et les Hommes: Lettrés, saints et sorciers au Maghreb (17e siècle)* (Paris: Les Éditions de l'École des Hautes études en Sciences Sociales, 1994).

Tweed, Thomas. *Crossing and Dwelling: A Theory of Religion* (Cambridge: Harvard University Press, 2008).

Ullmann, Manfred. *Islamic Medicine* (Edinburgh: Edinburgh University Press, 1978).

Die Natur und Geheimwissenschaften im Islam (Leiden: Brill, 1972).

Die Medizin im Islam (Leiden: Brill, 1970).

Valensi, Lucette. *Fables de la Mémoire: La glorieuse bataille des Trois Rois (1578).* (Paris: Chandeigne, 2009 [1992]).

van Bladel, Kevin. *The Arabic Hermes: From Pagan Sage to Prophet of Science* (Oxford: Oxford University Press, 2009).

van Dalen, Dorrit. "This Filthy Plant: The Inspiration of a Central Sudanic Scholar in the Debate on Tobacco," *Islamic Africa* 3 (2012), 227–47.

Varisco, Daniel Martin. *Reading Orientalism: Said and the Unsaid* (Seattle, WA: University of Washington Press, 2007).

Vásquez de Benito, Concepción. "La materia médica de Ibn al-Jaṭīb," *Boletín de la Asociación Española de Orientalistas* 15 (1979), 139–50.

Wainscott, Ann. "Islamic Modernism, Political Reform and the Arabisation of Education: The Relationship between Moroccan Nationalists and al-Azhar University." In Masooda Bano and Keiko Sakurai (eds.) *Shaping Global Islamic Discourses: The Role of al-Azhar, Al-Madinah and Al-Mustafa* (Edinburgh: Edinburgh University Press, 2015), 142–64.

Warscheid, Ismail. *Droit musulman et société au Sahara prémoderne: La justice islamique dans les oasis du Grand Touat (Algérie) aux XVIIe-XIXe siècles* (Brill: Leiden, 2017).

Weinberg, Steven. *To Explain the World: The Discovery of Modern Science* (New York: Harper, 2015).

White, Andrew D. *A History of the Warfare of Science with Theology in Christendom* (Buffalo, NY: Prometheus Books, 1993

White, Hayden. *Metahistory: The Historical Imagination of the Nineteenth Century* (Baltimore, MD: Johns Hopkins University Press, 1973).

Wiegers, Gerald. "A Life Between Europe and the Maghreb: The Writings and Travels of Aḥmad b. Qâsim ibn Aḥmad ibn al-faqîh Qâsim ibn al-shaykh al-Ḥajarî al-Andalusî (born c. 977/1569–70)." In Ed De Moor and Geert Jan van Gelder (eds.) *The Middle East and Europe: encounters and exchanges* (Leiden: Brill, 1992), 87–115.

Wisnovsky, Robert. "The Nature and Scope of Arabic Philosophical Commentary in Post-Classical (ca. 1100–1900 AD) Islamic Intellectual History: Some Preliminary Observations," *Bulletin of the Institute of Classical Studies* (University of London) 47 (2004), 149–91.

Witkam, J. J. *De egyptische arts Ibn al-Akfānī (gest. 749/1348) en zijn indeling van de wetenschappen* (Leiden: Ter Lugt Pers, 1989).

Wyrtzen, Jonathan. *Making Morocco: Colonial Intervention and the Politics of Identity* (Ithaca, NY: Cornell University Press, 2015).

Yalcinkaya, M. Alper. *Learned Patriots: Debating Science, State, and Society in the Nineteenth-Century Ottoman Empire* (Chicago, IL: The University of Chicago Press, 2015).

"Science as an Ally of Religion: A Muslim Appropriation of 'the Conflict Thesis'," *British Journal for the History of Science* 44 (2011), 161–81.

Yeo, Richard. *Defining Science: William Whewall, Natural Knowledge and Public Debate in Early Victorian England* (Cambridge: Cambridge University Press, 2003).

Ze'evi, Dror. "Back to Napoleon? Thoughts on the Beginning of the Modern Era in the Middle East," *Mediterranean Historical Review* 19 (2004), 73–94.

Zemon Davis, Natalie. *Trickster Travels: A Sixteenth-Century Muslim between Worlds* (New York: Hill and Wang, 2006).

Ziadat, Adel A. *Western Science in the Arab World: The Impact of Darwinism, 1860–1930*. (New York: Palgrave Macmillan, 1986).

Zilfi, Madeline. "The Kadizadelis: Discordant Revivalism in Seventeenth-Century Istanbul," *Journal of Near Eastern Studies* 45 (1986), 251–69.

Zirikili, *al-A'lām* (Beirut: Dār al-'ilm li l-malāyīn, 2003), 8 vols.

Zosimos of Panopolis, *The Book of the Keys of the Work Kitāb Mafātīḥ aṣ-ṣan'a* (Zurich: Living Human Heritage Publications, 2016).

Zouber, Mahmoud. *Aḥmad Bābā de Tombouctou (1556–1627)* (Paris: G.-P. Maisonneuve et Larose, 1997).

Zysow, Aron. "The Economy of Certainty: An Introduction to the Typology of Islamic Legal Theory." PhD Dissertation, Harvard University, 1984.

Index

Other Titles in the Series

Agricultural Innovation in the Early Islamic World: The Diffusion of Crops and Farming Techniques, 700–1100, Andrew M. Watson
Muslim Tradition: Studies in Chronology, Provenance, and Authorship of Early Hadith, G. H. A. Juynboll
Social History of Timbuktu: The Role of Muslim Scholars and Notables 1400–1900, Elias N. Saad
Sex and Society in Islam: Birth Control before the Nineteenth Century, B. F. Musallam
Towns and Townsmen of Ottoman Anatolia: Trade, Crafts, and Food Production in an Urban Setting 1520–1650, Suraiya Faroqhi
Unlawful Gain and Legitimate Profit in Islamic Law: Riba, Gharar and Islamic Banking, Nabil A. Saleh
Men of Modest Substance: House Owners and House Property in Seventeenth-Century Ankara and Kayseri, Suraiya Faroqhi
Roman, Provincial and Islamic Law: The Origins of the Islamic Patronate, Patricia Crone
Economic Life in Ottoman Jerusalem, Amnon Cohen
Mannerism in Arabic Poetry: A Structural Analysis of Selected Texts (3rd Century AH/9th Century AD – 5th Century AH/11th Century AD), Stefan Sperl
The Rise and Rule of Tamerlane, Beatrice Forbes Manz
Popular Culture in Medieval Cairo, Boaz Shoshan
Early Philosophical Shiism: The Ismaili Neoplatonism of Abu Yaʿqub Al-Sijistani, Paul E. Walker
Indian Merchants and Eurasian Trade, 1600–1750, Stephen Frederic Dale
Palestinian Peasants and Ottoman Officials: Rural Administration around Sixteenth-Century Jerusalem, Amy Singer
Arabic Historical Thought in the Classical Period, Tarif Khalidi
Mongols and Mamluks: The Mamluk-Ilkhanid War, 1260–1281, Reuven Amitai-Preiss
Knowledge and Social Practice in Medieval Damascus, 1190–1350, Michael Chamberlain
The Politics of Households in Ottoman Egypt: The Rise of the Qazdağlis, Jane Hathaway
Hierarchy and Egalitarianism in Islamic Thought, Louise Marlow
Commodity and Exchange in the Mongol Empire: A Cultural History of Islamic Textiles, Thomas T. Allsen
State and Provincial Society in the Ottoman Empire: Mosul, 1540–1834, Dina Rizk Khoury
The Mamluks in Egyptian Politics and Society, Thomas Philipp and Ulrich Haarmann (eds.)
The Delhi Sultanate: A Political and Military History, Peter Jackson
European and Islamic Trade in the Early Ottoman State: The Merchants of Genoa and Turkey, Kate Fleet
The Ottoman City between East and West: Aleppo, Izmir, and Istanbul, Edhem Eldem, Daniel Goffman, and Bruce Masters
The Politics of Trade in Safavid Iran: Silk for Silver, 1600–1730, Rudolph P. Matthee

The Idea of Idolatry and the Emergence of Islam: From Polemic to History, G. R. Hawting
A Monetary History of the Ottoman Empire, Şevket Pamuk
Classical Arabic Biography: The Heirs of the Prophets in the Age of Al-Ma'mun, Michael Cooperson
Empire and Elites after the Muslim Conquest: The Transformation of Northern Mesopotamia, Chase F. Robinson
Poverty and Charity in Medieval Islam: Mamluk Egypt, 1250–1517, Adam Sabra
Culture and Conquest in Mongol Eurasia, Thomas T. Allsen
Christians and Jews in the Ottoman Arab World: The Roots of Sectarianism, Bruce Masters
Arabic Administration in Norman Sicily: The Royal Diwan, Jeremy Johns
Law, Society and Culture in the Maghrib, 1300–1500, David S. Powers
Revival and Reform in Islam: The Legacy of Muhammad al-Shawkani, Bernard Haykel
Tolerance and Coercion in Islam: Interfaith Relations in the Muslim Tradition, Yohanan Friedmann
Guns for the Sultan: Military Power and the Weapons Industry in the Ottoman Empire, Gábor Ágoston
Marriage, Money and Divorce in Medieval Islamic Society, Yossef Rapoport
The Empire of the Qara Khitai in Eurasian History: Between China and the Islamic World, Michal Biran
Domesticity and Power in the Early Mughal World, Ruby Lal
Power, Politics, and Religion in Timurid Iran, Beatrice Forbes Manz
Postal Systems in the Pre-Modern Islamic World, Adam J. Silverstein
Kingship and Ideology in the Islamic and Mongol Worlds, Anne F. Broadbridge
Justice, Punishment, and the Medieval Muslim Imagination, Christian Lange
The Shiites of Lebanon under Ottoman Rule, 1516–1788, Stefan Winter
Women and Slavery in the Late Ottoman Empire, Madeline Zilfi
The Second Ottoman Empire: Political and Social Transformation in the Early Modern World, Baki Tezcan
The Legendary Biographies of Tamerlane: Islam and Heroic Apocrypha in Central Asia, Ron Sela
Non-Muslims in the Early Islamic Empire: From Surrender to Coexistence, Milka Levy-Rubin
The Origins of the Shi'a: Identity, Ritual, and Sacred Space in Eighth-Century Kufa, Najam Haider
Politics, Law, and Community in Islamic Thought: The Taymiyyan Moment, Ovamir Anjum
The Power of Oratory in the Medieval Muslim World, Linda G. Jones
Animals in the Qur'an, Sarra Tlili
The Logic of Law Making in Islam: Women and Prayer in the Legal Tradition, Behnam Sadeghi
Empire and Power in the Reign of Süleyman: Narrating the Sixteenth-Century Ottoman World, Kaya Şahin
Law and Piety in Medieval Islam, Megan H. Reid
Women and the Transmission of Religious Knowledge in Islam, Asma Sayeed

The New Muslims of Post-Conquest Iran: Tradition, Memory, and Conversion, Sarah Bowen Savant
The Mamluk City in the Middle East: History, Culture, and the Urban Landscape, Nimrod Luz
Disability in the Ottoman Arab World, 1500–1800, Sara Scalenghe
The Holy City of Medina: Sacred Space in Early Islamic Arabia, Harry Munt
Muslim Midwives: The Craft of Birthing in the Premodern Middle East, Avner Giladi
Doubt in Islamic Law: A History of Legal Maxims, Interpretation, and Islamic Criminal Law, Intisar A. Rabb
The Second Formation of Islamic Law: The Hanafi School in the Early Modern Ottoman Empire, Guy Burak
Sexual Violation in Islamic Law: Substance, Evidence, and Procedure, Hina Azam
Gender Hierarchy in the Qur'an: Medieval Interpretations, Modern Responses, Karen Bauer
Intellectual Networks in Timurid Iran: Sharaf al-Din 'Ali Yazdi and the Islamicate Republic of Letters, Ilker Evrim Binbaş
Authority and Identity in Medieval Islamic Historiography: Persian Histories from the Peripheries, Mimi Hanaoka
The Economics of Ottoman Justice: Settlement and Trial in the Sharia Courts, Metin Coşgel and Boğaç Ergene
The Mystics of al-Andalus: Ibn Barrajan and Islamic Thought in the Twelfth Century, Yousef Casewit
Muhammad's Heirs: The Rise of Muslim Scholarly Communities, 622–950, Jonathan E. Brockopp
The First of the Modern Ottomans: The Intellectual History of Ahmed Vasif, Ethan Menchinger
Non-Muslim Provinces under Early Islam: Islamic Rule and Iranian Legitimacy in Armenia and Caucasian Albania, Alison Vacca
Women and the Making of the Mongol Empire, Anne F. Broadbridge
Slavery and Empire in Central Asia, Jeff Eden
Christianity in Fifteenth-Century Iraq, Thomas A. Carlson
Child Custody in Islamic Law: Theory and Practice in Egypt since the Sixteenth Century, Ahmed Fekry Ibrahim
Ibadi Muslims of North Africa: Manuscripts, Mobilization, and the Making of a Written Tradition, Paul M. Love Jr.
Islamic Law of the Sea: Freedom of Navigation and Passage Rights in Islamic Thought, Hassan S. Khalilieh
Law and Politics under the Abassids: An Intellectual Portrait of al-Juwayni, Sohaira Z. M. Siddiqui
Friends of the Emir: Non-Muslim State Officials in Premodern Islamic Thought, Luke B. Yarbrough
The Crisis of Kingship in Late Medieval Islam: Persian Emigres and the Making of Ottoman Sovereignty, Christopher Markiewicz
Collective Liability in Islam: The 'Āqila and Blood Money Payments, Nurit Tsafrir
Arabic Poetics: Aesthetic Experience in Classical Arabic Literature, Lara Harb
The Sufi Saint of Jam: History, Religion, and Politics of a Sunni Shrine in Shi'i Iran, Shivan Mahendrarajah

Lightning Source UK Ltd.
Milton Keynes UK
UKHW051641020323
417905UK00017B/19

9 781107 638709